New Advances in Quantum Geometry

New Advances in Quantum Geometry

Editors

Shi-Dong Liang
Tiberiu Harko
Matthew J. Lake

MDPI

Basel • Beijing • Wuhan • Barcelona • Belgrade • Novi Sad • Cluj • Manchester

Editors

Shi-Dong Liang
Sun Yat-Sen University
Guangzhou, China

Tiberiu Harko
Babes-Bolyai University
Cluj-Napoca, Rumania

Matthew J. Lake
National Astronomical
Research Institute of Thailand
Chiang Mai, Thailand

Editorial Office
MDPI
St. Alban-Anlage 66
4052 Basel, Switzerland

This is a reprint of articles from the Special Issue published online in the open access journal *Physics* (ISSN 2624-8174) (available at: https://www.mdpi.com/journal/physics/special_issues/QuantumGeometry).

For citation purposes, cite each article independently as indicated on the article page online and as indicated below:

Lastname, A.A.; Lastname, B.B. Article Title. *Journal Name* **Year**, *Volume Number*, Page Range.

ISBN 978-3-0365-8770-7 (Hbk)
ISBN 978-3-0365-8771-4 (PDF)
doi.org/10.3390/books978-3-0365-8771-4

Contents

Preface

Dear Colleagues,

The expression "quantum geometry" means different things to different people. Although virtually all researchers agree that the quantum theory of gravity, whatever form it may take, must involve a description of quantised spacetime, there is no accepted definition of this term. In this Special Issue, we invited contributions exploring the multi-faceted meanings of these two deceptively simple words.

We aimed to collect, in one volume, a range of contributions representing the diverse meanings that have been attached to this innocent-sounding phrase over the past 95 years of research into quantum physics, information science, and gravity. Contributions from all fields were welcomed, with a particular emphasis on the following topics:

- Noncommutative geometry;
- Spin foams and loop quantum gravity;
- Nonlocal geometry;
- Generalised uncertainty relations and minimum-length scenarios;
- Quantum reference frames;
- Emergent geometry from quantum entanglement and the "it from bit" scenario;
- Information, stringy, and holographic geometry;
- Causal dynamical triangulations, asymptotically safe gravity, and fractal spacetimes;
- Weyl geometry in gravity and cosmology and Finsler geometry in physics.

Papers addressing other, less mainstream, approaches were also welcomed. The resulting collection includes both original research articles and review papers. A comprehensive summary of such a huge field is, of course, impossible within a single volume, but we hope that this issue will provide a valuable reference and starting point for dialogue between diverse approaches to a common theme: the problem of quantum geometry.

Shi-Dong Liang, Tiberiu Harko, and Matthew J. Lake

Editors

physics

MDPI

Editorial

New Advances in Quantum Geometry

Shi-Dong Liang [1,2,*], Tiberiu Harko [3,4,*] and Matthew J. Lake [1,3,5,6,7,*]

1 School of Physics, Sun Yat-sen University, Guangzhou 510275, China
2 State Key Laboratory of Optoelectronic Material and Technology, Guangdong Province Key Laboratory of Display Material and Technology, Sun Yat-sen University, Guangzhou 510275, China
3 Department of Physics, Babeș-Bolyai University, Mihail Kogălniceanu Street 1, 400084 Cluj-Napoca, Romania
4 Department of Theoretical Physics, National Institute of Physics and Nuclear Engineering (IFIN-HH), 077125 Bucharest, Romania
5 National Astronomical Research Institute of Thailand, 260 Moo 4, T. Donkaew, A. Maerim, Chiang Mai 50180, Thailand
6 Department of Physics and Materials Science, Faculty of Science, Chiang Mai University, 239 Huaykaew Road, T. Suthep, A. Muang, Chiang Mai 50200, Thailand
7 Office of Research Administration, Chiang Mai University, 239 Huaykaew Road, T. Suthep, A. Muang, Chiang Mai 50200, Thailand
* Correspondence: stslsd@mail.sysu.edu.cn (S.-D.L.); tiberiu.harko@aira.astro.ro (T.H.); matthewjlake@narit.or.th (M.J.L.)

Citation: Liang, S.-D.; Harko, T.; Lake, M.J. New Advances in Quantum Geometry. *Physics* **2023**, *5*, 688–689. https://doi.org/10.3390/physics5030045

Received: 26 June 2023
Accepted: 27 June 2023
Published: 30 June 2023

1. Introduction

Presently, we are in a period of rapid and intensive changes in our understanding of the gravitational interaction, triggered by the important observational findings of the late 1990s. We have witnessed the emergence of new cosmological paradigms and a better understanding of black hole properties, as well as a tremendous increase in the precision of observational data. However, in the understanding of one of the most fundamental questions of present day physics, the nature of quantum gravity, many unsolved questions still remain. Can gravity be quantized at all? Is the gravitational force a purely geometric effect, or is it field theoretic in nature? And, if gravity is pure geometry, what is the relation between geometry and the quantum? Is it possible to quantize geometry, to create a unified quantum geometric framework for gravity?

2. Scope and Aims of the Project

This Special Issue is focused on the fundamental question of quantum geometry, its various meanings, and its implications for the standard theoretical concepts in gravitation and cosmology. The Issue includes state-of-the-art research contributions in the following areas: the quantum geometry created by quantum matter [1], quantum metric fluctuations [2], cosmology in modified gravity models [3,4], de Sitter gauge theory [5], and matrix theory models of the gravitational interaction [6]. The nature of quantum configurations in phase space [7], momentum operators in intrinsically curved manifolds [8], uncertainty relations in the presence of a minimal length [9], generalized uncertainty black holes [10], and the effects of quantum gravity on mass scales at high energies [11] are also addressed. These fascinating topics, in which geometry, gravity and quantum mechanics are brought together, provide deeper insights into the unsolved mysteries of the gravitational interaction, at the smallest possible scales.

Review papers also play an essential role, both in synthesizing the available knowledge in a given field of science and in providing the key information required to understand the most advanced topics in contemporary research. Two reviews, one on noncommutativity in physics [12] and another on the Barbero–Immirzi parameter in loop quantum gravity [13], provide good introductions to these subjects, and overviews of some important recent results in these fascinating fields of basic research.

As the Guest Editors of this Special Issue, we hope that this collection will serve as a standard reference for initiating and continuing state-of-the-art research in the fundamental fields of quantum geometry, quantum gravity, and geometric theories of gravitation. Moreover, we hope that the Issue opens up some new perspectives on the quantum-geometric aspects of the gravitational field and its applications in astrophysics and cosmology. Our sincere thanks to all the authors who contributed to this volume, and without whom it would not have been possible, for their time and efforts.

Funding: This work was supported by the Grant of Scientific and Technological Projection of Guangdong Province (China), no. 2021A1515010036.

Conflicts of Interest: The authors declare no conflict of interest.

1. Ashtekar, A. Exploring quantum geometry created by quantum matter. *Physics* **2022**, *4*, 1384–1402. [CrossRef]
2. Haghani, Z.; Harko, T. Effects of quantum metric fluctuations on the cosmological evolution in Friedmann-Lemaitre-Robertson-Walker geometries. *Physics* **2021**, *3*, 689–714. [CrossRef]
3. Gadbail, G.N.; Mandal, S.; Sahoo, P.K. Parametrization of deceleration parameter in $f(Q)$ gravity. *Physics* **2022**, *4*, 1403–1412. [CrossRef]
4. Sofuoğlu, D.; Tiwari, R.K.; Abebe, A.; Alfedeel, A.H.A.; Hassan, E.I. The cosmology of a non-minimally coupled $f(R,T)$ gravitation. *Physics* **2022**, *4*, 1348–1358. [CrossRef]
5. Lu, J.-A. Cosmology of a polynomial model for de Sitter gauge theory sourced by a fluid. *Physics* **2022**, *4*, 1168–1179. [CrossRef]
6. Brahma, S.; Brandenberger, R.; Laliberte, S. BFSS matrix model cosmology: Progress and challenges. *Physics* **2023**, *5*, 1–10. [CrossRef]
7. Albuquerque, S.; Bezerra, V.B.; Lobo, I.P.; Macedo, G.; Morais, P.H.; Rodrigues, E.; Santos, L.C.N.; Varão, G. Quantum configuration and phase spaces: Finsler and Hamilton geometries. *Physics* **2023**, *5*, 90–115. [CrossRef]
8. Schürmann, T. On momentum operators given by Killing vectors whose integral curves are geodesics. *Physics* **2022**, *4*, 1440–1452. [CrossRef]
9. Rastegin, A.E. On Majorization uncertainty relations in the presence of a minimal length. *Physics* **2022**, *4*, 1413–1425. [CrossRef]
10. Lobos, N.J.L.S.; Pantig, R.C. Generalized extended uncertainty principle black holes: Shadow and lensing in the macro- and microscopic realms. *Physics* **2022**, *4*, 1318–1330. [CrossRef]
11. Singh, T.P. Why do elementary particles have such strange mass ratios?—The importance of quantum gravity at low energies. *Physics* **2022**, *4*, 948–969. [CrossRef]
12. Liang, S.-D.; Lake, M.J. An introduction to noncommutative physics. *Physics* **2023**, *5*, 436–460. [CrossRef]
13. Vyas, R.P.; Joshi, M.J. The Barbero–Immirzi parameter: An enigmatic parameter of loop quantum gravity. *Physics* **2022**, *4*, 1094–1116. [CrossRef]

physics

MDPI

Article

Effects of Quantum Metric Fluctuations on the Cosmological Evolution in Friedmann-Lemaitre-Robertson-Walker Geometries

Zahra Haghani [1,*] and Tiberiu Harko [2,3,4,5]

1 School of Physics, Damghan University, Damghan 36716-41167, Iran
2 Department of Theoretical Physics, National Institute of Physics and Nuclear Engineering (IFIN-HH), 077125 Bucharest, Romania; tiberiu.harko@aira.astro.ro
3 Astronomical Observatory, 400487 Cluj-Napoca, Romania
4 Department of Physics, Babes-Bolyai University, 400084 Cluj-Napoca, Romania
5 School of Physics, Sun Yat-Sen University, Guangzhou 510275, China
* Correspondence: z.haghani@du.ac.ir

Abstract: In this paper, the effects of the quantum metric fluctuations on the background cosmological dynamics of the universe are considered. To describe the quantum effects, the metric is assumed to be given by the sum of a classical component and a fluctuating component of quantum origin . At the classical level, the Einstein gravitational field equations are equivalent to a modified gravity theory, containing a non-minimal coupling between matter and geometry. The gravitational dynamics is determined by the expectation value of the fluctuating quantum correction term, which can be expressed in terms of an arbitrary tensor $K_{\mu\nu}$. To fix the functional form of the fluctuation tensor, the Newtonian limit of the theory is considered, from which the generalized Poisson equation is derived. The compatibility of the Newtonian limit with the Solar System tests allows us to fix the form of $K_{\mu\nu}$. Using these observationally consistent forms of $K_{\mu\nu}$, the generalized Friedmann equations are obtained in the presence of quantum fluctuations of the metric for the case of a flat homogeneous and isotropic geometry. The corresponding cosmological models are analyzed using both analytical and numerical method. One finds that a large variety of cosmological models can be formulated. Depending on the numerical values of the model parameters, both accelerating and decelerating behaviors can be obtained. The obtained results are compared with the standard ΛCDM (Λ Cold Dark Matter) model.

Keywords: semiclassical gravity; quantum metric fluctuations; cosmological models

Citation: Haghani, Z.; Harko, T. Effects of Quantum Metric Fluctuations on the Cosmological Evolution in Friedmann-Lemaitre-Robertson-Walker Geometries. *Physics* **2021**, *3*, 689–714. https://doi.org/10.3390/physics3030042

Received: 9 July 2021
Accepted: 11 August 2021
Published: 24 August 2021

1. Introduction

General relativity and quantum mechanics are the basic, and widely accepted, branches of theoretical physics, confirmed by a large number of experiments and observations. In particular, general relativity, a theory of gravity [1–3] is a typical example of a physical theory with a very beautiful geometric structure. Moreover, it is one of the very successful existing physical theories, with its predictions being confirmed in the past one hundred years with a high degree of accuracy [4–6]. In addition to the classical tests of general relativity performed at the Solar System level, recently two other fundamental predictions of general relativity, the existence of the gravitational waves, and the existence of black holes, have also been confirmed [7,8].

Despite the remarkable achievements of both quantum mechanics and general relativity, it had been known for a long time that these two fundamental theories of physics cannot be unified, and they seem to be incompatible with each other. The first in depth analysis of the possibility of the unification of quantum mechanics and general relativity was performed by Bronstein [9], whose analysis indicated the existence of an essential difference between quantum theory and the quantum theory of the gravitational field based on general relativity. These early results already pointed out to the fundamental difficulty of unifying quantum theory and general relativity. Even after almost ninety years of intensive effort the task of building a quantum theory of gravity is still unfulfilled, remaining an open assignment for modern theoretical physics. There are many approaches to quantum gravity, and for an introduction to the field, see, e.g., [10,11].

One possible avenue for the quantization of gravity would be its reformulation as a gauge theory, an approach pioneered in [12], and further developed in [13–17]. Another attempt to a quantum theory of gravity was initiated in [18–20], and it is based on the rewriting of the Hilbert-Einstein action in terms of a spin connection and a set of tetrads. Since both the tetrads and the spin connection are vector fields, the theory of gravitation can be reformulated as a vector gauge theory.

However, from a fundamental physical point of view, such an approach is not satisfactory, since in the case of gravitation the principles of equivalence and of the general covariance are more important than the gauge principle, on which the standard model of particle physics and its extensions are based [21]. Moreover, the theory is still based on the standard Hilbert-Einstein action, and it is not clear at this moment if it can be quantized. Various other approaches have been proposed for the quantization of the gravitational field in [22–29].

A possible way of dealing with the problem of the quantization of gravity, and a first step in this direction, is to assume that the matter fields are quantized, and that evolve in a classical spacetime, described by a metric $g_{\mu\nu}$, where the indices, denoted by Greek letters, take on the values 0, 1, 2, 3. There is an important difference in this case as compared to the evolution in a Minkowski spacetime, in the sense that in general there is no preferred vacuum state for the fields. Consequently, particle creation effects naturally take place. In semiclassical gravity the gravitational field is described classically, using the Hilbert-Einstein action, $S = \int \left(-R/2\kappa^2\right)\sqrt{-g}d^4x$, where R denotes the Ricci scalar, κ is the gravitational coupling constant, g is the determinant of the metric tensor, and x represents the time (x^0) and space (x^1, x^2, x^3) coordinates. Hence, in semiclassical gravity quantum matter is coupled to the gravitational field via the semiclassical Einstein equations,

$$R_{\mu\nu} - \frac{1}{2}g_{\mu\nu}R = \frac{8\pi G}{c^4}\langle\Psi|\hat{T}_{\mu\nu}|\Psi\rangle, \quad (1)$$

where G is the Newtonian constant of gravitation, c is the speed of light, and $\hat{T}_{\mu\nu}$ is the quantum operator associated with $T_{\mu\nu}$.

The above equations are obtained by replacing in the Einstein gravitational field equations $T_{\mu\nu}$, the classical matter energy-momentum tensor by the expectation value, $\langle\cdots\rangle$, in an arbitrary quantum state Ψ of the quantum operator associated with $T_{\mu\nu}$. The semiclassical approach to quantum gravity was proposed initially in [30,31], and it has been further developed and discussed in [32–44].

From Equation (1) it follows that the matter energy-momentum tensor $T_{\mu\nu}$ is obtained in the classical limit by assuming $\langle\Psi|\hat{T}_{\mu\nu}|\Psi\rangle = T_{\mu\nu}$. The semiclassical Einstein Equation (1) can also be derived from the variational principle $\delta\left(S_g + S_\psi\right) = 0$ [35], where S_g is the standard Hilbert-Einstein general relativistic classical action of the gravitational field, while the second component of the total action, generated by quantum effects, is given by:

$$S_\Psi = \int \left[\mathrm{Im}\langle\dot{\Psi}|\Psi\rangle - \langle\Psi|\hat{H}|\Psi\rangle + \alpha(\langle\Psi|\Psi\rangle - 1)\right]dt,$$

where α is a Lagrange multiplier, while $\hat{H}$ is the Hamilton operator of matter.

A more general pathway to semiclassical gravity was introduced in [35]. It is essentially based on the idea of the introduction of a nonminimal coupling between the classical Ricci scalar R, and the quantized matter fields. Specifically, one can introduce in the total action a term containing the quantum matter-geometry coupling with the simple form: $\int RF(\langle f(\phi)\rangle)_\Psi\sqrt{-g}d^4x$. In the coupling term, F and f denote some arbitrary functions, while $(\langle f(\phi)\rangle)_\Psi = \langle\Psi(t)|f[\phi(x)]|\Psi(t)\rangle$ denotes the average value of a function over the quantum fields. Then, the effective semiclassical Einstein equations are given by [35]:

$$R_{\mu\nu} - \frac{1}{2}Rg_{\mu\nu} = 16\pi G\left[\langle\hat{T}_{\mu\nu}\rangle_\Psi + G_{\mu\nu}F - \nabla_\mu\nabla_\nu F + g_{\mu\nu}\Box F\right]. \quad (2)$$

The semi-classical effective gravitational model described by Equation (2) has one important consequence, namely, that the mean value of the matter energy-momentum tensor $\langle\hat{T}^{\mu\nu}\rangle_\Psi$ is not conserved directly, since generally $\nabla_\mu\langle\hat{T}^{\mu\nu}\rangle_\Psi \neq 0$. Thus, the theoretical model described by Equation (7) can be interpreted theoretically as describing a process of particle production that is the direct result of an energy removal from geometry to matter.

An alternative method for the quantization of classical physical structures is the stochastic quantization method, introduced initially in [45]. In this approach to quantization of the classical systems the quantum fluctuations are characterized with the help of the stochastic Langevin equation. The stochastic quantization method goes back to the study initiated in [46], in which the Schrödinger equation was from the classical dynamics obtained using stochastic methods. However, the rigorous formulation of the stochastic quantization method was presented in [45]. In stochastic quantization the quantum mechanical picture of physical processes is constructed through the limit, with respect to a fictitious time variable t, of a hypothetical higher-dimensional stochastic process, assumed to be described by a Langevin type equation. For the early advancements in stochastic quantization theory see [47,48]. For the gravitational field the stochastic quantization procedure was introduced in [49,50], by assuming that the classical metric tensor obeys the covariant stochastic Langevin equation, given by [49]:

$$\dot{g}_{\mu\nu} = -2i\left[R_{\mu\nu} - \frac{\lambda+1}{2(2\lambda+1)}g_{\mu\nu}R\right] + \xi_{\mu\nu},$$

where $\xi_{\mu\nu}$ is a stochastic source term, λ is a parameter, and a dot denotes the derivative with respect to λ. For recent discussions on the stochastic quantization of gravity see [35]. For alternative Einstein-Langevin type equations, see, e.g., [51,52].

The quantum fluctuations of the space time are assumed to play an important role in the quantum description of gravity. In fact, long time ago, it was suggested that due to the Heisenberg uncertainty principle over extremely small distances and sufficiently small intervals of time, the geometry of spacetime may fluctuate [53,54]. The quantum fluctuations of the spacetime could be large enough to induce important deviations from the smooth spacetime one experiences at macroscopic scales, and giving spacetime a "foamy" character [53,54].

Quantum fluctuations play a central role in the alternative semi-classical description of quantum gravity, introduced in [55]. In this approach to quantum gravity, the quantized metric is assumed to have two components, and is given as the sum of classical and quantum terms. After performing this decomposition, the quantized Einstein equations can be approximated at the classical level by a modified gravity theory that includes a nonminimal coupling between matter and geometry. After introducing some natural hypotheses for the two-points expectation value of the product of the fluctuating quantum metric, one can obtain the effective semiclassical gravitational and scalar field Lagrangians [56]. For a vanishing expectation value of the first-order terms of the metric, the second order corrections can also be calculated. The second order quantum corrections also lead to a modified gravity theory.

The gravitational field equations and the modified conservation laws were obtained within the framework of the fluctuating metric approach in [57]. It was also shown that due to the quantum fluctuations a bouncing universe model can be constructed. Moreover, in a dark energy dominated phase, a decelerated expansionary cosmological evolution is also possible. Gravitational models with fluctuating metric were studied in [58–60]. In particular, after expressing generally the expectation value of the quantum correction in the form of a general second order tensor, in [61] the effective gravitational field equations at the classical level have been derived in a general formulation. Cosmological models with the quantum correction tensor given by the coupling of a scalar function and of a scalar field to the metric tensor, as well as by a quantum term proportional to the ordinary matter energy-momentum tensor, were analyzed. These models can describe the present day accelerated expansion of the universe.

One of the important implications of the gravitational theories with fluctuating quantum type metric is that they typically lead to modified gravity models in the presence of a geometry-matter coupling. Such theoretical models have been proposed already in the framework of standard classical general relativity as viable explanations of the recent cosmological observations that have determined a fundamental change in our comprehension of the universe. Very precise and detailed astronomical and astrophysical observations indicate that recently the universe experienced a transition from deceleration to an accelerating, de Sitter type regime [62–67]. These cosmological observations are usually interpreted through the postulation of the existence of a dominant constituent of the universe, called dark energy, and whose presence can give a reason for all recent cosmological observations [68,69]. However, a second major constituent, called dark matter, is also required to fully explain the observations [70,71].

On the other hand, one cannot reject a priori the possibility that the two major constituents of the universe, dark energy (usually modelled as a cosmological constant [72,73]), and dark matter, could be explained as a common and basic property of a generalized gravity theory that goes ahead of standard general relativity, and its Hilbert-Einstein variational formulation. Many extended gravity theories, modifying and generalizing Einstein's general relativity have been suggested recently. One of the first extensions of general relativity is represented by the $f(R)$ gravity theory, with gravitational action of the form $S = \left(1/2\kappa^2\right) \int f(R)\sqrt{-g}d^4x + \int L_m\sqrt{-g}d^4x$ [74–79], where L_m denotes the matter Lagrangian density. $f(R)$ gravity generalizes only the geometric part of the gravitational action, and thus it ignores the profound role the matter Lagrangian could have [80]. Moreover, $f(R)$ theory is still based on a minimal coupling between geometry and matter.

Extended gravity theories with arbitrary matter-geometry couplings were introduced initially in [81–84] in the form of the $f(R, L_m)$-modified gravity theory, with the gravitational action given by $S = \frac{1}{2\kappa^2} \int f(R, L_m)\sqrt{-g}d^4x$. In this approach, geometry becomes equivalent with matter, and thus matter plays a more important role in describing the properties of space-time as the one ascribed to it in standard general relativity.

The $f(R, T)$ gravity theory introduces an other type of matter-geometry coupling, with the gravitational action given by $S = \int \left[f(R, T)/2\kappa^2 + L_m\right]\sqrt{-g}d^4x$ [85,86]. Hence, in $f(R, T)$ theory matter and geometry are coupled through the trace T of the energy-momentum tensor. Many other gravitational theories with geometry-matter couplings have been proposed and studied widely up to now. Among them are the $f\left(R, T, R_{\mu\nu}T^{\mu\nu}\right)$ gravity theory [87,88], the $f(R, \mathcal{R})$ hybrid metric-Palatini gravity theory, with $\mathcal{R}$ representing the Ricci scalar, formed with the help of a connection not depending on the metric, such as in the case of the Levi-Civita connection [89–91], the Weyl-Cartan-Weitzenböck (WCW) theory [92], and the $f(Q, T)$ modified gravity theory [93,94], where Q is the non-metricity.

Modified gravity theories in which the torsion scalar $\tilde{T}$ couples nonminimally to the trace T of the matter energy-momentum tensor are called $f(\tilde{T}, \mathcal{T})$ gravity theories. These types of theories have also been extensively investigated [95]. In [96] theories with higher derivative matter fields were considered in detail. Extensive reviews of the $f(R, L_m)$, $f(R, T)$, and hybrid-metric-Palatini type gravity theories can be found in [97,98].

All gravitational theories with matter-geometry coupling have the curious particularity implying that the four-divergence of the matter energy-momentum tensor does not vanish generally, so that $\nabla_\mu T^{\mu\nu} \neq 0$. This non-conservation of $T_{\mu\nu}$ can be understood from a physical point of view using the formalism of the thermodynamics of open systems [86,99,100]. Hence, in these gravitational theories, one can presume that the energy and momentum balance equations describe irreversible particle creation processes. Thus, the non-conservation of $T_{\mu\nu}$ indicates an irreversible matter and energy transfer from the gravitational field to the newly produced particles.

The creation of particles from the cosmological vacuum is one of the significant predictions of the quantum field theory in curved space-times [101–105]. Quantum field theoretical approaches to gravity lead naturally to particles creation processes, and they play an essential role in the understanding of the foundations of the theory. In the expanding Friedmann-Lemaitre-Robertson-Walker geometry, quantum field theory in curved spacetimes predict that quanta of the minimally coupled scalar field are produced permanently from the cosmological vacuum [105–107].

Hence, the presence at the theoretical level of the particle creation processes in both quantum theories of gravity in curved space-times and in modified gravity theories with geometry-matter coupling suggests that a deep relationship may exist between these two, seemingly distinct physical theories. In fact, such a relationship was already obtained in [57], where it was found that in the nonperturbative approach for the quantization of the metric, as introduced in [55,56,58], as a consequence of the fluctuations of the spacetime, a specific $f(R,T)$ type gravitational model naturally emerges. The Lagrangian density of the theory is given by:

$$L = \left[(1-\alpha_0)\frac{R}{2\kappa^2} + \left(L_{\rm m} - \frac{\alpha_0}{2}T\right)\right]\sqrt{-g}, \tag{3}$$

where α_0 is a constant. This result suggests that a phenomenological description of quantum mechanical particle production processes may be possible in the $f(R, L_{\rm m})$ or $f(R,T)$ type theories. Such a semiclassical approach could lead to a better understanding of the quantum processes describing matter creation through an equivalent semi-classical description essentially involving the coupling between geometry and matter.

It is the main goal of the present study to further investigate the physical, astrophysical and cosmological implications of the effective modified gravitational theories induced by the quantum fluctuations of the space-time metric, as developed in [55–58], and further considered in [61]. Let us start the analysis by assuming that within a semiclassical approximation the quantized gravitational field can be described by a quantum metric, which can be decomposed into two terms. They are the classical, and a stochastic fluctuating component of quantum origin. Hence, the metric is obtained as the sum of these two components. As a result of this decomposition, the Einstein quantum gravity leads to an effective gravitational theory, analogous to the modified gravity models with a nonminimal coupling between geometry and matter, which have been already analyzed in [81,85,87]. Hence, it is proposed that a quantum gravitational theory can be illustrated within a semiclassical approximation.

To obtain some specific predictions from the effective gravitational theory obtained from the fluctuating quantum metric, one has to introduce the assumption that the expectation value of the quantum correction tensor $K_{\mu\nu}$ can be constructed from the metric, and from the thermodynamic quantities describing the ordinary matter content of the universe. In the present approach, the functional form of $K_{\mu\nu}$ is fixed using the Newtonian limit of the theory. By assuming that $K_{\mu\nu}$ can be represented as a linear combination of the metric, the Ricci tensor and the matter energy-momentum tensor, with the coefficients depending on the Ricci scalar R and on T, one derives first the Poisson equation in the presence of quantum fluctuations. Then, the functional form of $K_{\mu\nu}$ is determined by requiring compatibility with the Solar System observations. Hence, from the Newtonian limit, one obtains the form of $K_{\mu\nu}$ that satisfies all the Solar System constraints. Then, the cosmological implications of the obtained forms of $K_{\mu\nu}$ is investigated by considering four distinct cosmological scenarios.

The present paper is organized as follows. In Section 2, the field equations, induced by the quantum fluctuations of the metric, are obtained in general form using the variational principle. The modified Poisson equation for this modified gravity theory is also derived, and a set of constraints on the model parameters are obtained. The general cosmological implications of the modified gravity theories in the presence of quantum metric fluctuations are discussed in Section 3. Several cosmological models, obtained for different choices of the fluctuation tensor are investigated in detail in Section 4, using numerical methods. The results are discussed and concluded in Section 5. The full form of the generalized Friedmann equations obtained in the presence of a fluctuation tensor satisfying the Solar System tests are presented in Appendix A. In the present paper, a system of units with the speed of light $c = 1$ is used.

2. Quantum Metric Fluctuations Induced Gravitational Field Equations

Quantum mechanics is a very successful fundamental theory of physics, providing an excellent description of atoms, molecules, elementary particles, and classical fields, excluding gravitation. The quantum mechanical approach requires that all physical quantities must be described by operators acting in a Hilbert space. If gravity can also be described quantum mechanically, then it follows that all geometrical quantities characterizing the gravitational field must be quantized by identifying them with some suitably chosen operators. Hence, in one would like to construct a proper quantum theory of the gravitational field, Einstein's gravitational field equations must take an operator form, given by [55,56,58]:

$$\hat{R}_{\mu\nu} - \frac{1}{2}\hat{R}\hat{g}_{\mu\nu} = \frac{8\pi G}{c^4}\hat{T}_{\mu\nu}. \tag{4}$$

This formal representation corresponds to the non-perturbative quantum approach. Useful physical information should be extracted from the quantum Einstein operator equations by taking their average values over all possible products of the quantum metric operators $\hat{g}(x_1)\ldots\hat{g}(x_n)$ [55,56,58]. By introducing the Green functions, $\hat{G}_{\mu\nu}$, of the quantized gravitational field, the exact quantum approach to gravity implies to obtain the solutions of the infinite system of operator equations,

$$\begin{aligned}
\langle\Psi|\hat{g}(x_1)\hat{G}_{\mu\nu}|\Psi\rangle &= \langle\Psi|\hat{g}(x_1)\hat{T}_{\mu\nu}|\Psi\rangle, \\
\langle\Psi|\hat{g}(x_1)\hat{g}(x_2)\hat{G}_{\mu\nu}|\Psi\rangle &= \langle\Psi|\hat{g}(x_1)\hat{g}(x_2)\hat{T}_{\mu\nu}|\Psi\rangle. \\
\ldots &= \ldots,
\end{aligned}$$

In the above equations, $|\Psi\rangle$ represents the quantum state associated with the gravitational field. As this moment it is important to point out that $|\Psi\rangle$ may not necessarily represent the ordinary vacuum state of the standard quantum field theory in curved spacetimes. Unfortunately, no exact analytical solutions of the operator equations for the gravitational Green functions are known so far, and it seems that it may not be possible to obtain their solutions analytically. Therefore, the investigation of the physical implications of the quantum gravity models needs to use approximate methods [55–58]. A possible suggestion for the study of quantum gravity was presented in [55]. This approximations is based on the decomposition of the quantum metric operator, $\hat{g}_{\mu\nu}$, into the sum of two components. The first one is the average of the classical metric $g_{\mu\nu}$, while the second one corresponds to the fluctuating component, $\delta\hat{g}_{\mu\nu}$. Hence, in this approach, the quantum metric reads:

$$\hat{g}_{\mu\nu} = g_{\mu\nu} + \delta\hat{g}_{\mu\nu}. \tag{5}$$

Moreover, at this moment another approximation is introduced. Let us suppose that the average value of the fluctuating part of the metric, which is typically of a quantum nature, can be represented with the help of a classical tensor quantity $K_{\mu\nu}$, so that:

$$\langle\delta\hat{g}_{\mu\nu}\rangle = K_{\mu\nu} \neq 0. \tag{6}$$

Hence, in the present approach, the classical and quantum degrees of freedom are coupled using an expectation value. When such a coupling occurs there will be no effects of quantum fluctuations on the classical system [108]. Generally, in its Copenhagen interpretation, for a microscopic physical system quantum mechanics gives the amplitudes for many different states at a given time t. In the presence of quantum fluctuations, due to the interactions at later times, one cannot obtain a unique set of values of the physical or geometric quantities, but only a probability distribution for the different states. On the other hand, in the present semiclassical approach to quantum gravity, there is a unique solution for the metric, once the initial data for the metric and wave function are known. Hence, in this sense, quantum fluctuations do not influence the evolution of the metric. Moreover, if the metric and the initial state are homogeneous and isotropic, it follows that these symmetries are preserved by the dynamics of the gravitating system. Hence, one arrives at the important physical result that quantum fluctuations do not generate spatial variations in the energy-momentum tensor, or in the gravitational field itself [108]. Therefore, the character of the cosmological evolution is not influenced by the presence of the quantum fluctuations.

Then, by ignoring higher order fluctuations, the Lagrangian of the gravitational theories that also considers the consequences of the quantum fluctuations can be obtained as [55]:

$$\begin{aligned} L &= \sqrt{-\hat{g}}\mathcal{L}_g(\hat{g}_{\mu\nu}) + \sqrt{-\hat{g}}\mathcal{L}_m(\hat{g}_{\mu\nu}) \\ &\approx \sqrt{-g}(\mathcal{L}_g + \mathcal{L}_m) + \left[\frac{\delta(\sqrt{-g}\mathcal{L}_g)}{\delta g^{\mu\nu}} + \frac{\delta(\sqrt{-g}\mathcal{L}_m)}{\delta g^{\mu\nu}}\right]\delta\hat{g}^{\mu\nu} \\ &= \sqrt{-g}\left[\frac{1}{2\kappa^2}(R + G_{\mu\nu}\delta\hat{g}^{\mu\nu}) + \mathcal{L}_m - \frac{1}{2}T_{\mu\nu}\delta\hat{g}^{\mu\nu}\right], \end{aligned} \tag{7}$$

where $\kappa^2 = 8\pi G/c^4$, $\mathcal{L}_g(\hat{g}_{\mu\nu})$ denotes the (quantized) Lagrangian of the gravitational field, $\mathcal{L}_m(\hat{g}_{\mu\nu})$ is the matter Lagrangian, while $T_{\mu\nu}$ is the energy-momentum tensor of the classical matter, defined as:

$$T_{\mu\nu} = -\frac{2}{\sqrt{-g}}\frac{\delta(\sqrt{-g}\mathcal{L}_m)}{\delta g^{\mu\nu}}. \tag{8}$$

Therefore, in the present approach to quantum gravity, the analysis starts with the full system of the gravitational field equations in operator form. As a next step, the metric is decomposed into two terms, the first being the classical part, while the second term a stochastic fluctuating part. Thus, one obtains an effective semiclassical theory of gravity, completely described in terms of classical concepts and quantities. However, in this approach one cannot obtain the functional form of $K^{\mu\nu}$, the important quantum perturbation tensor, from the first principles. Therefore, the form of $K_{\mu\nu}$ must be chosen from physical considerations.

The first-order corrected quantum Lagrangian (7) leads to the following general gravitational field equations:

$$\begin{aligned} G_{\mu\nu} \;=\; & \kappa^2\Big(T_{\mu\nu} + \gamma^{\alpha\beta}_{\mu\nu}T_{\alpha\beta} - \frac{1}{2}g_{\mu\nu}T_{\alpha\beta}K^{\alpha\beta} \\ & + 2\frac{\delta^2\mathcal{L}_m}{\delta g^{\alpha\beta}\delta g^{\mu\nu}}K^{\alpha\beta} - \frac{1}{2}\mathcal{L}_m K g_{\mu\nu} + \frac{1}{2}KT_{\mu\nu} \\ & + \mathcal{L}_m K_{\mu\nu}\Big) - \frac{1}{2}\Big(2\gamma^{\alpha\beta}_{\mu\nu}G_{\alpha\beta} - g_{\mu\nu}G_{\alpha\beta}K^{\alpha\beta} \\ & + g_{\mu\nu}\nabla_\alpha\nabla_\beta K^{\alpha\beta} - KR_{\mu\nu} + RK_{\mu\nu}\Big) \\ & + \frac{1}{2}\left[\nabla_\alpha\nabla_{(\nu}K^\alpha_{\mu)} + \Box K g_{\mu\nu} - \Box K_{\mu\nu} - \nabla_\nu\nabla_\mu K\right], \end{aligned} \tag{9}$$

where $K = g_{\mu\nu}K^{\mu\nu}$and $A_{\alpha\beta}\delta K^{\alpha\beta} = \delta g^{\mu\nu}(\gamma^{\alpha\beta}_{\mu\nu}A_{\alpha\beta})$. Moreover, $A_{\alpha\beta}$ is either $R_{\alpha\beta}$ or $T_{\alpha\beta}$. On the other hand, $\gamma^{\alpha\beta}_{\mu\nu}$ may represent an operator, an algebraic tensor, or their combination. One can also obtain the conservation of the energy-momentum tensor:

$$\begin{aligned}(K-2)\nabla^{\mu}T_{\mu\nu} &= 2\nabla^{\mu}\left(2K_{\alpha(\nu}T^{\alpha}_{\mu)} - K_{\mu\nu}\mathcal{L}_{\mathrm{m}}\right) \\ &\quad +\nabla_{\nu}\left(K\mathcal{L}_{\mathrm{m}} - T^{\alpha\beta}K_{\alpha\beta}\right) - T_{\mu\nu}\nabla^{\mu}K \\ &\quad +\frac{1}{2\kappa^2}\Big\{2G^{\alpha\mu}\left(\nabla_{\nu}K_{\alpha\mu} - 2\nabla_{\mu}K_{\alpha\nu}\right) \\ &\quad +\nabla^{\mu}\left[\gamma^{\alpha\beta}_{\mu\nu}\left(\kappa^2T_{\alpha\beta} - G_{\alpha\beta}\right)\right]\Big\}.\end{aligned} \tag{10}$$

One can see immediately that in the case of a vanishing $K_{\mu\nu}$, the matter energy-momentum tensor becomes a conserved quantity.

The Modified Poisson Equation

The tensor $K_{\mu\nu}$ is a second order symmetric tensor, and it is responsible for the quantum corrections of the quantum metric tensor. In general, it can be proportional to any classical second rank tensor existing in general relativity. However, one may assume that it is a function of the geometric and thermodynamic quantities describing a gravitational system. Moreover, in the following it is assumed that $K_{\mu\nu}$ has a linear dependence on these quantities, and hence, a general form of the tensor $K_{\mu\nu}$ is:

$$K_{\mu\nu} = A_1\, g_{\mu\nu} + A_2\, R_{\mu\nu} + A_3\, T_{\mu\nu}, \tag{11}$$

where $A_i = A_i(R,T)$, $i = 1,2,3$, are general functions of the classical Ricci scalar R, and of the trace T of the energy-momentum tensor. We would like to emphasize that here the Latin letters do not denote the spatial coordinates. The background solution for the Minkowski space time of Equation (9) is $K_{\mu\nu} = \alpha\eta_{\mu\nu}$, where α is a constant.

To obtain the Newtonian limit of this model, one perturbs the field equations around the Minkowski space time up to first order in perturbed quantities. Then, the perturbed metric is represented in the Newtonian gauge:

$$ds^2 = -c^2\left(1+\frac{2\phi}{c^2}\right)dt^2 + \left(1-\frac{2\psi}{c^2}\right)\delta_{ij}dx^i dx^j, \tag{12}$$

where ϕ and ψ are general functions of spatial coordinates and the indices, denoted by Latin letters, stay for spatial coordinates. In the first order of perturbations the Lagrangian of the matter field and its energy-momentum tensor takes the form:

$$\mathcal{L}_{\mathrm{m}} = -c^2\rho, \quad T^{\mu}_{\nu} = \mathrm{diag}\left(-c^2\rho, 0, 0, 0\right), \tag{13}$$

where ρ is the matter energy-density.

One should note that since the Newtonian limit of the model is considered, the matter is assumed to be non-relativistic with the thermodynamic pressure $p = 0$. In the first order of approximation of Equation (11), the coefficients A_i, appearing in $K_{\mu\nu}$, contribute up to the linear order in R and T which gives:

$$A_i(R,T) = \alpha_i + \beta_i R + \gamma_i T, \quad i = 1,2,3, \tag{14}$$

where α_i, β_i and γ_i are constants. The α_1 term reproduce the Einstein-Hilbert action. The term containing β_1 is redundant because the same terms are generated by α_2 and γ_1, respectively. Hence, in the following, Equation (14) is considered with $\alpha_1 = 0 = \beta_1$.

With these assumptions, the first-order off-diagonal components of Equation (9) yield:

$$\left(\alpha_2\vec{\nabla}^2+1\right)(\phi-\psi)+\frac{c^4}{2}(\alpha_3+2\gamma_1)\rho=0. \tag{15}$$

Using the above equation, the (ii) components of Equation (9) will be satisfied in first order, while the (00) component becomes:

$$\alpha_2\vec{\nabla}^2\left(\vec{\nabla}^2\phi\right)+\frac{c^4}{4}\left(2\gamma_1+2\alpha_3-\kappa^2\alpha_2\right)\vec{\nabla}^2\rho + \vec{\nabla}^2\phi=c^4\kappa^2\rho. \tag{16}$$

The above equation differs from the Poisson equation for the gravitational potential ϕ due to the presence of the first and second terms in the left-hand side. The generalized Poisson equation provides a very powerful theoretical tool for investigating the consistency of modified gravity models. For example, in [109] the modified Poisson equations for $f(R)$-gravity were obtained in the form of the system:

$$\nabla^2\phi+\nabla^2\psi-2f''(0)\nabla^4\phi+2f''(0)\nabla^4\psi=2\mathcal{X}\rho, \tag{17}$$
$$\nabla^2\phi-\nabla^2\psi+3f''(0)\nabla^4\phi-3f''(0)\nabla^4\psi=-\mathcal{X}\rho, \tag{18}$$

where $\mathcal{X}=8\pi G/c^4$, and ϕ and ψ are the two metric potentials, as introduced in Equation (12). By a prime we have denoted the derivative with respect to the argument of the function.

After eliminating the higher-order terms, one can recover the standard Poisson equation of general relativity. As an application of the modified Poisson equation, used together with the collisionless Boltzmann equation, the Jeans stability criterion in $f(R)$ gravity was investigated in [109], by considering a small perturbation from the equilibrium and linearizing the field equations. From the performed analysis, unstable modes, not present in the standard Jeans analysis, were obtained.

To be compatible with the Solar System observations, the coefficients of these two terms should be very small. In the following, we set $\alpha_2=0$. Additionally, one should have $\gamma_1+\alpha_3\ll 1$, a condition that can be safely satisfied for $\gamma_1=-\alpha_3$.

As a result, the form of the tensor $K_{\mu\nu}$ up to the linear order is obtained:

$$K_{\mu\nu} = (\beta_2R+\gamma_2T)R_{\mu\nu}+(\gamma_1+\beta_3R+\gamma_3T)T_{\mu\nu}-\gamma_1Tg_{\mu\nu}. \tag{19}$$

In Section 3, several classes of cosmological solutions of the quantum metric fluctuations, induced modified gravity theory with the above form for the tensor $K_{\mu\nu}$, are considered.

3. Cosmological Models with Quantum Metric Fluctuations

In this Section, the cosmological implications of the extended gravity models, obtained from the effective approach to quantized gravity introduced in the previous sections, are investigated. After presenting the basic geometrical and physical assumptions, defining the basic parameters used for the characterization of the cosmological models, one considers four specific models of the universe, obtained by adopting some specific functional forms for the fluctuation tensor $K_{\mu\nu}$.

3.1. Metric and Field Equations

In the following, for the metric of the universe, we adopt the homogeneous and isotropic flat Friedmann-Lemaitre-Robertson-Walker line element [110],

$$ds^2=-dt^2+a(t)^2d\vec{x}^2, \tag{20}$$

where the scale factor $a(t)$ is a function of the cosmological time only. At this moment the Hubble function H, defined as $H(t)=\dot{a}/a$, is introduced. Moreover, we assume that the matter content of the universe consists of a perfect fluid, characterized by two thermodynamic quantities only, the energy density ρ, and the thermodynamic pressure p. Then the matter energy-momentum tensor is given by:

$$T^{\mu\nu} = (\rho + p)u^{\mu}u^{\nu} + pg^{\mu\nu}. \tag{21}$$

For the matter equation of state, let us adopt the linear barotropic equation of state $p = \omega\rho$, where ω is a constant. For the choice of $K_{\mu\nu}$, as given by Equation (19), the cosmological field Equation (9), representing the generalized Friedmann equations, are:

$$3H^2 = \kappa^2(\rho + \rho_{\rm eff}), \tag{22}$$

and

$$2\dot{H} + 3H^2 = -\kappa^2(p + p_{\rm eff}), \tag{23}$$

where $\rho_{\rm eff}$ and $p_{\rm eff}$ are given in Appendix A.

To specify the accelerating/decelerating type of the cosmological expansion, one uses the deceleration parameter q defined as:

$$q = \frac{d}{dt}\frac{1}{H} - 1 = -\frac{\dot{H}}{H^2} - 1. \tag{24}$$

Using Equations (22) and (23):

$$q = \frac{1}{2} + \frac{3}{2}\frac{p + p_{\rm eff}}{\rho + \rho_{\rm eff}} = \frac{1}{2}\left[1 + \frac{3(\omega + \omega_{\rm eff})}{1 + \Omega_{\rm eff}}\right], \tag{25}$$

where $\omega_{\rm eff} = p_{\rm eff}/\rho$ and $\Omega_{\rm eff} = \rho_{\rm eff}/\rho$. A dust universe reaches the marginally accelerating state with $q = 0$ once the condition $1 + \Omega_{\rm eff} = 3\omega_{\rm eff}$ is satisfied. The general condition for accelerating expansion can be formulated as $\omega_{\rm eff}/(1 + \Omega_{\rm eff}) < -1/3$.

To simplify the mathematical formalism, a set of dimensionless parameters $(\Omega_{\rm m}, \tau, h)$ is introduced:

$$\rho = \frac{3H^2}{\kappa^2}\Omega_{\rm m}, \quad \tau = H_0 t, \quad H = H_0 h, \tag{26}$$

where H_0 is the current value of the Hubble parameter.

To expedite the testing of the theoretical predictions of the model with the cosmological observations, one introduces, instead of the cosmological time variable t, as independent variable the redshift z, defined as:

$$1 + z = \frac{1}{a}, \tag{27}$$

where one normalizes the scale factor a by imposing the condition that its present-day value is one, $a(0) = 1$. Hence, one replaces in all cosmological evolution equations the derivatives with respect to the cosmological time t with the derivatives with respect to the redshift z, so that

$$\frac{d}{dt} = \frac{dz}{dt}\frac{d}{dz} = -(1 + z)H(z)\frac{d}{dz}. \tag{28}$$

As a function of the cosmological redshift, the deceleration parameter is obtained as:

$$q(z) = (1 + z)\frac{1}{H(z)}\frac{dH(z)}{dz} - 1. \tag{29}$$

In what follows, the cosmological evolution of the universe filled with non-relativistic matter with $\omega = 0$ for four independent choices of the fluctuation tensor $K_{\mu\nu}$ is considered.

3.2. The Standard ΛCDM Model

In this investigation, we also include a comparison of the behavior of the physical and geometric cosmological quantities obtained from the present version of the modified gravity induced by the quantum metric fluctuations with the standard ΛCDM (Cold Dark Matter) model, where Λ is the cosmological constant.

The recent results of the study of the cosmic microwave background radiation by the Planck satellite has provided high precision cosmological data [62,111–117]. In the following, the simplifying assumption that the present day universe consists mostly of dust matter, with negligible pressure, is adopted. Hence, the energy conservation equation, $\dot{\rho}+3H\rho=0$, of standard general relativity gives for the variation of the matter energy density the expression, $\rho=\rho_0/a^3=\rho_0(1+z)^3$, where $\rho_0=\rho(0)$ is the present-day matter density. As a function of the scale factor the time variation of the Hubble function is obtained in the form [115]:

$$H=H_0\sqrt{(\Omega_b+\Omega_{\rm DM})a^{-3}+\Omega_\Lambda}, \tag{30}$$

where Ω_b, $\Omega_{\rm DM}$, and Ω_Λ denote the density parameters of the baryonic matter, the cold (pressureless) dark matter, and the dark energy (described by a cosmological constant), respectively. The three density parameters satisfy the important relation $\Omega_b+\Omega_{\rm DM}+\Omega_\Lambda=1$, indicating that the geometry of the universe is flat.

As a function of the redshift the dimensionless form of the Hubble function $H(z)=H_0h(z)$ is obtained as:

$$h(z)=\sqrt{(\Omega_{\rm DM}+\Omega_b)(1+z)^3+\Omega_\Lambda}. \tag{31}$$

For the deceleration parameter, as a function of the redshift, then one finds

$$q(z)=\frac{3(1+z)^3(\Omega_{\rm DM}+\Omega_b)}{2[\Omega_\Lambda+(1+z)^3(\Omega_{\rm DM}+\Omega_b)]}-1. \tag{32}$$

In this study, for the density parameters, the numerical values $\Omega_{\rm DM}=0.2589$, $\Omega_b=0.0486$, and $\Omega_\Lambda=0.6911$ [115] are are adopted, as obtained from the Planck data. For the total matter density parameter, $\Omega_{\rm m}=\Omega_{\rm DM}+\Omega_b$, the value $\Omega_{\rm m}=0.3089$ is taken. The present day value of the deceleration parameter is given by $q(0)=-0.5381$, corresponding to an accelerating expansion of the universe [115,116]. The dependence of the dimensionless matter density on the redshift is given, in the standard ΛCDM cosmological model, by $r(z)=\Omega_{\rm m}(1+z)^3=0.3089(1+z)^3$ [115,116].

4. Specific Cosmological Models

Here, a few particular cosmological models are investigated, in which for the fluctuation tensor some particular forms are adopted, which follow from the general representation (19). These specific forms of $K_{\mu\nu}$ are obtained by fixing the values of the arbitrary coefficients γ_i, and β_i, $i=2,3$.

4.1. $K_{\mu\nu}=(\beta_2\,R+\gamma_2\,T)R_{\mu\nu}$

As a first example of a cosmological model in the modified gravity theory induced by the metric fluctuations, one considers that the tensor $K_{\mu\nu}$ is proportional to the Ricci tensor:

$$K_{\mu\nu}=(\beta_2\,R+\gamma_2\,T)R_{\mu\nu}. \tag{33}$$

This form of $K_{\mu\nu}$ is obtained by taking $\gamma_1 = \gamma_3 = \beta_3 = 0$ in Equation (19). To obtain a dimensionless form of the cosmological evolution equations, the dimensionless parameters b_2 and g_2 are introduced:

$$b_2 = \beta_2 H_0^4, \qquad g_2 = \frac{H_0^4 \gamma_2}{\kappa^2}. \tag{34}$$

In terms of redshift z, the field equations (A1) and (A2) in this case take the form:

$$\begin{aligned}
1-\Omega_{\rm m} &= 9g_2h^4\Omega_{\rm m}^2 - 9b_2h^4(3\Omega_{\rm m}+4) \\
&+ (z+1)h^3\left(3b_2(30h'\Omega_{\rm m} - 64h' + 9h\Omega_{\rm m}') \right.\\
&\left.+ 3g_2\left(-6h'\Omega_{\rm m}^2 + \Omega_{\rm m}(16h' - 3h\Omega_{\rm m}') + 6h\Omega_{\rm m}'\right)\right) \\
&+ (z+1)^2\Big(- 3b_2h^2(2hh''(3\Omega_{\rm m} - 16) + 3h'^2(5\Omega_{\rm m} - 18) \\
&+ 6hh'\Omega_{\rm m}') - 3g_2h^2\left(2hh'\Omega_{\rm m}' + \Omega_{\rm m}\left(2hh'' + 5h'^2\right)\right)\Big) \\
&- 12b_2h(z+1)^3h'\left(3hh'' + 2(h')^2\right),
\end{aligned} \tag{35}$$

$$\begin{aligned}
1-\frac{2(z+1)h'}{3h} &= 9(\Omega_{\rm m} - 4)(b_2 - g_2\Omega_{\rm m}) \\
&+ (z+1)\left(b_2h^3(60h'\Omega_{\rm m} - 56h' + 21h\Omega_{\rm m}')\right. \\
&\left.+ g_2h^3(3h(2-3\Omega_{\rm m})\Omega_{\rm m}' - h'\Omega_{\rm m}(9\Omega_{\rm m}+16))\right) \\
&+ (z+1)^2\Big(g_2h^2(3h\left(h\Omega'^2 - 2h\Omega_{\rm m}'' - 16h'\Omega_{\rm m}'\right) \\
&+ 6\Omega_{\rm m}^2\left(hh'' + 4h'^2\right) + \Omega_{\rm m}(3h(h\Omega_{\rm m}'' - 6h'' + 9h'\Omega_{\rm m}') \\
&- 67h'^2)) - b_2h^2(3h(4h''(3\Omega_{\rm m} - 8) + 3h\Omega_{\rm m}'') \\
&+ h'^2(135\Omega_{\rm m} - 374) + 81hh'\Omega_{\rm m}')\Big) + (z+1)^3 \\
&\times \Big(2b_2h(27h'^3(\Omega_{\rm m} - 4) + 21hh'^2\Omega_{\rm m}' + h^2(h'''(3\Omega_{\rm m} - 16) \\
&+6h''\Omega_{\rm m}') + hh'(2h''(15\Omega_{\rm m} - 67) + 3h\Omega_{\rm m}'')) + 2g_2h\Big(h(hh'\Omega_{\rm m}'' \\
&+ \left(2hh'' + 7h'^2\right)\Omega_{\rm m}') + \Omega_{\rm m}(9h'^3 + h^2h''' + 10hh'h'')\Big)\Big) \\
&+ 12b_2(z+1)^4\left(h^2h''^2 + 2h'^4 + 6hh'^2h'' + h^2h'h'''\right),
\end{aligned} \tag{36}$$

where a prime denotes the derivative with respect to the independent redshift variable z.

The non-conservation equation of the energy-momentum tensor (see Equation (10)) can be written as:

$$\begin{aligned}
(z+1)(4h'\Omega_{\rm m} + 2h\Omega_{\rm m}') - 6h\Omega_{\rm m} &= -216b_2h^5\Omega_{\rm m} \\
&+ 108g_2h^5(\Omega_{\rm m} - 2)\Omega_{\rm m} + (z+1)(36b_2h^4(9h'\Omega_{\rm m} + h\Omega_{\rm m}') \\
&+ 18g_2h^4(h'(16 - 7\Omega_{\rm m})\Omega_{\rm m} - h(\Omega_{\rm m} - 2)\Omega_{\rm m}')) \\
&+ (z+1)^2(18g_2h^3h'(2h'(\Omega_{\rm m} - 3)\Omega_{\rm m} + h(\Omega_{\rm m} - 2)\Omega_{\rm m}') \\
&- 18b_2h^3h'(10h'\Omega_{\rm m} + 3h\Omega_{\rm m}')) + (z+1)^3(18b_2h^2(h')^2 \\
&\times (2h'\Omega_{\rm m} + h\Omega_{\rm m}') + 6g_2h^2(h')^2(2h'\Omega_{\rm m} + h\Omega_{\rm m}')).
\end{aligned} \tag{37}$$

Figures 1–3 show the cosmological parameters $h(z)$, $\Omega_m(z)$ and $q(z)$, obtained by numerically solving the generalized Friedmann equations, for different values of b_2 and g_2. One can see that there are ascending and descending curves for the density parameter, depending on the values of the pair of the parameters (b_2, g_2). Additionally, an accelerated phase is present at late times, while, for earlier times, a decelerating phase is found.

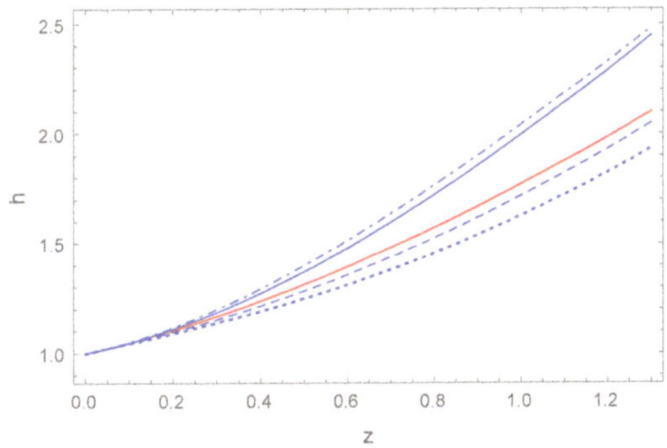

Figure 1. Variation of the dimensionless Hubble function, h, as a function of the redshift z for the tensor $K_{\mu\nu} = (\beta_2 R + \gamma_2 T)R_{\mu\nu}$ with different values of the parameters b_2 and g_2: $b_2 = 0.044$ and $g_2 = -0.01$ (solid curve), $b_2 = -0.0076$ and $g_2 = 0.001$ (dotted curve), $b_2 = -0.013$ and $g_2 = -0.0007$ (dashed curve), and $b_2 = 0.013$ and $g_2 = 0.03$ (dot-dashed curve). The evolution of the Hubble function in the standard ΛCDM model is represented by the red solid curve.

As shown in Figure 1, the Hubble function is an increasing function of the redshift z (and a decreasing function of time), indicating an expansionary evolution of the universe. The evolution of h depends strongly on the model parameters b_2 and g_2. For low redshifts $z \leq 0.3$, the evolution is basically independent of b_2 and g_2. The model can reproduce well the evolution of h in the standard ΛCDM model. The matter density, displayed in Figure 2, shows significant differences with respect to standard cosmology. For the chosen set of parameters, two different behaviors can be observed.

The matter energy density is either increasing or decreasing function of z (decreasing or increasing function of time). The latter case, implying a matter density that increases in time, may be considered unphysical. A scenario with an almost constant density is also possible. For the adopted set of parameters, the evolution of the matter density is significantly different from the evolution of the matter density in the ΛCDM model.

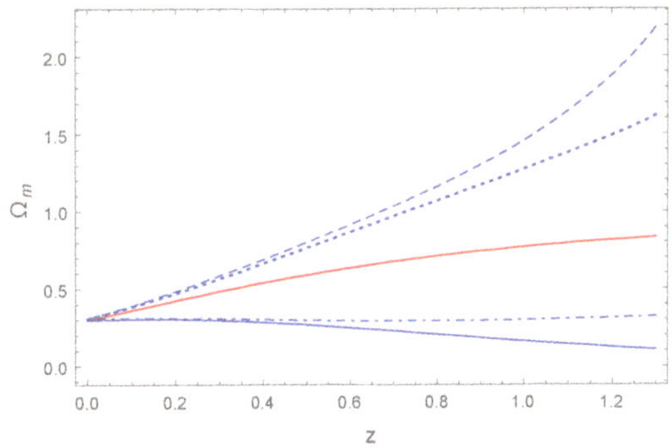

Figure 2. Variation of the dimensionless matter density, Ω_m, as a function of z for $K_{\mu\nu} = (\beta_2 R + \gamma_2 T)R_{\mu\nu}$ with different values of the parameters b_2 and g_2: $b_2 = 0.044$ and $g_2 = -0.01$ (solid curve), $b_2 = -0.0076$ and $g_2 = 0.001$ (dotted curve), $b_2 = -0.013$ and $g_2 = -0.0007$ (dashed curve), and $b_2 = 0.013$ and $g_2 = 0.03$ (dot-dashed curve). The evolution of the matter density in the standard ΛCDM model is shown by the red solid curve.

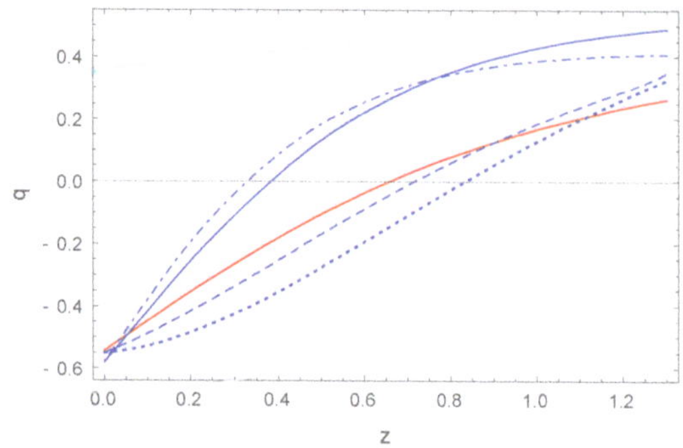

Figure 3. Variation of the deceleration parameter, q, as a function of z for $K_{\mu\nu} = (\beta_2 R + \gamma_2 T)R_{\mu\nu}$ with different values of the parameters b_2 and g_2: $b_2 = 0.044$ and $g_2 = -0.01$ (solid curve), $b_2 = -0.0076$ and $g_2 = 0.001$ (dotted curve), $b_2 = -0.013$ and $g_2 = -0.0007$ (dashed curve), and $b_2 = 0.013$ and $g_2 = 0.03$ (dot-dashed curve) The evolution of the deceleration parameter in the standard ΛCDM model is depicted by the red solid curve.

The deceleration parameter, q, shown in Figure 3, indicates the existence of a transition from the decelerating to an accelerating expansion. The evolution of q is strongly dependent on the model parameters, and a wide variety of cosmological behaviors can be constructed. The evolution of q in standard cosmology can also be reproduced.

4.2. $K_{\mu\nu} = (\beta_3 R + \gamma_3 T)T_{\mu\nu}$

In this Subsection, the expression of the fluctuation tensor $K_{\mu\nu}$, containing only the energy-momentum tensor, is considered. Hence, the γ_1 term is ignored, since it corresponds to a term proportional to the metric tensor. Therefore, the tensor $K_{\mu\nu}$ is given in this case by:

$$K_{\mu\nu} = (\beta_3 R + \gamma_3 T)T_{\mu\nu}. \tag{38}$$

The dimensionless parameters b_3 and g_3 are defined as:

$$b_3 = \frac{H_0^4}{\kappa^2}\beta_3, \qquad g_3 = \frac{H_0^4}{\kappa^4}\gamma_3. \tag{39}$$

Then, the field Equations (A1) and (A2) read for this case:

$$\begin{aligned} 1 - \Omega_{\rm m} = {} & \frac{9}{2}h^4\Omega_{\rm m}(2b_3(\Omega_{\rm m}+3) - g_3\Omega_{\rm m}(\Omega_{\rm m}+1)) \\ & - 9b_3h^3(z+1)(2\Omega_{\rm m}-1)(2h'\Omega_{\rm m} + h\Omega'_{\rm m}), \end{aligned} \tag{40}$$

$$\begin{aligned} 1 - \frac{2(z+1)h'}{3h} = {} & \frac{9}{2}h^4\Omega_{\rm m}(14b_3 - 6\Omega_{\rm m}b_3 + 3g_3\Omega_{\rm m}^2 \\ & - 5g_3\Omega_{\rm m}) + (z+1)h^3\Big(6g_3\Omega_{\rm m}(3h'\Omega_{\rm m} + h\Omega'_{\rm m}) \\ & - b_3(3h(2\Omega_{\rm m}+3)\Omega'_{\rm m} + 84h'\Omega_{\rm m})\Big) + 3b_3h^2(z+1)^2 \\ & \times \Big(h\Big(2h\Omega'^2_{\rm m} - 7h'\Omega'_{\rm m} - h\Omega''_{\rm m}\Big) + 4\Omega^2_{\rm m}\Big(hh'' + 4h'^2\Big) \\ & + 2\Omega_{\rm m}(h^2\Omega''_{\rm m} - hh'' + 9hh'\Omega'_{\rm m} - 3h'^2)\Big). \end{aligned} \tag{41}$$

The non-zero component of Equation (10) is:

$$\begin{aligned}&(z+1)\left(4h'\Omega_{\rm m}+2h\Omega_{\rm m}'\right)-6h\Omega_{\rm m}=108h^5(1-\Omega_{\rm m})\Omega_{\rm m}\\&\times(g_3\Omega_{\rm m}-2b_3)+(z+1)\left(36b_3h^4(h(1-2\Omega_{\rm m})\Omega_{\rm m}'\right.\\&-7h'(\Omega_{\rm m}-1)\Omega_{\rm m})+9g_3h^4\Omega_{\rm m}(6h'(\Omega_{\rm m}-1)\Omega_{\rm m}\\&\left.+h(3\Omega_{\rm m}-2)\Omega_{\rm m}')\right)-18b_3h^3(z+1)^2h'\\&\times\left(h(1-2\Omega_{\rm m})\Omega_{\rm m}'-4h'(\Omega_{\rm m}-1)\Omega_{\rm m}\right)\end{aligned} \tag{42}$$

In Figures 4–6 the behaviors of the Hubble function, the matter density and the deceleration parameters are shown for different values of the parameters b_3 and g_3.

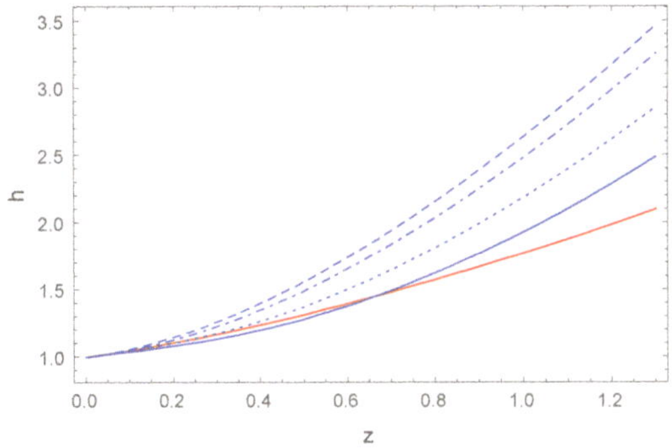

Figure 4. Variation of the dimensionless Hubble function h as a function of z for $K_{\mu\nu} = (\beta_3 R + \gamma_3 T)T_{\mu\nu}$ with different values of the parameters b_3 and g_3: $b_3 = 0.073$ and $g_3 = 0.1$ (solid curve), $b_3 = 0.08$ and $g_3 = 0.097$ (dotted curve), $b_3 = 0.15$ and $g_3 = 0.30$ (dashed curve), and $b_3 = 0.21$ and $g_3 = 0.12$ (dot-dashed curve). The evolution of the Hubble function in the standard ΛCDM model is described by the red solid curve.

The Hubble function, $h(z)$, represented in Figures 4, is an increasing function of z, and it can reproduce the evolution of the Hubble function in the standard ΛCDM model.

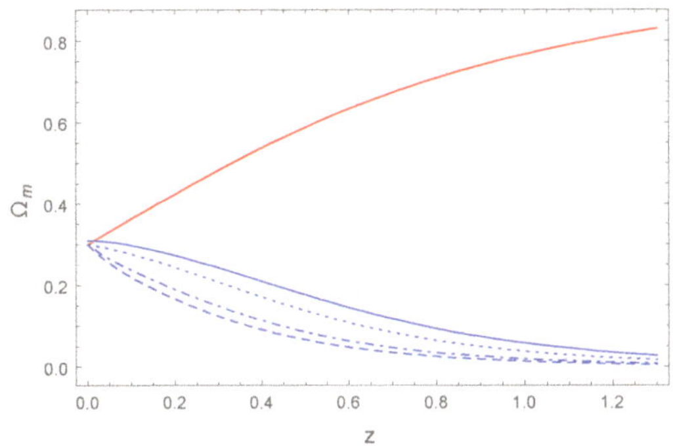

Figure 5. Variation of the dimensionless matter density, $\Omega_{\rm m}$, as a function of z for $K_{\mu\nu} = (\beta_3 R + \gamma_3 T)T_{\mu\nu}$ with different values of the parameters b_3 and g_3: $b_3 = 0.073$ and $g_3 = 0.1$ (solid curve), $b_3 = 0.08$ and $g_3 = 0.097$ (dotted curve), $b_3 = 0.15$ and $g_3 = 0.30$ (dashed curve), and $b_3 = 0.21$ and $g_3 = 0.12$ (dot-dashed curve). The evolution of the matter density in the standard ΛCDM model corresponds to the red solid curve.

However, significant differences do appear in the behavior of the matter density, as shown in Figure 5. As opposed to the matter behavior in standard cosmology, in this model, the matter energy density is an increasing function of time. Moreover, its the evolution is significantly different compared to the matter density evolution in the standard ΛCDM model for all adopted values of the model parameters.

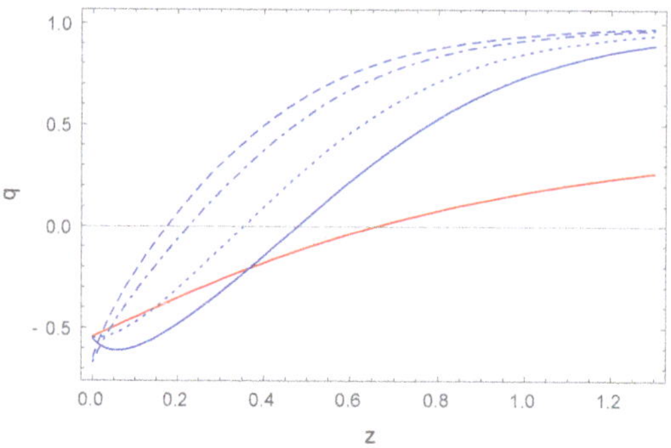

Figure 6. Variation of the deceleration parameter, q, as a function of z for $K_{\mu\nu} = (\beta_3\, R + \gamma_3\, T) T_{\mu\nu}$ with different values of the parameters b_3 and g_3: $b_3 = 0.073$ and $g_3 = 0.1$ (solid curve), $b_3 = 0.08$ and $g_3 = 0.097$ (dotted curve), $b_3 = 0.15$ and $g_3 = 0.30$ (dashed curve), and $b_3 = 0.21$ and $g_3 = 0.12$ (dot-dashed curve). The evolution of the deceleration parameter in the standard ΛCDM model is described by the red solid curve.

However, the deceleration parameter q, shown in Figure 6, indicates a transition from a decelerating to an accelerating phase, which can reproduce the present day value of q.

4.3. $K_{\mu\nu} = (\beta_2 R_{\mu\nu} + \beta_3 T_{\mu\nu}) R$

Now, let us consider the case where $\gamma_i = 0$, $i = 1,2,3$. Hence, the expression of $K_{\mu\nu}$,

$$K_{\mu\nu} = R(\beta_2 R_{\mu\nu} + \beta_3 T_{\mu\nu}), \tag{43}$$

is obtained.

In this case, the field equations are:

$$\begin{aligned}
1 - \Omega_{\rm m} = {} & 9h^4(b_3\Omega_{\rm m}(\Omega_{\rm m}+3) - b_2(3\Omega_{\rm m}+4)) \\
& + 3h^3(z+1)(b_2(30h'\Omega_{\rm m} - 64h' + 9h\Omega'_{\rm m}) \\
& - 3b_3(2\Omega_{\rm m}-1)(2h'\Omega_{\rm m} + h\Omega'_{\rm m})) - 3b_2h^2(z+1)^2 \\
& \times \left(2hh''(3\Omega_{\rm m}-16) + 3h'^2(5\Omega_{\rm m}-18) + 6hh'\Omega'_{\rm m}\right) \\
& - 12\, b_2\, h\, h'(z+1)^3\left(3hh'' + 2h'^2\right),
\end{aligned} \tag{44}$$

and

$$
\begin{aligned}
1-\frac{2(z+1)h'}{3h} &= 9h^4(b_2(\Omega_{\rm m}-4)+b_3(7-3\Omega_{\rm m})\Omega_{\rm m})\\
&+h^3(z+1)(4h'(15b_2\Omega_{\rm m}-14\,b_2-21b_3\Omega_{\rm m})\\
&+3h(7b_2-2b_3\Omega_{\rm m}-3b_3)\Omega'_{\rm m})-h^2(z+1)^2\\
&\times\left(b_2(12hh''(3\Omega_{\rm m}-8)+9h^2\Omega''_{\rm m}+h'^2(135\Omega_{\rm m}-374)\right.\\
&+81hh'\Omega'_{\rm m})+3b_3(7h\,h'\Omega'_{\rm m}+h^2\left(\Omega''_{\rm m}-2\Omega'^2_{\rm m}\right)\\
&-4\Omega^2_{\rm m}(hh''+4h'^2)+2\Omega_{\rm m}(hh''-h^2\Omega''_{\rm m}-9hh'\Omega'_{\rm m}\\
&+3h'^2)))+2\,b_2\,h(z+1)^3(27h'^3(\Omega_{\rm m}-4)+21hh'^2\Omega'_{\rm m}\\
&+h^2(h'''(3\Omega_{\rm m}-16)+6h''\Omega'_{\rm m})+hh'(2h''(15\Omega_{\rm m}-67)\\
&+3h\Omega''_{\rm m}))+12\,b_2\,(z+1)^4(h^2h''^2+2h'^4+6hh'^2h''\\
&+h^2h'h'''),
\end{aligned} \tag{45}
$$

and the temporal component of Equation (10) in terms of redshift is:

$$
\begin{aligned}
(z+1)\left(4h'\Omega_{\rm m}+2h\Omega'_{\rm m}\right)-6h\Omega_{\rm m} &= 216h^5\Omega_{\rm m}\\
&\times(b_3(\Omega_{\rm m}-1)-b_2)+36h^4(z+1)\left(h'\Omega_{\rm m}(9b_2\right.\\
&-7b_3(\Omega_{\rm m}-1))+h(b_3(1-2\Omega_{\rm m})+b_2)\Omega'_{\rm m})\\
&+18h^3(z+1)^2h'(2h'\Omega_{\rm m}(2b_3(\Omega_{\rm m}-1)-5b_2)\\
&-h(b_3(1-2\Omega_{\rm m})+3b_2)\Omega'_{\rm m})+18\,b_2\,h^2\,h'^2(z+1)^3\\
&\times\left(2h'\Omega_{\rm m}+h\Omega'_{\rm m}\right).
\end{aligned} \tag{46}
$$

The evolution in terms of redshift z of the cosmological parameters h, $\Omega_{\rm m}$ and q are shown in Figures 7–9 for different values of the parameters b_2 and b_3. In this case, as well as in all other cases the parameters were chosen in such a way to obtain the closest possible approximation of the ΛCDM model.

The Hubble function, h, presented in Figure 7, is an increasing function of z and reproduces well the standard ΛCDM model. For low redshifts, the behavior of h is basically independent on the model parameters.

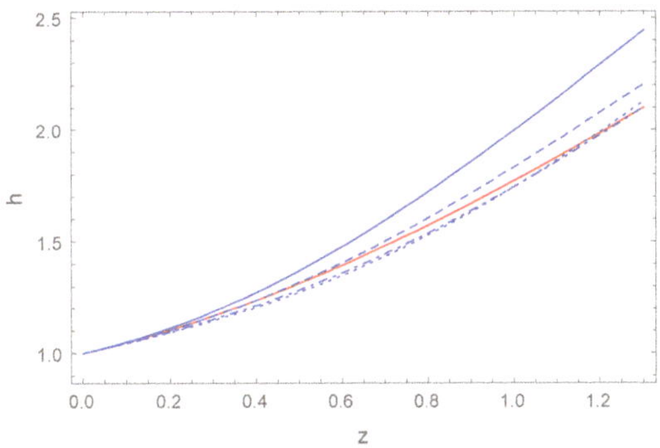

Figure 7. Variation of the Hubble function h as a function of z for $K_{\mu\nu}=R(\beta_2R_{\mu\nu}+\beta_3T_{\mu\nu})$ with different values of the parameters b_2 and b_3: $b_2=-.007$ and $b_3=0.011$ (solid curve), $b_2=-0.08$ and $b_3=0.098$ (dotted curve), $b_2=-0.02$ and $b_3=0.009$ (dashed curve), and $b_2=-0.01$ and $b_3=0.006$ (dot-dashed curve). The evolution of the Hubble function in the standard ΛCDM model is described by the red solid curve.

The ordinary matter density, Ω_m, shown in Figure 8, is found to be a monotonically decreasing function of time, whose evolution at high redshifts is strongly dependent on the model parameters. For a particular set of values of b_2 and b_3, the variation of Ω_m in the standard cosmological model is reproduced almost exactly.

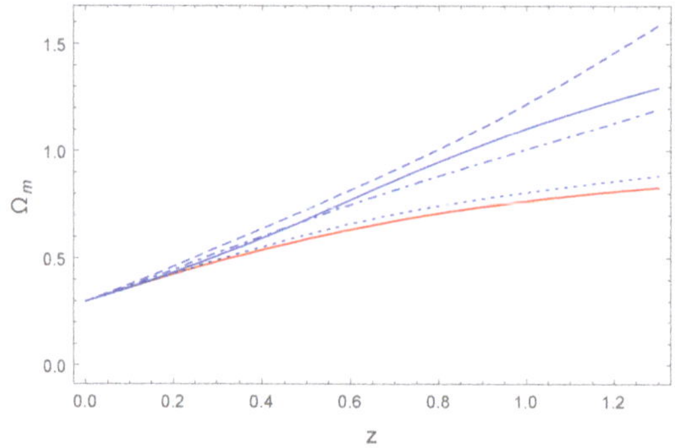

Figure 8. Variation of the matter density Ω_m as a function of z for $K_{\mu\nu} = R(\beta_2 R_{\mu\nu} + \beta_3 T_{\mu\nu})$ with different values of the parameters b_2 and b_3: $b_2 = -.007$ and $b_3 = 0.011$ (solid curve), $b_2 = -0.08$ and $b_3 = 0.098$ (dotted curve), $b_2 = -0.02$ and $b_3 = 0.009$ (dashed curve), and $b_2 = -0.01$ and $b_3 = 0.006$ (dot-dashed curve). The evolution of the matter density in the standard ΛCDM model is described by the red solid curve.

The deceleration parameter, q, represented in Figures 9, shows a transition from a decelerating to an accelerating cosmological phase. The evolution of the deceleration parameter depends strongly on the adopted numerical values of the model parameters, b_2 and b_3. For a range of these parameters, the present model can reproduce quite well the behavior of q at low redshifts.

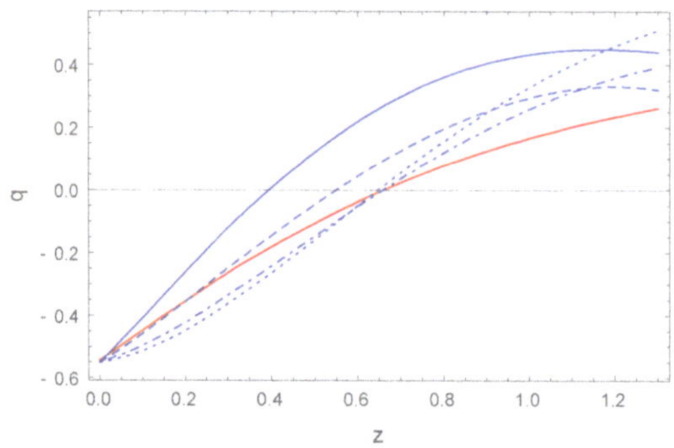

Figure 9. Variation of the deceleration parameter, q, as a function of z for $K_{\mu\nu} = R(\beta_2 R_{\mu\nu} + \beta_3 T_{\mu\nu})$ with different values of the parameters b_2 and b_3: $b_2 = -.007$ and $b_3 = 0.011$ (solid curve), $b_2 = -0.08$ and $b_3 = 0.098$ (dotted curve), $b_2 = -0.02$ and $b_3 = 0.009$ (dashed curve), and $b_2 = -0.01$ and $b_3 = 0.006$ (dot-dashed curve). The evolution of the deceleration parameter in the standard ΛCDM model is represented by the red solid curve.

4.4. $K_{\mu\nu} = -\gamma_1(T g_{\mu\nu} - T_{\mu\nu})$

Finally, let us consider the case when only the coefficient γ_1 is kept in Equation (19), with all the other coefficients set to zero. Hence,

$$K_{\mu\nu} = -\gamma_1(T g_{\mu\nu} - T_{\mu\nu}) \tag{47}$$

is obtained.

Let us also introduce the dimensionless parameter g_1 defined as:

$$g_1 = \frac{H_0^2}{\kappa^2}\gamma_1. \tag{48}$$

With the use of Equations (A1) and (A2), one can obtain the Friedmann and Raychaudhuri equations in terms of redshift for this case:

$$1-\Omega_{\rm m} = \frac{3}{2}g_1 h\left(\Omega_{\rm m}(3h-4(z+1)h') - 2h(z+1)\Omega_{\rm m}'\right), \tag{49}$$

$$\begin{aligned} 1-\frac{2(z+1)h'}{3h} &= g_1\left(\frac{3}{2}h^2\Omega_{\rm m}(2\Omega_{\rm m}-3)\right. \\ &+ (z+1)^2\left(5hh'\Omega_{\rm m}' + h^2\Omega_{\rm m}'' + 2\left(hh''+2h'^2\right)\Omega_{\rm m}\right) \\ &\left. - 2(z+1)\left(hh'\Omega_{\rm m} + h^2\Omega_{\rm m}'\right)\right). \end{aligned} \tag{50}$$

The divergence of the energy-momentum tensor (10) takes the form:

$$\begin{aligned} (z+1)(4h'\Omega_{\rm m} + 2h\Omega_{\rm m}') - 6h\Omega_{\rm m} &= 18g_1h^3(3-\Omega_{\rm m})\Omega_{\rm m} \\ &+ 6g_1(z+1)^2\left(2hh'^2\Omega_{\rm m} + h^2h'\Omega_{\rm m}'\right) \\ &+ g_1(z+1)\left(-9h^3\Omega_{\rm m}' - 48h^2h'\Omega_{\rm m}\right). \end{aligned} \tag{51}$$

The evolutions in terms of redshift z of the Hubble parameter, h, the density parameter, $\Omega_{\rm m}$, and the deceleration parameter, q, are shown in Figures 10–12. The curves are obtained for different values of the parameter $g_1 = -10, -1, 5, 15$.

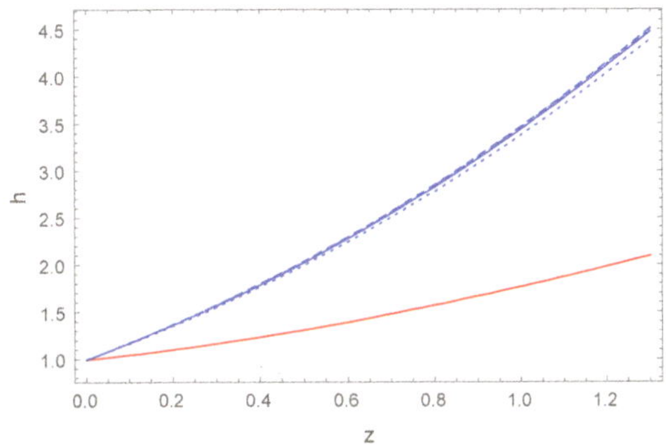

Figure 10. Variation of the Hubble function, h, as a function of z for $K_{\mu\nu} = -\gamma_1(Tg_{\mu\nu} - T_{\mu\nu})$ with different values of the parameter g_1: $g_1 = -10$ (solid curve), $g_1 = -1$ (dotted curve), $g_1 = 5$ (dashed curve), and $g_1 = 15$ (dot-dashed curve). The evolution of the Hubble function in the standard ΛCDM model is represented by the red solid curve.

For this choice of $K_{\mu\nu}$, no accelerating expansion at late times is found. The evolution of the universe is still expansionary, as shown by the evolution of the Hubble function in Figure 10. The matter density, shown in Figure 11, is a decreasing function of the redshift, in contrast with the ΛCDM model, indicating an increase of the matter density in time. This effect is due to the non-conservation of the energy-momentum tensor of Equation (51). As one can see from Figure 12, the deceleration parameter is positive, and is roughly constant in the considered range of z. The variation of the numerical values of the parameter γ_1 has a little effect on the behavior of the cosmological parameters $\Omega_{\rm m}$, q and h.

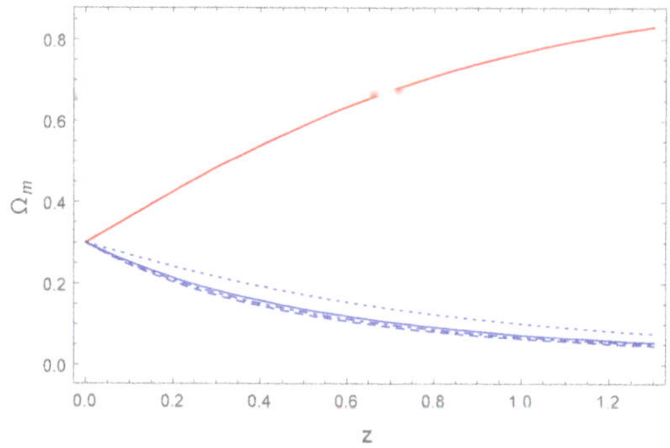

Figure 11. Variation of the matter density, Ω_m, as a function of z for $K_{\mu\nu} = -\gamma_1 (T g_{\mu\nu} - T_{\mu\nu})$ with different values of the parameter g_1: $g_1 = -10$ (solid curve), $g_1 = -1$ (dotted curve), $g_1 = 5$ (dashed curve), and $g_1 = 15$ (dot-dashed curve). The evolution of the matter density in the standard ΛCDM model is represented by the red solid curve.

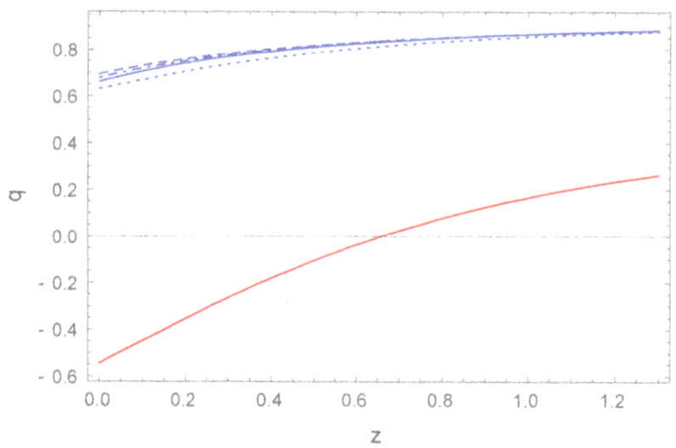

Figure 12. Variation of the deceleration parameter, q, as a function of z for $K_{\mu\nu} = -\gamma_1 (T g_{\mu\nu} - T_{\mu\nu})$ with different values of the parameter g_1: $g_1 = -10$ (solid curve), $g_1 = -1$ (dotted curve), $g_1 = 5$ (dashed curve), and $g_1 = 15$ (dot-dashed curve). The evolution of the deceleration parameter in the standard ΛCDM model is represented by the red solid curve.

5. Discussions and Final Remarks

The search for quantum gravity is one of the major topics of interest in nowadays theoretical physics. However, despite the intensive effort invested in this direction, the quantum properties of gravity are still elusive, and we presently lack a theory fully unifying the two basic branches of physics. Hence, in order to have at least a basic understanding of the quantum properties of gravity, we need to resort to some mathematical approximations, or to some qualitative approaches. One of such semiclassical directions of research was proposed in [55,57], and it is based on the idea of the decomposition of the quantum metric into two components, one being the classical metric tensor, while the second one is a fluctuating tensor, of quantum origin. By adopting another semiclassical approximation, one can substitute the quantum fluctuating part by the average value of the tensor $K_{\mu\nu}$, representing an effective classical term to be added to the standard metric of general relativity.

A first interesting consequence of this approach is that it leads, in the first order of approximation and on a classical level, to several classes of modified gravity theories, with geometry-matter coupling. For example, by assuming that$K_{\mu\nu} \propto g_{\mu\nu}$, a particular class of the modified $f(R, T)$ gravity theory [98] as a function of the Ricci scalar, R, and the trace of the energy-momentum tensor T, with geometry-matter coupling is obtained. These extensions of standard general relativity have been intensively studied in their different versions [98], but the possible relation with effective semiclassical theories of gravity has not been pointed out. Modified gravity theories with quantum metric fluctuations, even formulated in a semiclassical and effective form may give some insights into the quantum nature of gravity, and its manifestations at the level of the classical world. An important property of all modified theories with geometry-matter coupling is the non-conservation of the matter energy-momentum tensor. This is also a basic property of quantum field theories in curved space-times. Hence, the particle creation processes present in modified gravity theories with geometry-matter coupling may point towards a possible relation between these classes of theories and quantum effects in gravity.

However, in the general formulation of the modified gravity theories in the presence of quantum metric fluctuations, the mathematical form of the fluctuation tensor $K_{\mu\nu}$ is arbitrary. In the present paper, the functional form of $K_{\mu\nu}$ is fixed by considering the Newtonian limit of the theory. This leads to the derivation of the generalized Poisson equation, which contains several correction terms, with respect to its standard form. By requiring that the model passes the standard tests of gravity at the level of the Solar System, we can fix the form of the fluctuation tensor as given by Equation (19). Generally, $K_{\mu\nu}$ can be obtained as a linear combination of the Ricci tensor, the energy-momentum tensor, and the metric tensor, with the coefficients functions of the Ricci scalar R and of the trace of the energy-momentum tensor T.

As a next step, in this investigation, the quantum corrected classical Lagrangian is obtained along with the general effective Einstein equations, corresponding to an arbitrary $K_{\mu\nu}$ [61]. Then, using the general form of $K_{\mu\nu}$, a few classes of cosmological models can be constructed that are consistent with the Solar System tests. More exactly, four classes of models are studied. In the first two models, $K_{\mu\nu}$ is proportional to the Ricci tensor, and the matter energy-momentum tensor, respectively, with the proportionality coefficients given by linear combinations of R and T. In the third model, we have assumed that $K_{\mu\nu}$ can be obtained as a linear combination of the Ricci tensor and the energy-momentum tensor. In the fourth model, $K_{\mu\nu}$ is determined by the energy-momentum tensor and its trace only, i.e., by the properties of the matter filling the universe.

A detailed investigation of the cosmological properties of the models, obtained for these specific functional forms of $K_{\mu\nu}$, was performed using numerical methods. Exact solutions of the field equations seem to be impossible to be found, due to the extreme mathematical complexity of the generalized Friedmann equations. The behavior of the Hubble function, of the matter density and of the deceleration parameter were investigated, and for each case, the model predictions are compared with the cosmological results, obtained in the framework of the standard ΛCDM model. Overall, the cosmological evolution strongly depends on the choice of the function $K_{\mu\nu}$, and of the parameters of the specific models. Generally, the models can reproduce the predictions of the standard ΛCDM model, and, thus, describe both decelerating and accelerating phases. To test these models, a detailed comparison with the observational data is necessary. In the present approach, a phenomenological approach is used, by adopting, for the quantum perturbation tensor, some specific functional forms that passes the Solar System tests. However, even this strong criterion cannot uniquely fix the form of $K_{\mu\nu}$, and, thus, for different choices of the quantum perturbation tensor drastically different astrophysical and cosmological behaviors may emerge.

In most of the considered models, in the large time (small redshift) limit, the universe enters into an accelerating phase. The present day value of the deceleration parameter $q \approx -0.53$ can be obtained for large range of parameter values for the first three considered forms of $K_{\mu\nu}$. By varying the model parameters, a wide variety of cosmological evolutions can be constructed, with some of them reproducing almost exactly the results of the standard ΛCDM model. However, other models do show significant deviations from it. If the variations of the Hubble function and of the deceleration parameter are, for the first three models, qualitatively consistent with observations, some significant differences do appear in the evolution of the matter density. In the first, the second and the fourth here considered cosmological models, for some specific values of parameters the matter energy-density is increasing in time. This unusual behavior is a direct consequence of the non-conservation of the energy-momentum tensor, which can also be interpreted as related to the creation of ordinary matter by the gravitational field. Such processes may play an important role in the early stages of the evolution of the universe, as possible alternatives to the reheating phase of the post-inflationary era. We would also like to point out is that purely decelerating cosmological evolution over a large range of redshifts can also be obtained.

The ultimate challenge present day theoretical physics faces is the problem of the quantization of the gravitational field. Unfortunately no exact solutions for this problem are known. Hence to answer the question of the existence of quantum effects in gravity one should resort to approximate, semiclassical methods. A promising way in this direction could be represented by the consideration, in an additive way, of some tensor fluctuating terms in the metric. The quantum mechanical origin of these terms can be well motivated physically. Such approaches lead to classical gravity models with geometry-matter coupling, and to the non-conservation of the matter energy-momentum tensor. Consequently, the particle production processes, specific to these classes of theories, may be an indication of their deep relation with effective descriptions of quantum gravity. On the other hand, the investigations of the gravitational models with fluctuating quantum metrics could lead to a better understanding of the physical basis of the modified gravity theories with geometry-matter coupling. In the present paper, we considered some of the cosmological implications of the modified gravity models, induced by the quantum metric fluctuations, and some basic theoretical and mathematical tools were introduced that might be used for further investigations of the quantum mechanical effects in gravity and in the geometry of the spacetime.

Author Contributions: Conceptualization, Z.H. and T.H.; Formal analysis, Z.H. and T.H.; Investigation, Z.H. and T.H.; Software, Z.H.; Writing—original draft, T.H.; Writing—review and editing, Z.H. and T.H. All authors have read and agreed to the published version of the manuscript.

Funding: This research received no external funding.

Acknowledgments: We would like to thank the two anonymous reviewers for their useful comments and suggestions that helped to improve the paper. T.H. would like to thank the Yat Sen School of the Sun Yat-Sen University in Guangzhou, China, for kind hospitality during the preparation of this paper.

Conflicts of Interest: The authors declare no conflict of interest.

Appendix A. The Generalized Friedmann Equations

The generalized Friedmann and Raychaudhuri equations, obtained by adopting for the fluctuation tensor the expression given by Equation (19), can be written as:

$$\begin{aligned}\kappa^2\rho_{\rm eff} = &\frac{3}{2}\gamma_1 H(\mathcal{A}_3+\dot{\rho}) + \gamma_2\left(3\kappa^2 H\mathcal{A}_1\rho - 18\dot{\rho}H^3\right.\\ &\left.-6\rho H\ddot{H} - 6H\dot{H}\mathcal{A}_3 + 3\rho\dot{H}^2\right) - \frac{1}{2}\gamma_3\rho^2\left(3H^2+\kappa^2\rho\right)\\ &+9\beta_2\Big\{\kappa^2\left[\rho(\dot{H}^2-2H\ddot{H}) - 3H^3\mathcal{A}_1 - 2H\dot{H}\mathcal{A}_3\right]\\ &-12H^4(H^2-8\dot{H}) + 2\dot{H}^2(17H^2-2\dot{H})\\ &+4H\ddot{H}\left(8H^2+3\dot{H}\right)\Big\} + 3\beta_3 H\left[3H^2(\mathcal{A}_3-2\dot{\rho})\right.\\ &\left.+\kappa^2\rho(\mathcal{A}_1+\dot{\rho})\right],\end{aligned} \tag{A1}$$

and

$$
\begin{aligned}
-\kappa^2 p_{\text{eff}} = \gamma_1\Big[\ddot{\rho} - 2\rho\dot{H} + 3H\dot{\rho} + \rho\,(\kappa^2\rho - \frac{9}{2}H^2)\Big] \\
+\frac{1}{2}\gamma_3\rho\Big[3\rho\Big(\kappa^2\rho - 5H^2\Big) - 4\dot{H}\rho - 4H\dot{\rho}\Big] + \gamma_2\Big[18\rho H^2 \times \\
\Big(\dot{H} + 2H^2\Big) - 12H\dot{\rho}\,\Big(H^2 + 2\dot{H}\Big) - 2\ddot{\rho}\,\Big(\dot{H} + 3H^2\Big) \\
-6\dot{H}\mathcal{A}_2 - \rho\Big(2\dddot{H} + 3\dot{H}^2\Big) + \kappa^2\Big(\rho\ddot{\rho} + \dot{\rho}\,\mathcal{A}_4 \\
-3\rho^2(\dot{H} + H^2)\Big)\Big] + \beta_2\Big\{12\dddot{H}\Big(8H^2 + 3\dot{H}\Big) + 12\dot{H} \times \\
\Big(48H^3 + 3\ddot{H} + 50H\dot{H}\Big) + 6\dot{H}^2\Big(243H^2 + 28\dot{H}\Big) - 108H^4 \times \\
(H^2 - 6\dot{H}) + \kappa^2\Big[\rho\,(3H^4 - 2\dddot{H}) - \rho\dot{H}\Big(5\dot{H} + 16H^2\Big) \\
-2\dot{\rho}H(9\dot{H} + 5H^2) - \ddot{\rho}\,(2\dot{H} + 3H^2) - 4\dot{H}\mathcal{A}_3\Big]\Big\} \\
+\beta_3\Big\{3H^2(21H^2\rho + 2H\dot{\rho} - \ddot{\rho}) + 6H\dot{H}(11H\rho - \dot{\rho}) \\
+6\rho\dot{H}^2 - 2\kappa^2\Big[\rho^2\Big(4\dot{H} + 9H^2\Big) - 2\dot{\rho}\,\mathcal{A}_2 - 2\rho\ddot{\rho}\Big]\Big\},
\end{aligned} \tag{A2}
$$

with

$$
\mathcal{A}_n = \dot{\rho} + n\,H\rho, \quad n = 1, 2, 3, 4. \tag{A3}
$$

References

1. Hilbert, D. Die Grundlagen der Physik. In *Nachrichten von der Gesellschaft der Wissenschaften zu Göttingen–Mathematische-Physikalische Klasse*; Vandenhoeck & Ruprecht: Göttingen, Germany, 1915; Volume 1915, pp. 395–407.
2. Einstein, A. Die Feldgleichungen der Gravitation. In *A. Königlich Preussische Akademie der Wissenschaften*; Vandenhoeck & Ruprecht: Göttingen, Germany, 1915; Volume 25, pp. 844–847.
3. Einstein, A. Die Grundlage der allgemeinen Relativitätstheorie. *Annalen Phys.* **1916**, *49*, 769. [CrossRef]
4. Turyshev, S.G. Experimental Tests of General Relativity. *Ann. Rev. Nucl. Part. Sci.* **2008**, *58*, 207. [CrossRef]
5. Will, C.M. The Confrontation between General Relativity and Experiment. *Living Rev. Rel.* **2014**, *17*, 4. [CrossRef]
6. Marchi, F.D.; Cascioli, G. Testing General Relativity in the Solar System: Present and future perspectives. *arXiv* **2019**, arXiv1911.05561.
7. Abbott, B.P.; Abbott, R.; Abbott, T.D.; Abernathy, M.R.; Acernese, F.; Ackley, K.; Adams, C.; Adams, T.; Addesso, P.; Adhikari, R.X.; et al. LIGO Scientific and Virgo Collaborations. Observation of Gravitational Waves from a Binary Black Hole Merger. *Phys. Rev. Lett.* **2016**, *116*, 061102. [CrossRef]
8. Akiyama, K.; Alberdi, A; Alef, W.; Asada, K.; Azulay, R.; Baczko, A.-K.; Ball, D; Baloković, M.; Barrett, J.; Bintley, D.; et al. Event Horizon Telescope Collaboration, First M87 Event Horizon Telescope Results. I. The Shadow of the Supermassive Black Hole. *Astrophys. J. Lett.* **2019**, *875*, L1. [CrossRef]
9. Bronstein, M. Quantum theory of weak gravitational fields. *Phys. Z. Der Sowjetunion* **1936**, *9*, 140; republished as Bronstein, M. Republication of: Quantum theory of weak gravitational fields. *Gen. Relativ. Gravit.* **2012**, *44*, 267. [CrossRef]
10. Mukhanov, V.; Winitzki, S. *Introduction to Quantum Effects in Gravity*; Cambridge University Press: Cambridge, UK, 2007.
11. Kiefer, C. *Quantum Gravity*, 3rd ed.; Oxford University Press: Oxford, UK, 2012.
12. Utiyama, R. Invariant Theoretical Interpretation of Interaction. *Phys. Rev.* **1956**, *101*, 1597. [CrossRef]
13. Kibble, T.W.B. Lorentz invariance and the gravitational field. *J. Math. Phys.* **1961**, *2*, 212. [CrossRef]
14. Moller, C. Conservation laws and absolute parallelism in general relativity. *K. Dan. Vidensk. Selsk. Mat. Fys. Skr.* **1961**, *1*, 1.
15. Pellegrini, C.; Plebanski, J. Tetrad fields and gravitational fields. *K. Dan. Vidensk. Selsk. Mat. Fys. Skr.* **1962**, *2*, 1.
16. Hayashi, K.; Nakano, T. Extended translation invariance and associated gauge fields. *Prog. Theor. Phys.* **1967**, 38, 491. [CrossRef]
17. Blagojevic, M. *Gravitation and Gauge Symmetries*; IOP Publishing: Bristol, UK, 2002.
18. Ashtekar, A. New Variables for Classical and Quantum Gravity. *Phys. Rev. Lett.* **1986**, 57, 2244. [CrossRef] [PubMed]
19. Ashtekar, A. New Hamiltonian formulation of general relativity. *Phys. Rev. D* **1987**, *36*, 1587. [CrossRef] [PubMed]
20. Ashtekar, A. *Lectures on Non-perturbative Canonical Gravity. Notes Prepared in Collaboration with R.S. Tate*; World Scientific: Singapore, 1991.
21. Funai, S.S.; Sugawara, H. Current Algebra Formulation of Quantum Gravity and Its Application to Cosmology. *arXiv* **2020**, arXiv:2004.02151.

22. Jacobson, T. Thermodynamics of Spacetime: The Einstein Equation of State. *Phys. Rev. Lett.* **1995**, *75*, 1260. [CrossRef]
23. Connes, A. Gravity coupled with matter and the foundation of non-commutative geometry. *Commun. Math. Phys.* **1996**, *182*, 155. [CrossRef]
24. Reuter, M. Nonperturbative evolution equation for quantum gravity. *Phys. Rev. D* **1998**, *57*, 971. [CrossRef]
25. Rovelli, C. Loop Quantum Gravity. *Living Rev. Rel.* **1998**, *1*, 1. [CrossRef]
26. Gambin, R.; Pullin, J. Consistent Discretization and Loop Quantum Geometry. *Phys. Rev. Lett.* **2005**, *94*, 101302. [CrossRef] [PubMed]
27. Ashtekar, A. Gravity and the quantum. *New J. Phys.* **2005**, *7*, 198. [CrossRef]
28. Horava, P. Quantum gravity at a Lifshitz point. *Phys. Rev. D* **2009**, *79*, 084008. [CrossRef]
29. Verlinde, E.P. On the Origin of Gravity and the Laws of Newton. *JHEP* **2011**, *29*, 1104.
30. Møller, C. *Les théories Relativistes de la Gravitation*; Colloques Internationaux CNRS vol 91; Tonnelat, A., Ed.; CNRS: Paris, France, 1962; pp. 1–96.
31. Rosenfeld, L. On quantization of fields. *Nucl. Phys.* **1963**, *40*, 353. [CrossRef]
32. Davies, P.C.W. Singularity avoidance and quantum conformal anomalies. *Phys. Lett. B* **1977**, *6*, 402. [CrossRef]
33. Fischetti, M.V.; Hartle, J.B.; Hu, B.L. Quantum Effects in the early Universe. I. Influence of Trace Anomalies on Homogeneous, Isotropic, Classical Geometries. *Phys. Rev. D* **1979**, *20*, 1757. [CrossRef]
34. Starobinsky, A.A. A new type of isotropic cosmological models without singularity. *Phys. Lett. B* **1980**, *91*, 99. [CrossRef]
35. Kibble, T.W.B.; Randjbar-Daemi, S. Non-linear coupling of quantum theory and classical gravity. *J. Phys. A Math. Gen.* **1980**, *13*, 141. [CrossRef]
36. Nojiri, S.; Odintsov, S.D. Quantum deSitter cosmology and phantom matter. *Phys. Lett. B* **2003**, *562*, 147. [CrossRef]
37. Nojiri, S.; Odintsov, S.D. Quantum escape of sudden future singularity. *Phys. Lett. B* **2004**, *595*, 1.
38. Carlip, S. Is Quantum Gravity Necessary? *Class. Quant. Grav.* **2008**, *25*, 154010. [CrossRef]
39. Ho, P.-M.; Kawai, H.; Matsuo, Y.; Yokokura, Y. Back reaction of 4D conformal fields on static black-hole geometry. *JHEP* **2018**, *56*, 11. [CrossRef]
40. Han, M. Einstein equation from covariant loop quantum gravity in semiclassical continuum limit. *Phys. Rev. D* **2017**, *96*, 024047. [CrossRef]
41. dos Reis, E.A.; Krein, G.; de Paula Netto, T.; Shapiro, I.L. Stochastic quantization of a self-interacting nonminimal scalar field in semiclassical gravity. *Phys. Lett. B* **2019**, *798*, 134925. [CrossRef]
42. Juárez-Aubry, A. Semi-classical gravity in de Sitter spacetime and the cosmological constant. *Phys. Lett. B* **2019**, *797*, 134912. [CrossRef]
43. Satin, S. Correspondences of matter fluctuations in semiclassical and classical gravity for cosmological spacetime. *Phys. Rev. D* **2019**, *100*, 044032. [CrossRef]
44. Matsui, H.; Watamura, N. Quantum spacetime instability and breakdown of semiclassical gravity. *Phys. Rev. D* **2020**, *101*, 025014. [CrossRef]
45. Parisi, G.; Wu, Y.S. Perturbation Theory Without Gauge Fixing. *Sci. Sin.* **1981**, *24*, 483.
46. Nelson, E. Derivation of the Schrödinger Equation from Newtonian Mechanics. *Phys. Rev.* **1966**, *150*, 1079. [CrossRef]
47. Damgaard, P.H.; Huffel, H. *Stochastic Quantization*; World Scientific: Singapore, 1988.
48. Namiki, M. Basic Ideas of Stochastic Quantization. *Prog. Theor. Phys. Suppl.* **1993**, *1*, 111. [CrossRef]
49. Rumpf, H. Stochastic quantization of Einstein gravity. *Phys. Rev. D* **1986**, *942*, 33. [CrossRef]
50. Rumpf, H. Stochastic Quantum Gravity in D Dimension. *Prog. Theor. Phys. Suppl.* **1993**, *111*, 63. [CrossRef]
51. Hu, B.L.; Roura, A.; Verdaguer, E. Induced quantum metric fluctuations and the validity of semiclassical gravity. *Phys. Rev. D* **2004**, *70*, 044002. [CrossRef]
52. Satin, S.; Cho, H.T.; Hu, B.L. Conformally-related Einstein-Langevin equations for metric fluctuations in stochastic gravity. *Phys. Rev. D* **2016**, *94*, 064019. [CrossRef]
53. Wheeler, J.A. On the nature of quantum geometrodynamics. *Ann. Phys.* **1957**, *2*, 604. [CrossRef]
54. Misner, C.W.; Thorne, K.S.; Wheeler, J.A. *Gravitation*; W. H. Freeman: San Francisco, CA, USA, 1973.
55. Dzhunushaliev, V.; Folomeev, V.; Kleihaus, B.; Kunz, J. Modified gravity from the quantum part of the metric. *Eur. Phys. J. C* **2014**, *74*, 2743. [CrossRef]
56. Dzhunushaliev, V.; Folomeev, V.; Kleihaus, B.; Kunz, J. Modified gravity from the nonperturbative quantization of a metric. *Eur. Phys. J. C* **2015**, *75*, 157. [CrossRef] [PubMed]
57. Yang, R.-J. Effects of quantum fluctuations of metric on the universe. *Phys. Dark Univ.* **2016**, *13*, 87. [CrossRef]
58. Dzhunushaliev, V. Nonperturbative quantization: Ideas, perspectives, and applications. *arXiv* **2015**, arXiv:1505.02747.
59. Dzhunushaliev, V.; Quevedo, H. Einstein equations with fluctuating volume. *Gravit. Cosmol.* **2017**, *23*, 280. [CrossRef]
60. Dzhunushaliev, V.; Folomeev, V.; Quevedo, H. Nonperturbative Quantization à La Heisenberg: Modified Gravities, Wheeler-DeWitt Equations, and Monopoles in QCD. *Gravit. Cosmol.* **2019**, *25*, 1. [CrossRef]
61. Liu, X.; Harko, T.; Liang, S.-D. Cosmological implications of modified gravity induced by quantum metric fluctuations. *Eur. Phys. J. C* **2016**, *76*, 420. [CrossRef]
62. Perlmutter, S.; Aldering, G.; Goldhaber, G.; Knop, R.A.; Nugent, P.; Castro, P.G.; Deustua, S.; Fabbro, S.; Goobar, A.; Groom, D.E.; et al. Measurements of Ω andΛ from 42 High-Redshift Supernovae. *Astrophys. J.* **1999**, *517*, 565. [CrossRef]

63. Knop, R.A.; Aldering, G.; Amanullah, R.; Astier, P.; Blanc, G.; Burns, M.S.; Conley, A.; Deustua, S.E.; Doi, M.; Ellis, R.; et al. New Constraints on Ω_m, ω_Λ, and w from an Independent Set of Eleven High-Redshift Supernovae Observed with HST. *Astrophys. J.* **2003**, *598*, 102. [CrossRef]
64. Riess, A.G.; Strolger, L.-G.; Casertano, S.; Ferguson, H.C.; Mobasher, B.; Gold, B.; Challis, P.J.; Filippenko, A.V.; Jha, S.; Li, W.; et al. New Hubble Space Telescope Discoveries of Type Ia Supernovae at z > 1: Narrowing Constraints on the Early Behavior of Dark Energy. *Astrophys. J.* **2007**, *659*, 98. [CrossRef]
65. Amanullah, R.; Lidman, C.; Rubin, D.; Aldering, G.; Astier, P.; Barbary, K.; Burns, M.S.; Conley, A.; Dawson, K.S.; Deustua, S.E.; et al. Spectra and Light Curves of Six Type Ia Supernovae at $0.511 < z < 1.12$ and the Union2 Compilation. *Astrophys. J.* **2010**, *716*, 712. [CrossRef]
66. Weinberg, D.H.; Mortonson, M.J.; Eisenstein, D.J.; Hirata, C.; Riess, A.G.; Rozo, E. Observational probes of cosmic acceleration. *Phys. Rep.* **2013**, *530*, 87. [CrossRef]
67. Aad, G.; Abbott, B.; Abdallah, J.; Abdinov, O.; Aben, R.; Abolins, M.; Abouzeid, O.S.; Abramowicz, H.; Abreu, H.; Abreu, R.; et al. Search for Dark Matter in Events with Missing Transverse Momentum and a Higgs Boson Decaying to Two Photons in pp Collisions at $\sqrt{s} = 8$ TeV with the ATLAS Detector. *Phys. Rev. Lett.* **2015**, *115*, 131801. [CrossRef]
68. Peebles, P.J.E.; Ratra, B. The cosmological constant and dark energy. *Rev. Mod. Phys.* **2003**, *75*, 559. [CrossRef]
69. Padmanabhan, T. Cosmological Constant—The Weight of the Vacuum. *Phys. Repts.* **2003**, *380*, 235. [CrossRef]
70. Overduin, J.M.; Wesson, P.S. Dark Matter and Background Light. *Phys. Rep.* **2004**, *402*, 267. [CrossRef]
71. Baer, H.; Choi, K.-Y.; Kim, J.E.; Roszkowski, L. Dark matter production in the early Universe: Beyond the thermal WIMP paradigm. *Phys. Rep.***2015**, *555*, 1. [CrossRef]
72. Weinberg, S. The cosmological constant problem. *Rev. Mod. Phys.* **1989**, *61*, 1. [CrossRef]
73. Weinberg, S. The Cosmological Constant Problems (Talk given at Dark Matter 2000, February, 2000). *arXiv* **2000**, arXiv:astro-ph/0005265.
74. Buchdahl, H.A. Non-Linear Lagrangians and Cosmological Theory. *Mon. Not. Roy. Astron. Soc.* **1970**, *150*, 1. [CrossRef]
75. Silvestri, A.; Trodden, M. Approaches to Understanding Cosmic Acceleration. *Rept. Prog. Phys.* **2009**, *72*, 096901. [CrossRef]
76. De Felice, A.; Tsujikawa, S. $f(R)$ theories. *Living Rev. Rel.* **2010**, *13*, 3. [CrossRef] [PubMed]
77. Sotiriou, T.P.; Faraoni, V. $f(R)$ theories of gravity. *Rev. Mod. Phys.* **2010**, *82*, 451. [CrossRef]
78. Capozziello, S.; De Laurentis, M. Extended Theories of Gravity. *Phys. Rept.* **2011**, *509*, 167. [CrossRef]
79. Nojiri, S.; Odintsov, S.D. Unified cosmic history in modified gravity: From $F(R)$ theory to Lorentz non-invariant models. *Phys. Rept.* **2011**, *505*, 59. [CrossRef]
80. Haghani, Z.; Harko, T.; Sepangi, H.R.; Shahidi, S. Matter may matter. *Int. J. Mod. Phys. D* **2014**, *23*, 1442016. [CrossRef]
81. Bertolami, O.; Boehmer, C.G.; Harko, T.; Lobo, F.S.N. Extra force in $f(R)$ modified theories of gravity. *Phys. Rev. D* **2007**, *75*, 104016. [CrossRef]
82. Harko, T. Modified gravity with arbitrary coupling between matter and geometry. *Phys. Lett. B* **2008**, *669*, 376. [CrossRef]
83. Harko, T.; Lobo, F.S.N. $f(R, L_m)$ gravity. *Eur. Phys. J. C* **2010**, *70*, 373. [CrossRef]
84. Harko, T.; Lobo, F.S.N.; Minazzoli, O. Extended $f(R, L_m)$ gravity with generalized scalar field and kinetic term dependences. *Phys. Rev. D* **2013**, *87*, 047501. [CrossRef]
85. Harko, T.; Lobo, F.S.N.; Nojiri, S.; Odintsov, S.D. $f(R, T)$ gravity. *Phys. Rev. D* **2011**, *84*, 024020. [CrossRef]
86. Harko, T. Thermodynamic interpretation of the generalized gravity models with geometry-matter coupling. *Phys. Rev. D* **2014**, *90*, 044067. [CrossRef]
87. Haghani, Z.; Harko, T.; Lobo, F.S.N.; Sepangi, H.R.; Shahidi, S. Further matters in space-time geometry: $f(R, T, R_{\mu\nu}T^{\mu\nu})$ gravity. *Phys. Rev. D* **2013** *88*, 044023. [CrossRef]
88. Odintsov, S.D.; Sáez-Gómez, D. $f(R, T, R_{\mu\nu}T^{\mu\nu})$ gravity phenomenology and ΛCDM universe. *Phys. Lett. B* **2013**, *725*, 437. [CrossRef]
89. Harko, T.; Koivisto,T.S.; Lobo, F.S.N.; Olmo, G.J. Metric-Palatini gravity unifying local constraints and late-time cosmic acceleration. *Phys. Rev. D* **2012**, *85*, 084016. [CrossRef]
90. Tamanini, N.; Böhmer, C.G. Generalized hybrid metric-Palatini gravity. *Phys. Rev. D* **2013**, *87*, 084031. [CrossRef]
91. Capozziello, S.; Harko, T.; Koivisto, T.S.; Lobo, F.S.N.; Olmo, G.J. Hybrid metric-Palatini gravity. *Universe* **2015**, *1*, 199. [CrossRef]
92. Haghani, Z.; Harko, T.; Sepangi, H.R.; Shahidi, S. Weyl-Cartan-Weitzenböck gravity as a generalization of teleparallel gravity. *JCAP* **2012**, *10*, 061. [CrossRef]
93. Harko, T.; Koivisto,T.S.; Lobo, F.S.N.; Olmo, G.J.; Rubiera-Garcia, D. Coupling matter in modified Q gravity. *Phys. Rev. D* **2018**, *98*, 084043. [CrossRef]
94. Xu, Y.; Li, G.; Harko, T.; Liang, S.-D. $f(Q, T)$ gravity. *Eur. Phys. J. C* **2019**, *79*, 708. [CrossRef]
95. Harko, T.; Lobo, F.S.N.; Otalora, G.; Saridakis, E.N. $f(T, \mathcal{T})$ gravity and cosmology. *JCAP* **2014**, *21*, 12. [CrossRef]
96. Harko, T.; Lobo, F.S.N.; Saridakis, E.N. Cosmology with higher-derivative matter fields. *Int. J. Geom. Meth. Mod. Phys.* **2016**, *13*, 1650102. [CrossRef]
97. Harko, T.; Lobo, F.S.N. Generalized Curvature-Matter Couplings in Modified Gravity. *Galaxies* **2014**, *2*, 410. [CrossRef]
98. Harko, T.; Lobo, F.S.N. *Extensions of $f(R)$ Gravity: Curvature-Matter Couplings and Hybrid Metric-Palatini Theory;* Cambridge University Press: Cambridge, UK, 2018.

99. Harko, T.; Lobo, F.S.N. Irreversible thermodynamic description of interacting dark energy—Dark matter cosmological models. *Phys. Rev. D* **2013**, *87*, 044018. [CrossRef]
100. Harko, T.; Lobo, F.S.N.; Mimoso, J.P.; Pavón, D. Gravitational induced particle production through a nonminimal curvature-matter coupling. *Eur. Phys. J. C* **2015**, *75*, 386. [CrossRef]
101. Parker, L. Particle Creation in Expanding Universes. *Phys. Rev. Lett.* **1968**, *21*, 562. [CrossRef]
102. Zeldovich, Y.B. Particle Production in Cosmology. *J. Exper. Theor. Phys. Lett.* **1970**, *12*, 307.
103. Parker, L. Quantized Fields and Particle Creation in Expanding Universes. II. *Phys. Rev. D* **1971**, *3*, 2546. [CrossRef]
104. Fulling, S.A. *Aspects of Quantum Field Theory in Curved Space-Time*; Cambridge University Press: Cambridge, UK, 1989.
105. Parker, L.E.; Toms, D.J. *Quantum Field Theory in Curved Spacetime-Quantized Fields and Gravity*; Cambridge University Press: Cambridge, UK, 2009.
106. Lee, J.-W. Are galaxies extending? *Phys. Lett. B* **2009**, *681*, 118. [CrossRef]
107. Park, C.-G.; Hwang, J.-C.; Noh, H. Axion as a cold dark matter candidate: Low-mass case. *Phys. Rev. D* **2012**, *86*, 083535. [CrossRef]
108. Boucher, W.; Traschen, J. Semiclassical physics and quantum fluctuations. *Phys. Rev. D* **1988**, *37*, 3522. [CrossRef]
109. Capozziello, S.; De Laurentis, M.; De Martino, I.; Formisano, M.; Odintsov, S.D. Jeans analysis of self-gravitating systems in $f(R)$ gravity. *Phys. Rev. D* **2012**, *85*, 044022. [CrossRef]
110. Landau, L.D.; Lifshitz, E.M. *The Classical Theory of Fields*; Butterworth-Heinemann: Oxford, UK, 1998.
111. Riess, A.G.; Filippenko, A.V.; Challis, P.; Clocchiatti, A.; Diercks, A.; Garnavich, P.M.; Gilliland, R.L.; Hogan, C.J.; Jha, S.; Kirshner, R.P.; et al. Observational Evidence from Supernovae for an Accelerating Universe and a Cosmological Constant. *Astron. J.* **1998**, *116*, 1009. [CrossRef]
112. de Bernardis, P.; Ade, P.A.R.; Bock, J.J.; Bond, J.R.; Borrill, J.; Boscaleri, A.; Coble, K.; Crill, B.P.; De Gasperis, G.; Farese, P.C.; et al. A Flat Universe from High-Resolution Maps of the Cosmic Microwave Background Radiation. *Nature* **2000**, *404*, 955. [CrossRef]
113. Hanany, S.; Ade, P.; Balbi, A.; Bock, J.; Borrill, J.; Boscaleri, A.; de Bernardis, P.; Ferreira, P.G.; Hristov, V.V.; Jaffe, A.H.; et al. MAXIMA-1: A Measurement of the Cosmic Microwave Background Anisotropy on angular scales of 10 arcminutes to 5 degrees. *Astrophys. J.* **2000**, *545*, L5. [CrossRef]
114. Riess, A.G.; Macri, L.; Casertano, S.; Lampeitl, H.; Ferguson, H.C.; Filippenko, A.V.; Jha, S.W.; Li, W.; Chornock, R.A. 3% Solution: Determination of the Hubble Constant with the Hubble space Telescopr and Wide Field Camera 3. *Astrophys. J.* **2011**, *730*, 119. [CrossRef]
115. Ade, P.A.R.; Aghanim, N.; Arnaud, M.; Ashdown, M.; Aumont, J.; Baccigalupi, C.; Banday, A.J.; Barreiro, R.B.; Bartlett, J.G.; Bartolo, N.; et al. Planck collaboration, Planck 2015 results. XIII. Cosmological parameters. *Astron. Astrophys.* **2016**, *594*, A13. [CrossRef]
116. Aghanim, N.; Akrami, Y.; Arroja, F.; Ashdown, M.; Aumont, J.; Baccigalupi, C.; Ballardini, M.; Banday, A.J.; Barreiro, R.B.; Bartolo, N.; et al. Planck 2018 results. I. Overview and the cosmological legacy of Planck. *Astron. Astrophys.* **2020**, *641*, A1. [CrossRef]
117. Aghanim, N.; Akrami, Y.; Ashdown, M.; Aumont, J.; Baccigalupi, C.; Ballardini, M.; Banday, A.J.; Barreiro, R.B.; Bartolo, N.; Basak, S.; et al. Planck 2018 results. VI. Cosmological parameters. *Astron. Astrophys.* **2020**, *641*, A6. [CrossRef]

physics

MDPI

Article

Why Do Elementary Particles Have Such Strange Mass Ratios?—The Importance of Quantum Gravity at Low Energies †

Tejinder P. Singh

Tata Institute of Fundamental Research, Homi Bhabha Road, Mumbai 400005, India; tpsingh@tifr.res.in

† This article is an expanded version of an essay written for the Gravity Research Foundation 2022 Awards for Essays on Gravitation (USA). This essay received an Honorable Mention.

Abstract: When gravity is quantum, the point structure of space-time should be replaced by a non-commutative geometry. This is true even for quantum gravity in the infra-red. Using the octonions as space-time coordinates, we construct pre-spacetime, pre-quantum Lagrangian dynamics. We show that the symmetries of this non-commutative space unify the standard model of particle physics with $SU(2)_R$ chiral gravity. The algebra of the octonionic space yields spinor states which can be identified with three generations of quarks and leptons. The geometry of the space implies quantisation of electric charge, and leads to a theoretical derivation of the mysterious mass ratios of quarks and the charged leptons. Quantum gravity is quantisation not only of the gravitational field, but also of the point structure of space-time.

Keywords: quantum gravity; quantum foundations; trace dynamics; noncommutative geometry; octonions; unification; standard model; mass ratios

Citation: Singh, T.P. Why Do Elementary Particles Have Such Strange Mass Ratios?—The Importance of Quantum Gravity at Low Energies. *Physics* **2022**, *4*, 948–969. https://doi.org/10.3390/physics4030063

Received: 12 May 2022
Accepted: 11 August 2022
Published: 25 August 2022

1. When Is Quantum Gravity Necessary?

Consider a massive object in a quantum superposition of its two different classical position states A and B. The resulting gravitational field is also then in a superposition of the field corresponding to position A and the field corresponding to position B. A clock kept at a field point C will not register a definite value of time, nor will a measurement of the metric yield a well-defined result [1]. Let us now imagine a thought experiment in which every object in today's universe is in a superposition of its two different position states. The space-time metric will then undergo quantum fluctuations. Now, the Einstein hole argument shows that, in order for space-time points to be operationally distinguishable, the manifold must be overlaid by a (classical) metric [2]. Therefore, in our thought experiment, the point structure of space-time is lost, even though the energy scales of interest are much smaller than Planck scale, and the gravitational fields are weak.

When we describe microscopic systems by the laws of quantum theory, we take it for granted that the universe is dominated by classical bodies, so that a background space-time can be achieved and is available for defining time evolution of quantum systems. However, if everything were to be quantised at once, in the sense of the aforementioned thought experiment, no classical time will be available, and yet we ought to be able to describe the dynamics. This is an example of quantum gravity in the infra-red (QG in IR): the action of the gravitational field is much larger than the reduced Planck's constant, $\hbar$ (unlike for Planck scale quantum gravity), and yet the point structure of space-time is lost. The manifold has to be replaced by something non-classical: quantum gravity is quantisation not only of the gravitational field, but also of the point structure of space-time.

Since the energy scale is not a relevant criterion for deciding whether gravity is classical or quantum, we propose that a gravitational field is quantum in nature when one or more of the following three (energy independent) criteria are satisfied: (i) the time scales of interest are of the order of Planck time, t_P; (ii) the length scales of interest are of the order of Planck length, L_P; and (iii) every sub-system has an action of the order $\hbar$ (and is hence quantum and obeys quantum superposition). If (iii) holds but (i) and (ii) do not, we have QG in IR If (iii) holds along with (i) and (ii), then we have quantum gravity in the UV (ultra-violet).

Put differently, there ought to exist a reformulation of quantum (field) theory, even at low energies, which does not depend on classical time. Such a reformulation is essential also for the standard model of particle physics. In fact, we show that it helps us understand why the standard model has the symmetries it does, and why its free parameters take the specific values they do, and also shows how to unify gravity with the other fundamental forces: electroweak (EW) and strong. We construct such a dynamics using t_P, L_P and $\hbar$ as the only three fundamental parameters in the theory. We note that in these units the low energy fine structure constant, $\alpha_f = e^2/\hbar c \equiv e^2\ t_P/\hbar L_P \sim 1/137$, where e is electron charge and c is the speed of light, is order unity and hence quantum gravitational in origin (QG in IR). On the other hand, particles masses, $m \sim \epsilon m_P \equiv \epsilon \hbar t_P/L_P^2$, are not because $\epsilon \ll 1$. However, mass ratios (at low energies) can be, and, in fact, are quantum gravitational in origin.

To achieve our goal, we build on Adler's pre-quantum theory, i.e., trace dynamics (TD) [3,4]. Starting from classical Lagrangian dynamics, TD retains the classical space-time manifold, but all configuration variables and their canonical momenta are raised to the status of matrices (equivalently operators). This step is the same as in quantum theory; however, the canonical Heisenberg commutation relations, $[q,p] = i\hbar$, are not imposed; here, q and p are canonical configuration variable and momentum, respectively. Instead, we have a matrix-valued Lagrangian dynamics, where the Lagrangian is the trace of a matrix polynomial made from matrix-valued configuration variables and their time derivatives (i.e., the velocities). A 'trace' derivative enables the derivation of Lagrange equations of motion, and a global unitary invariance of the trace Hamiltonian (this being an elementary consequence of invariance of the trace under cyclic permutations) implies the existence of the novel conserved Noether charge, $\tilde{C} \equiv \sum_i [q_i, p_i]$, where i stands for the various degrees of freedom. The Hamiltonian of the theory is in general not self-adjoint, and dynamical evolution is not restricted to be unitary. Assuming these dynamics hold on Planck time scale resolution, one asks what the averaged dynamics on lower energy scales will be, if one coarse-grains the dynamics on time scales much larger than Planck time. Using the techniques of statistical thermodynamics, it is shown that, if the anti-self-adjoint part of the Hamiltonian is negligible, the emergent dynamic is relativistic quantum (field) theory. The aforementioned Noether charge is equi-partitioned over all bosonic and fermionic degrees of freedom, and canonical commutation and anti-commutation relations emerge for the statistically averaged canonical variables, which obey the Heisenberg equations of motion. If the anti-self-adjoint part of the Hamiltonian becomes significant (this is enabled by large-scale quantum entanglement), spontaneous localisation results, leading to the quantum-to-classical transition and emergence of classical dynamics. For a detailed explanation of the emergence of the classical universe, the reader is referred to Section XIII of [5].

2. Replacing the Point Structure of Spacetime by the Non-Commutative Geometry of the Octonions

Next, TD is generalised, so as to replace the four-dimensional (4D) Minkowski space-time manifold by a higher dimensional non-commutative space-time, and incorporate matrix-valued pre-gravitation, thus taking TD to a pre-space-time, pre-quantum theory. Let us recall that, in special relativity, given the four-vector $V^\mu = dt\ \hat{t} + dx\ \hat{x} + dy\ \hat{y} + dz\ \hat{z}$ connecting two neighbouring space-time points having a separation (dt, dx, dy, dz), one can define the line element, $ds^2 = \eta_{\mu\nu} V^\mu V^\nu$, and the four-velocity, dq^μ/ds, of a particle having the configuration variable, $q^\mu = (q^t, q^x, q^y, q^z)$. Here, Greek letters denote the time (0) and space components, and $\eta_{\mu\nu}$ is the flat spacetime metric. The action for the particle is $mc \int ds$, where m is the particle mass, and the transition to curved space-time and general relativity (GR) is made by introducing the metric $g_{\mu\nu}$, i.e., $ds^2 = g_{\mu\nu}dx^\mu dx^\nu$, and writing down the action,

$$S = \frac{c^4}{16\pi G}\int d^4x\,\sqrt{-g}\,R + \sum_i m_i c\int ds + S_{YM} \tag{1}$$

Here, the first term is the Einstein–Hilbert action, S_{YM} stands for the action of Yang–Mills fields and also includes their current sources, G is the Newtonian constant of gravity, g is the deteminant of $g_{\mu\nu}$, and R stands for scalar curvature.

We now generalise this action to construct a pre-spacetime, pre-quantum action principle [6], from which the sought for quantum theory without classical time emerges, and whose symmetries imply the standard model of particle physics and fix its free parameters. The space-time coordinates (t,x,y,z) are replaced by a set $\{e_i,\ i=0,1,2,\ldots,m-1\}$ of m non-commuting coordinates, to be specified later in this Section. The configuration variable q^μ for a particle is replaced by a matrix q_F, whose entries are odd-grade Grassmann elements over the field of complex numbers (so as to represent fermions). q_F has m components q_F^i, one for each of the coordinates e_i, i.e., $q_F = (q_F^0\,e_0 + \cdots + q_F^{(m-1)}\,e_{m-1})$. The point structure of space-time is lost; instead, we have a non-commutative geometry, and the matrix-valued velocity is defined as $dq_F/d\tau \equiv \dot{q}_F$. Here, the newly introduced Connes time, τ, is a unique property of a non-commutative geometry; it is an absolute real-valued time parameter distinct from the non-commuting coordinates e_i, and is used to describe evolution [7].

To introduce pre-gravitation into trace dynamics, we recall the spectral action principle of Chamseddine and Connes, according to which the Einstein–Hilbert action can be cast in terms of the eigenvalues of the square of the (regularised) Dirac operator, D_B, on a Riemannian manifold, by making use of a truncated heat kernel expansion [8]:

$$Tr\,[L_P^2\,D_B^2] \sim \int d^4x\sqrt{g}\,\frac{R}{L_P^2} + \mathcal{O}(L_P^0) \sim L_P^2\sum_n \lambda_n^2\,. \tag{2}$$

Here, the eigenvalues λ_n of the Dirac operator play the role of dynamical variables of general relativity [9]. Following trace dynamics, each eigenvalue λ_n is raised to the status of a canonical matrix momentum: $\lambda_n \to p_{Bn} \propto q_{Bn}/d\tau \equiv D_B$, and the bosonic matrix q_B (with even grade Grassmann elements as entries) is now the configuration variable, and it has m matrix components q_B^m over the non-commuting coordinates e_i. Therefore, we have N copies of the Dirac operator (n runs from 1 to N, with $N\to\infty$). The trace Lagrangian (space-time part) of the matrix dynamics for the n-th degree of freedom is given by $L_P^2\ Tr\ (dq_{Bn}/d\tau)^2$. The full action for the total matrix dynamics (space-time part) is $S \sim \sum_n \int d\tau\ L_P^2\ Tr\ (dq_{Bn}/d\tau)^2$. Yang–Mills fields are expressed by the matrices q_{Bn}, pre-gravitation by the $\dot{q}_{Bn}$, the fermionic degrees of freedom by fermionic matrices q_{Fn} and by their 'velocities', $\dot{q}_{Fn}$. Each of the n degrees of freedom has a fundamental action, which is given by [5,10]

$$\frac{S}{\hbar} = \frac{a_0}{2}\int\frac{d\tau}{\tau_P}\,Tr\left[\dot{q}_B^\dagger + i\frac{\alpha}{L}q_B^\dagger + a_0\beta_1\left(\dot{q}_F^\dagger + i\frac{\alpha}{L}q_F^\dagger\right)\right]\times\left[\dot{q}_B + i\frac{\alpha}{L}q_B + a_0\beta_2\left(\dot{q}_F + i\frac{\alpha}{L}q_F\right)\right], \tag{3}$$

where $a_0 \equiv L_P^2/L^2$, and τ_P is Planck time. The net action of this generalised trace dynamics is therefore the sum over n of N copies of the above action, one copy for each degree of freedom, and this new action replaces action (1) in the pre-theory. This full action defines the pre-spacetime, pre-quantum theory, with each degree of freedom (defined by the above action) considered as an 'atom' of space-time-matter (an STM atom). L is a length parameter (scaled with respect to L_P; q_B and q_F have dimensions of length), which characterise the STM atom, and α is the dimensionless Yang–Mills coupling constant. β_1 and β_2 are two unequal complex Grassmann numbers [5].

The subsequent analysis of this pre-space-time, pre-quantum theory is carried out analogously to the pre-quantum trace dynamics. Equations of motion are derived, and there is again a conserved Noether charge. Assuming that the theory is valid at the Planck time scale, the coarse-grained emergent low-energy approximation obeys quantum commutation rules and Heisenberg equations of motion, and this is also the sought for reformulation of quantum theory without classical time. The emergent dynamics is also the desired quantum theory of QR in IR. If a sufficient number of STM atoms get entangled, the anti-self-adjoint part of the Hamiltonian becomes important, and spontaneous localisation results; the fermionic part of the entangled STM atoms gets localised. There hence emerges a 4D classical space-time manifold (labelled by the positions of collapsed fermions), which is sourced by point masses and by gauge fields, and whose geometry obeys the laws of GR given above by Equation (1). Those STM atoms which are not sufficiently entangled, continue to remain quantum; their dynamics are described by the low energy pre-theory itself, or approximately by quantum field theory on the 4D space-time background generated by the entangled and collapsed fermions (these being the macroscopic bodies of the universe).

Note that the non-commutative coordinate system $\{e_i\ ;\ i = 1,2,\ldots,n\}$ is not impacted by the coarse-graining. The averaging takes place only over the time-scale τ and hence over energy; therefore, the non-commuting coordinates e_i remain valid at low energies as well. What, then, should we choose as our e_i, in place of the four real numbers (t,x,y,z) which label the 4D space-time manifold in classical physics? We take clue from the normed division algebras, i.e., number systems, in which the four operations of addition, subtraction, multiplication and division can be defined. There are only four such number systems: real numbers $\mathbb{R}$, complex numbers $\mathbb{C}$, quaternions $\mathbb{H}$, and the octonions $\mathbb{O}$. A quaternion, $H = (a_0e_0 + a_1\hat{\imath} + a_2\hat{\jmath} + a_3\hat{k})$, is a generalisation of complex numbers, such that the a_i here are reals, $e_0^2 = 1$ and

$$\hat{\imath}^2 = \hat{\jmath}^2 = \hat{k}^2 = -1\ ; \quad \hat{\imath}\hat{\jmath} = -\hat{\jmath}\hat{\imath} = \hat{k}; \quad \hat{\jmath}\hat{k} = -\hat{k}\hat{\jmath} = \hat{\imath}; \quad \hat{k}\hat{\imath} = -\hat{\imath}\hat{k} = \hat{\jmath}\,. \tag{4}$$

Quaternions are used to describe rotations in three dimensions, and the automorphism group formed by the three imaginary directions is $SO(3)$. Complex quaternions $\mathbb{C}\times\mathbb{H}$ generate the Lorentz algebra $SL(2,\mathbb{C}) \sim SO(1,3)$ and the Clifford algebra $Cl(2)$, if one of the three quaternionic imaginary directions is kept fixed. If no direction is kept fixed, they generate the Lorentz algebra in 6D: $SL(2,\mathbb{H}) \sim SO(1,5)$ and the Clifford algebra $Cl(3)$. However, they are not a big enough number system for unifying all the standard model symmetries with the Lorentz symmetry, whereas the octonions seem to be just right for that purpose.

An octonion is defined as $O = a_0e_0 + a_1e_1 + a_2e_2 + a_3e_3 + a_4e_4 + a_5e_5 + a_6e_6 + a_7e_7$ such that the a_i are reals, $e_0^2 = 1$, each of the seven imaginary directions $(e_1, e_2, \ldots, e_7)$ squares to -1, these directions anti-commute with each other, and their multiplication rule is given by the so-called Fano plane. Octonionic multiplication is non-associative. The imaginary directions form the automorphism group G_2, which is the smallest of the five exceptional Lie groups G_2, F_4, E_6, E_7, E_8, all of which have to do with the symmetries of the octonion algebra. F_4 is the automorphism group of the exceptional Jordan algebra: the algebra of 3×3 Hermitean matrices with octonionic entries, and E_6 is the automorphism group of the complexified exceptional Jordan algebra [11]. The octonions are what is thought for non-commuting coordinates e_i, on which the action principle (3) is constructed. They generate 10D space-time: $SL(2,\mathbb{O}) \sim SO(1,9)$. The coordinate geometry of the octonions dictates the allowed symmetry groups, and definition and properties of fermions such as quantisation of electric charge [12], value of the low energy fine structure constant [13], and mass-ratios [14]. The parameter L and the coupling constant α in (3) are determined by the algebra of the octonions, not by the dynamics of q_F and q_B. This way, not only does the geometry tell matter how to move, it also tells matter what to be. The dynamical variables (q_B, q_F) curve the flat geometry $\{e_i\}$; however, even before the dynamics are switched on, the low-energy standard model of particle physics is fixed by the e_i, unlike when space-time is $\mathbb{R}^4$. The transition $e_i \rightarrow q_F^ie_i + q_B^ie_i$ is akin to the transition $\eta_{\mu\nu}x^\mu x^\nu \rightarrow g_{\mu\nu}x^\mu x^\nu$, with the important difference that the former transition takes place at the 'square-root of metric' level, as if for tetrads, and the matrices q_B and q_F incorporate standard model forces besides gravity, and also fermionic matter. In fact, with the redefinition, $\breve{\dot{Q}}_B \equiv (i\alpha q_B + L\dot{q}_B)/L$ and $\breve{\dot{Q}}_F \equiv (i\alpha q_F + L\dot{q}_F)/L$, the Lagrangian in Equation (3) can be brought to the elegant and revealing form, as if describing a two-dimensional (because $\beta_1 \neq \beta_2$) free particle:

$$\frac{S}{\hbar} = \frac{a_0}{2}\int \frac{d\tau}{\tau_{Pl}}\, Tr\left(\breve{\dot{Q}}_B^{\dagger} + \frac{L_P^2}{L^2}\beta_1 \breve{\dot{Q}}_F^{\dagger}\right)\left(\breve{\dot{Q}}_B + \frac{L_P^2}{L^2}\beta_2 \breve{\dot{Q}}_F\right). \tag{5}$$

In this fundamental form of the action, the coupling constant α is not present. Actually, α, along with mass ratios, emerges only after (left-right) symmetry breaking segregates the unifying dynamical variable $\breve{\dot{Q}}_B$ into its gravitational part $\dot{q}_B$ and Yang–Mills part q_B.

3. Spinor States for Quarks and Leptons from the Algebra of the Complex Octonions

The automorphism group G_2 of the octonions has two maximal sub-groups $SU(3)$ and $SO(4) \sim SU(2) \times SU(2)$, the first of which is the element preserver group of the octonions, and the second is the stabiliser group of the quaternions inside the octonions [15]. The two groups have a $SU(2)$ intersection. Keeping one of the seven imaginary directions, say e_7, fixed, the remaining six directions can be used to form an MTIS (maximally totally isotropic subspace) and the following generators (along with their adjoints) for the Clifford algebra $Cl(6)$:

$$\alpha_1 = \frac{-e_5 + ie_4}{2}, \quad \alpha_2 = \frac{-e_3 + ie_1}{2}, \quad \alpha_3 = \frac{-e_6 + ie_2}{2}. \tag{6}$$

(This is a covariant choice as all the imaginary directions are equivalent and interchanging any of them does not change the analysis or results.) From here, one can construct spinors as minimum left ideals of the algebra, by first constructing the idempotent $\Omega\Omega^{\dagger}$, where $\Omega = \alpha_1\alpha_2\alpha_3$. The eight resulting spinors are

$$\begin{aligned} &\mathcal{V} = \Omega\Omega^{\dagger}; \quad V_{ad1} = \alpha_1^{\dagger}\mathcal{V}; \quad V_{ad2} = \alpha_2^{\dagger}\mathcal{V}; \quad V_{ad3} = \alpha_3^{\dagger}\mathcal{V}; \\ &V_{u1} = \alpha_3^{\dagger}\alpha_2^{\dagger}\mathcal{V}; \quad V_{u2} = \alpha_1^{\dagger}\alpha_3^{\dagger}\mathcal{V}; \quad V_{u3} = \alpha_2^{\dagger}\alpha_1^{\dagger}\mathcal{V}; \quad V_{e+} = \alpha_3^{\dagger}\alpha_2^{\dagger}\alpha_1^{\dagger}\mathcal{V}. \end{aligned} \tag{7}$$

After defining the operator $Q = (\alpha_1^{\dagger}\alpha_1 + \alpha_2^{\dagger}\alpha_2 + \alpha_3^{\dagger}\alpha_3)/3$ as one-third of the $U(1)$ number operator, we find that the states $\mathcal{V}$ and V_{e+} are singlets under $SU(3)$ and respectively have the eigenvalues $Q = 0$ and $Q = 1$. The states $V_{ad1}, V_{ad2}, V_{ad3}$ are anti-triplets under $SU(3)$ and have $Q = 1/3$ each, whereas the states V_{u1}, V_{u2}, V_{u3} are triplets under $SU(3)$ and each have $Q = 2/3$. These results allow Q to be interpreted as electric charge, and the eight states represent a neutrino, three anti-down quarks, three up quarks and the positron having the standard model symmetries $SU(3)_c \times U(1)_{em}$ (subscripts 'c' and 'em' stand for 'color' and 'electromagnetic', respectively). Anti-particle states are obtained by complex conjugation. The eight $SU(3)$ generators can also be expressed in terms of the octonions and represent the eight gluons, whereas the $U(1)$ generator is for the photon. We hence see the standard model of particle physics emerging from the symmetries of the physical octonionic space, and the quantisation of electric charge is a consequence of the coordinate geometry of the octonions [12].

To see how the weak force (and electroweak) and chiral gravity emerge from the other maximal sub-group $SO(4) \sim SU(2) \times SU(2)$, we must consider three fermion generations and the larger exceptional Lie group E_6 because these symmetries are shared pair-wise across fermion generations, as shown in Figure 1. Furthermore, the neutrino is assumed to be Majorana because only then the correct values of mass ratios are obtained [14]. In addition, notably, E_6 is the only one of the exceptional groups having complex representations.

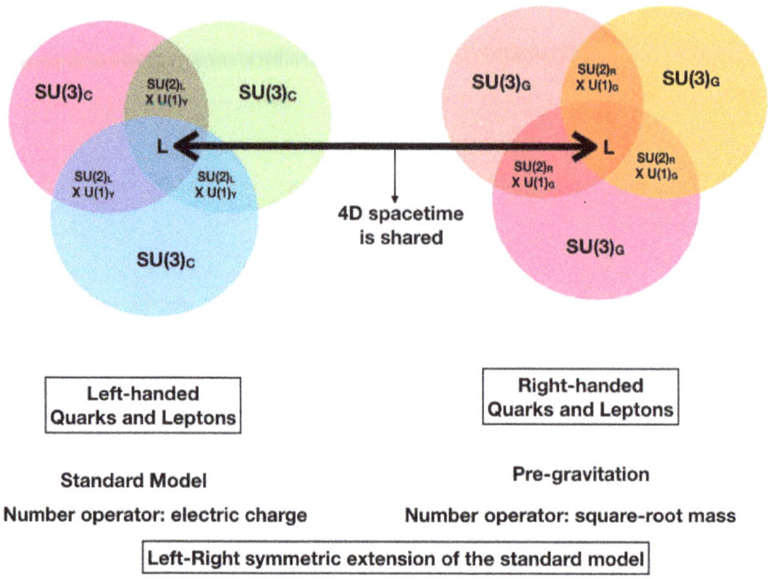

Figure 1. Unification from symmetries of E_6. See text for details.

4. Pre-Gravitation and Mass Ratios from a Left-Right Symmetric Extension of the Standard Model of Particle Physics

The 78-dimensional exceptional Lie group E_6 is the automorphism group of the complexified Jordan algebra, and admits the sub-group structure shown in Figure 1, as motivated by the discussion in [16]. E_6 contains three intersecting copies of $Spin(9,1) \sim SL(2,\mathbb{O})$, which have an $SO(8)$ intersection, and the triality property of $SO(8)$ motivates that there are exactly three fermion generations. In order to account for the symmetries of E_6 and to obtain chiral fermions, we now work with split bioctonions (instead of octonions, which are used in Equation (5), i.e., before symmetry breaking) [17].

Embedded in the three $Spin(9,1)$ are three copies of $SU(3)$, one of which is $SU(3)_c$ and one is for generational symmetry—this is shown in the left part of Figure 1. There is a pairwise intersection amongst a pair of generations, which is the electroweak group $SU(2)_L \times U(1)_Y$ (subscript 'L' denotes 'left' and subscript 'Y' stays for 'hypercarge') from which the weak interaction and electromagnetism (EM) can be obtained. There is a three way intersection marked L which is the 4D Lorentz group $SO(3,1)$. The $Cl(4)$ generators of $SU(2)_L$ are constructed from the $Cl(6)$ of and $Cl(2)$ of the Lorentz algebra, and it can be shown that $SU(2)_L$ acts only on left handed (LH) fermions. The spinor states for the LH quarks and leptons of one generation are constructed analogously to those in spinors (7), by using the LH active Majorana neutrino as the idempotent, with complex conjugation giving the corresponding antiparticles. The spinor states for the second and third generation are respectively obtained by applying two successive $2\pi/3$ rotations on the eight states of the first generation, while staying in the plane defined by the form $(e_i + ie_j)$ of a given first generation particle ($SO(8)$ symmetry implies eight independent great circles on an 8-sphere, one for each of the eight particles, and three particles of three generations). There are a total of 24 LH fermions and their 24 anti-particles and 12 gauge bosons. The unified symmetry group of Lagrangian (5) is E_6.

Similarly, three generations of right-handed (RH) fermions are obtained by using split octonions and the three RH sterile Majorana neutrinos as the idempotent. We identify the associated $SU(3)$ with $SU(3)_G$—a newly introduced gravitational sector; and identify the $U(1)$ number operator with ($\pm$square-root) of the mass of a quark/lepton (in Planck mass units), and the eight respective spinor states of one generation are: the sterile neutrino, three positrons of three different gravi-colors, three RH up quarks of three colors same as $SU(3)_c$, and one down quark which is a singlet under $SU(3)_G$, along with the singlet sterile neutrino [18]. These obtain the respective square-root mass number $(0, 1/3, 2/3, 1)$ explaining why the down quark is nine times heavier than the electron. The $SU(2)_R$ is RH chiral gravity (LQG, loop quantum gravity) [19], which reduces, in the classical limit induced by spontaneous localisation, to GR. The Lorentz group (whose Casimir invariant is the introduced mass number) is common with the LH particles, and its six dimensions, together with the 24 RH fermions and 12 new gauge bosons, when added to the LH sector, give the correct count of 78 for E_6. The split complex number gives a scalar field that acts as the Higgs mediating between the LH charge eigenstates, and RH mass eigenstates. Is $U(1)$ gravity the dark energy? This possibility is discussed further in Section 5.

The group E_6 is also the symmetry group for the Dirac equation in 10D [16] for three fermion generations (either LH or RH). The eigenvalue and eigenmatrix problem for the Dirac equation is in fact the same as $J_3(8)X = \lambda X$, where $J_3(8)$ is the exceptional Jordan algebra with symmetry group F_4. Substituting the above-mentioned spinor states of LH fermions (these being eigenstates of electric charge) and solving this eigenvalue problem expresses the LH charge eigenstates as superpositions of the RH mass eigenstates (thus fixing α and L in Equation (3)), and the ratios of the eigenvalues yield mass ratios of charged fermions as shown in Figure 2; these exhibit very good agreement with the mysterious mass ratios, as shown in the Table 1 below [14].

E_6 has three copies of 10D spacetime. We never compactify the extra six complex dimensions—these represent the standard model internal forces which determine the geometry of these extra dimensions. Quantum systems do not live in 4D spacetime. They live in E_6 and their true dynamics is the generalised trace dynamics, with evolution given by Connes time. Only classical systems live in 4D spacetime, where they descend as a result of spontaneous localisation of highly entangled fermions (compactification without compactification). This overcomes the troublesome non-unique compactification problem of string theory.

Mass ratios: Square root of the mass of a charged fermion with respect to the down quark

Down quark	Strange quark	Bottom quark
1	$\frac{1+\sqrt{3/8}}{1-\sqrt{3/8}}$	$\frac{1+\sqrt{3/8}}{1-\sqrt{3/8}} \times \frac{1+\sqrt{3/8}}{1-\sqrt{3/8}} \times \frac{1+\sqrt{3/8}}{1}$
Up quark	**Charm quark**	**Top quark**
$2/3$	$\frac{2}{3} \times \frac{\frac{2}{3}+\sqrt{3/8}}{\frac{2}{3}-\sqrt{3/8}}$	$\frac{2}{3} \times \frac{\frac{2}{3}+\sqrt{3/8}}{\frac{2}{3}-\sqrt{3/8}} \times \frac{\frac{2}{3}}{\frac{2}{3}-\sqrt{3/8}}$
Electron	**Muon**	**Tau lepton**
$1/3$	$\frac{1}{3} \times \frac{1+\sqrt{3/8}}{1-\sqrt{3/8}} \times \frac{1/3+\sqrt{3/8}}{\lvert 1/3-\sqrt{3/8}\rvert}$	$\frac{1}{3} \times \frac{1+\sqrt{3/8}}{1-\sqrt{3/8}} \times \frac{1+\sqrt{3/8}}{1-\sqrt{3/8}} \times \frac{1/3+\sqrt{3/8}}{\lvert 1/3-\sqrt{3/8}\rvert}$

Figure 2. Square-root mass ratios of charged elementary fermions [14].

Table 1. Comparison of theoretically predicted square-root mass ratios with experimentally known range [14].

Square Root Mass Ratios			
Particles	**Theoretical Mass Ratio**	**Minimum Experimental Value**	**Maximum Experimental Value**
muon/electron	14.10	14.37913078	14.37913090
taun/electron	58.64	58.9660	58.9700
charm/up	23.57	21.04	26.87
top/up	289.26	248.18	310.07
strange/down	4.16	4.21	4.86
bottom/down	28.44	28.25	30.97

Apart from the two mass ratios of charged leptons, other theoretical mass ratios lie within the experimental bounds [20]. On accounting for the so-called Karolyhazy correction [21], we might possibly get more accurate mass ratios for all particles including charged leptons. This will be investigated in future work.

In quantum theory, even at low energies, assuming a point structure for spacetime is an approximation; it is because of this approximation that the standard model of particle physics has so many unexplained free parameters. When we replace this approximate description by a non-commutative spacetime, we find evidence that these parameter values get fixed. In particular, we derive the low energy fine structure constant [13,21] and mass ratios of charged fermions [14] from first principles. We do not need experiments at ever higher energies to understand the low energy standard model. Instead, we need a better understanding of the quantum nature of spacetime at low energies, such that the quantum spacetime is consistent with the principle of quantum linear superposition.

5. Further Developments, Clarifying Remarks, and Current Status of the Present Unification Programme

The aforesaid essay is intended to give the reader a short overview of a new approach to quantum gravity and unification, details of which can be found in [5]. This Section reports on a few new insights not described in our earlier work, and provide clarifying details on some of the statements in the previous sections.

One further way to motivate the present theory is to recall that, when one takes the square root of the Klein–Gordon equation to arrive at the Dirac equation for spin-half fermions described by spinors, one does not take the corresponding square root of the four-dimensional Minkowski spacetime labelled by four real numbers. However, suppose one were to take the latter square root as well; then, one arrives at a spinor description of spacetime, i.e., Penrose's twistor space, labelled by complex numbers, and one notes that $SL(2,C)$ is the double cover of the Lorentz group $SO(3,1)$. We can now try to describe fermions on this twistor space (both the fermions as well as the space-time are now spinorial) and we can write down the Dirac equation on this twistor space. We can go even further and replace the complex numbers by one or the other of the only two additional division algebras, the quaternions (H) and the octonions (O). Quaternionic twistor space is equivalent to 6D spacetime because $SL(2,H)$ is the double cover of $SO(5,1)$, whereas octonion-valued twistor space is equivalent to 10D spacetime because $SL(2,O)$ is the double cover of $SO(9,1)$. When we describe fermions on octonionic twistor space, we begin to deduce remarkable results such as three generations of quarks and leptons, and quantisation of electric charge, just as in the standard model. We are now solving the Dirac equation on the spinor equivalent of 10D Minkowski space-time, and this implies mass quantisation, a derivation of the low energy fine structure constant and of mass ratios of quarks and charged leptons. The bosonic sector now includes pre-gravitation in unification with the standard model forces, and the theory is shown to obey an $E_8 \times E_8$ symmetry [18]. Indeed, the theory we arrive at is a revised string theory without the troublesome nonuniqueness problem of compactification, and the fundamental entities are 2-branes with an area of the order of Planck area.

It is important to note that, in going from real numbers to complex numbers to octonions, we have not changed the energy scale of the problem; rather, we have gone from 4D Minkowski spacetime to 8D twistor space. In so doing, we have simply found a new mathematical description of the standard model which explains the standard model origin and its unification with pre-gravitation. We therefore arrive at the inescapable conclusion that elementary particles live in a space with $E_8 \times E_8$ symmetry even at low energies (built on a 10D complex non-commutative spacetime). They do not live in ordinary spacetime, and definitely not in 4D classical space-time. The extra dimensions are not Planck size, but large extra dimensions, and have an absolute modulus of the order of the scale of the strong force and the weak force (10^{-18} m to 10^{-15} m). This can be approximately justified by recalling that, in this theory, the absolute modulus, l, of the extra dimensions goes as $l \sim R^{1/3}$ (because of the holographic principle [22]), where R is the cosmic length scale, and both l and R are expressed in units of Planck length. Further work remains to be done to make this result rigorous, and also to derive the Higgs mass and the masses of the W bosons from fundamental considerations.

The action principle in Equation (3) is motivated by the action principle for a free particle in Newton's mechanics, which of course is nothing but the integral of the kinetic energy over absolute time. Now, over octonionic space (more precisely, complex split bioctonionic space) we have written the equivalent expression for the kinetic energy of a 2-brane over Connes time. The undotted variables are related to the left-chiral sector, this being the gauge bosons of the standard model and three generations of left chiral fermions, defined over octonionic space. The dotted variables are related to the right chiral sector, this being pre-gravitation $SU(3)_G \times SU(2)_R \times U(1)_G$ and three generations of right chiral fermions, defined over the split part of bioctonionic space [18]. A detailed investigation of the Lagrangian as regards its particle content is currently in progress. $\dot{q}_B$ is the Dirac operator on octonionic space and q_B, the Yang–Mills field, is the correction to the Dirac operator, as in conventional quantum field theory. The spectral action principle tells us the classical limit of the trace of the Dirac operator squared, when Yang–Mills fields are present. This classical limit has been discussed briefly in our earlier study [10] and is given by the following equation, from [6], where $D_{B\text{new}} \equiv D_B + \alpha A$ is the corrected Dirac operator resulting after including the Yang–Mills potential A:

$$
\begin{aligned}
Tr\,[L_P^2\, D_{B\text{new}}^2] = \frac{N}{48\pi^2}\Bigg[& 12L_P^{-4} f_0 \int d^4x\sqrt{g} + L_P^{-2} f_2 \int d^4x\sqrt{g}R \\
& + f_4 \int d^4x\sqrt{g}\Big[-\frac{3}{20}C_{\mu\nu\rho\sigma}C^{\mu\nu\rho\sigma} + \frac{11}{20}R^*R^* + \frac{1}{10}R_{;\mu}{}^{\mu} \\
& + \frac{g^2}{N}F^i_{\mu\nu}F^{\mu\nu i}\Big] + \mathcal{O}\left(L_P^2\right)\Bigg],
\end{aligned}
$$

where

- $\dfrac{NL_P^{-2}f_2}{48\pi^2}\int d^4x\sqrt{g}R$ term is the Einstein–Hilbert action;
- $\dfrac{NL_P^{-4}f_0}{4\pi^2}\int d^4x\sqrt{g}$ term is responsible for the cosmological constant;
- $\dfrac{f_4 g^2}{48\pi^2}\int d^4x\sqrt{g}F^i_{\mu\nu}F^{\mu\nu i}$ term is the Yang–Mills action;
- $-\dfrac{Nf_4}{320\pi^2}\int d^4x\sqrt{g}C_{\mu\nu\rho\sigma}C^{\mu\nu\rho\sigma}$ term would be responsible for the conformal gravity;
- $\dfrac{11Nf_4}{960\pi^2}\int d^4x\sqrt{g}R^*R^*$ term would be responsible for the Gauss–Bonnet gravity.

This is the expansion of the squared Dirac operator when gauge fields are included alongside gravity. We do not yet take into account the volume term, and conformal gravity, and Gauss–Bonnet gravity in our present work. It is, however, an issue of great significance that in the classical limit, GR is being modified by conformal gravity, and encourages us to relate our theory, described to MOND (modified Newtonian dynamics) and RMOND (relativistic MOND), as an alternative to cold dark matter. Interestingly, there so far seems to be no cold dark matter candidate particle in our theory, and MOND and sterile neutrinos seem to arise naturally in our approach to unification; see Section XIV of [5]; further work is currently in progress in this direction.

The matrix-valued equations of motion are easily written down after first defining:

$$q_1 = q_B + a_0 \beta_1 q_F, \qquad q_2 = q_B + a_0 \beta_2 q_F, \tag{8}$$

where we have set $a_0 \equiv L_P^2/L^2$. In terms of these two variables, the above trace Lagrangian can be written as

$$Tr\mathcal{L} = \frac{1}{2} a_1 a_0 Tr\left[\dot{q}_1 \dot{q}_2 - \frac{\alpha^2 c^2}{L^2} q_1 q_2 + i\frac{\alpha c}{L}(\dot{q}_1 q_2 + q_1 \dot{q}_2)\right], \tag{9}$$

where $a_1 \equiv \hbar / c L_P$. The last term in the trace Lagrangian is a total time derivative, and hence does not contribute to the equations of motion, so that we can get dynamics from the Lagrangian:

$$Tr\mathcal{L} = \frac{1}{2} a_1 a_0 Tr\left[\dot{q}_1 \dot{q}_2 - \frac{\alpha^2 c^2}{L^2} q_1 q_2\right]. \tag{10}$$

The canonical momenta are given by (δ denotes variation with respect to the matrix variable)

$$p_1 = \frac{\delta Tr\mathcal{L}}{\delta \dot{q}_1} = \frac{1}{2} a_1 a_0 \dot{q}_2; \qquad p_2 = \frac{\delta Tr\mathcal{L}}{\delta \dot{q}_2} = \frac{1}{2} a_1 a_0 \dot{q}_1 . \tag{11}$$

The Euler–Lagrange equations of motion are

$$\ddot{q}_1 = -\frac{\alpha^2 c^2}{L^2} q_1; \qquad \ddot{q}_2 = -\frac{\alpha^2 c^2}{L^2} q_2 . \tag{12}$$

In terms of these two complex variables, the 2-brane behaves like two independent complex-valued oscillators. However, the degrees of freedom of the 2-brane couple with each other when expressed in terms of the self-adjoint variables q_B and q_F. This is because q_1 and q_2 both depend on q_B and q_F, the difference being that q_1 depends on β_1 and q_2 depends on β_2.

The trace Hamiltonian is

$$Tr\mathcal{H} = Tr[p_1 \dot{q}_1 + p_2 \dot{q}_2 - Tr\mathcal{L}] = \frac{a_1 a_0}{2} Tr\left[\frac{4}{a_1^2 a_0^2} p_1 p_2 + \frac{\alpha^2 c^2}{L^2} q_1 q_2\right], \tag{13}$$

and the Hamilton's equations of motion are

$$\dot{q}_1 = \frac{2}{a_1 a_0} p_2 \qquad \dot{q}_2 = \frac{2}{a_1 a_0} p_1; \qquad \dot{p}_1 = -\frac{a_1 a_0 \alpha^2 c^2}{2L^2} q_2; \qquad \dot{p}_2 = -\frac{a_1 a_0 \alpha^2 c^2}{2L^2} q_1 . \tag{14}$$

It is understood that these generalised trace dynamics are defined over complex bioctonionic space, and elementary particles and gauge bosons including those for pre-gravitation are special cases of these dynamical variables, reminiscent of the different vibrations of the string in string theory.

5.1. Octonions, Elementary Particle Physics, and Gravitation

Octonions have been associated with physics for a very long time. In fact, already in the 1920s, Jordan discovered the algebraic formulation of quantum mechanics, and the Jordan algebras, and, in particular, his work led to the discovery of the exceptional Jordan algebra $J_3(8)$ (also known as the Albert algebra). This is the algebra of 3×3 Hermitean matrices with octonionic entries. Since the 1970s, there have been extensive investigations of how the octonions bear a fundamental connection with elementary particles, quarks as well as leptons of the standard model [15,23–34].

How does our work compare with and differ from those earlier works on applications of octonions to quantum mechanics and particle physics? Our work is strongly inspired by those earlier investigations and is built on them, especially with regard to the definition of elementary particle states using Clifford algebras of octonionic chains, applications of the exceptional Lie groups and the exceptional Jordan algebra. However, our fundamental perspective is different from those earlier studies, which have been largely focused on relating octonions to the standard model and its extensions, such as Grand Unified Theories (GUTs), to the octonions. Our starting point is fully quantum foundational: to seek a reformulation of quantum theory, which does not depend on classical time. This turns out to be a pre-quantum, pre-spacetime theory, which is a matrix-valued Lagrangian dynamics on an octonionic twistor space. Remarkably, it is also a theory of quantum (pre-)gravity and a unification of the standard model of particle physics with pre-gravitation. The eigenvalue problem for the Dirac equation on octonionic space (equivalently 10D spacetime) is solved by the characteristic equation of the exceptional Jordan algebra. In combination with the Lagrangian of the theory, this leads to a derivation of the low energy fine structure constant, and mass ratios of quarks and charged leptons. Thus, by bringing in gravitation, and trace dynamics, our work significantly expands the scope of earlier related research. This new scenario is summarised in Figure 3. Below, we elaborate on some aspects of these new developments.

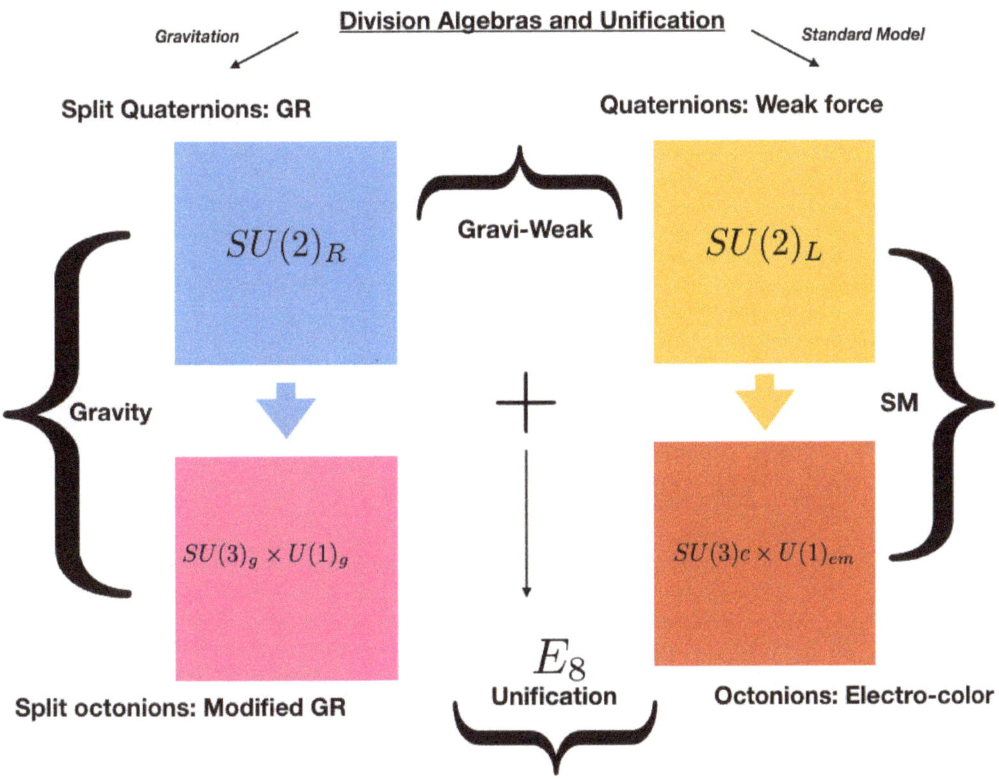

Figure 3. Octonions, and an $E_8 \times E_8$ unification of standard model and pre-gravitation [18]. Here, 'GR' stays for 'general relativity', 'SM' stays for 'standard model'.

5.2. Mass Quantisation from a Number Operator

The masses of the electron, the up quark, and the down quark, are in the ratio 1:4:9. This fact calls for a theoretical explanation. A few years ago, Cohl Furey proved the quantisation of electric charge as a consequence of constructing the states for quarks and leptons from the algebra of the octonions [12]. The complex octonions are used to construct a Clifford algebra $Cl(6)$, which is then used to make states for one generation of quarks and leptons. The automorphism group G_2 of the octonions has a sub-group $SU(3)$, and these particle states have the correct transformation properties as expected if this $SU(3)$ is $SU(3)_c$ of QCD (quantum chromodynamics). Furthermore, (one-third of) a number operator, built from the $Cl(6)$ generators, has the eigenvalues (0, 1/3, 2/3, 1) (with 0 and 1 for the $SU(3)$ singlets and 1/3, 2/3 for the triplets) allowing this to be identified with electric charge. This proves charge quantisation, and the $U(1)$ symmetry of the number operator is identified with $U(1)_{em}$. Anti-particle states obtained by complex conjugation of particle states are shown to have electric charge (0, −1/3, −2/3, −1). Thus, the algebra describes the electro-colour symmetry for the neutrino, down quark, up quark, electron, and their anti-particles. Note that it could instead be the second fermion generation, or the third generation. Each generation has the same charge ratio (0, 1/3, 2/3, 1).

This same analysis can now be used to show that the square-roots of the masses of electron, up and down are in ratio 1:2:3. All we have to do is to identify the eigenvalues of the number operator with the square-root of the mass of an elementary particle, instead of its electric charge. In addition, we also obtain a classification of matter and anti-matter, after noting that complex conjugation now sends matter to anti-matter, as follows:

Matter $\sqrt{\text{mass}}$	Anti–matter $\sqrt{\text{mass}}$
anti–neutrino 0	neutrino 0
electron 1/3	positron −1/3
up–quark 2/3	anti–up −2/3
down–quark 1	anti–down −1

Compared to the electric charge case above, the electron and down quark have switched places, and we already have our answer to the mass quantisation question, placed at the start of this Subsection. There is again an $SU(3)$ and a $U(1)$, but obviously this is no longer QCD and EM. We identify this symmetry is identified here with a newly proposed $SU(3)_G \times U(1)_G$, whose full physical implications remain to be unravelled (GR emerges from $SU(2)_R$ this being an analog of the weak force $SU(2)_L$). $E_6 \times E_6$ admits a sub-group structure with two copies each of $SU(3)$, $SU(2)$ and $U(1)$. Therefore, one set is identified with the standard model $SU(3) \times SU(3) \times U(1)$ (electric charge based) and the other with the newly introduced $SU(3)_G \times SU(2)_R \times U(1)_G$ (sqaure-root mass based). In the early universe, the separation of matter from anti-matter is the separation of particles with positive square-root mass from particles of negative square root mass. This separation effectively converts the vector-interaction of pre-gravitation into an attractive only emergent gravitation.

However, the second and third fermion generations do not have the mass ratios 1:4:9 unlike the electric charge ratios which are the same for all three generations. Why so? Because mass eigenstates are not the same as charge eigenstates. We make our measurements using eigenstates of electric charge; these have strange mass ratios, e.g., the muon is 206 times heavier than the electron. If we were to make our measurements using eigenstates of square-root mass, we would find that all three generations have the mass ratios 1:4:9 whereas this time around the electric charge ratios seems to be strange. There is a perfect duality between electric charge and square-root mass. A free electron in flight—is it in a charge eigenstate or a mass eigenstate? Neither. It is in a superposition of both, and collapses to one or the other, depending on what we choose to measure. In fact, the free electron in flight does not separately have a mass and a charge; it has a quantum number which could be called charge-sqrt mass, which is the quantum number for the unified force. Unification is broken by measurement: if we measure EM effect then we attribute electric charge to the source. If we measure inertia or gravity, we attribute mass to the source. These statements are independent of energy scale. A classical measuring apparatus emerges from its quantum constituents as a consequence of sufficient entanglement: the emergence of such classical apparatus is the prelude to the breaking of unification symmetry. In the early universe, sufficient entanglement is impossible above a certain energy (possibly the EW scale), and it appears as if symmetry breaking depends on energy. This is only an indirect dependence. The true dependence of symmetry breaking is on the amount of entanglement. In our current low energy universe, we have both low entanglement systems (quantum, unified) and high entanglement systems (classical, unification broken).

5.3. Some Further Insights into the Origin of Mass Ratios for Three Fermion Generations

Prior to the left-right symmetry breaking, the $E_8 \times E_8$ symmetry is intact. We have three identical fermion generations of lepto-quark states having an associated $U(1)$ quantum number, which we call 'electric-charge-square-root-mass' and which takes the values $(0, 1/3, 2/3, 1)$. These lepto-quarks are excitations of the Dirac neutrino (quantum number 0) and can be labelled as LHdownquark-RHelectron (1/3), LHupquark-RHupquark (2/3) and LHelectron-RHdownquark (1) with the understanding that all three generations are identical. The left-right symmetry breaking segregates these lepto-quark states into left chiral fermions with the associated $U(1)$ quantum number being electric charge, and the right chiral fermions with the associated $U(1)$ quantum number being square-root mass. The Dirac neutrino segregates into a LH active Majorana neutrino and a RH sterile Majorana neutrino. The LH chiral fermions are excitations of the active LH Majorana neutrino, and are the anti-down quark (1/3), up quark (2/3) and positron (1) and their antiparticles. The RH chiral fermions are excitations of the sterile LH Majorana neutrino, and are the electron (1/3), up quark (2/3) and the down quark (1); numbers in brackets now show square-root mass. A Higgs coming from the right sector gives masses to the LH fermions; and, remarkably, a charged Higgs coming from the left sector gives electric charge to the LH fermions. The gauge symmetry associated with the LH sector is the standard model symmetry $SU(3)_c \times SU(2)_L \times U(1)_Y$ and that associated with the RH sector is pre-gravitation $SU(3)_G \times SU(2)_R \times U(1)_G$.

The LH states, being eigenstates of electric charge, are different from the RH states, which are eigenstates of square-root mass. When we solve the Dirac equation for three generations of a family with a given electric charge, assuming the neutrino to be Majorana, it reveals mass quantisation, and the charge eigenstates are superpositions of square-root mass eigenstates and the corresponding eigenvalues carry information about the mass ratios across three generations. We called these eigenvalues 'Jordan eigenvalues' [21]; they are shown in the table in Figure 4.

For a given electric charge value $q \neq 0$, the three eigenvalues take the form $q \pm \epsilon\sqrt{3/8}$ where $\epsilon = (-1, 0, 1)$. $\sqrt{3/8}$ is also the magnitude of the octonion, which describes the state of the three generations of a family of charged fermions. For the neutrino, the $\sqrt{3/8}$ is replaced by $\sqrt{3/2}$. If the Jordan eigenvalues are calculated assuming the neutrino to be a Dirac particle then there is a subtle change in the values: the $\sqrt{3/8}$ is replaced by $\sqrt{3/2}$ for the charged leptons (no other change), and for the neutrino the eigenvalues are now very different: $(-1/2 - \sqrt{3/2}, 1, -1/2 + \sqrt{3/2})$. The shift from $\sqrt{3/2}$ to $\sqrt{3/8}$ results because, in going from the Dirac neutrino to the Majorana neutrino, the lepto-quark state splits into two halves—the LH charge eigenstate and the RH mass eigenstate—and hence the magnitude $\sqrt{3/2}$ is divided into two equal halves of $\sqrt{3/8}$ each. The Jordan eigenvalues hold for the anti-particles as well, with the charge q being replaced by its negative value $-q$. Obviously, the eigenvalues of the neutrino do not change from the values, shown in Figure 4; consistent with the neutrino being its own anti-particle.

The Jordan Eigenvalues

Neutrinos: Magntitude 3/4	$-\frac{\sqrt{3}}{2}$	**0**	$\frac{\sqrt{3}}{2}$
1/3 Quarks: Mag. 3/8	$\frac{1}{3}-\sqrt{\frac{3}{8}}$	$\frac{1}{3}$	$\frac{1}{3}+\sqrt{\frac{3}{8}}$
2/3 Quarks Mag. 3/8	$\frac{2}{3}-\sqrt{\frac{3}{8}}$	$\frac{2}{3}$	$\frac{2}{3}+\sqrt{\frac{3}{8}}$
Charged Leptons Mag. 3/8	$1-\sqrt{\frac{3}{8}}$	**1**	$1+\sqrt{\frac{3}{8}}$

These are all numbers in Base four !!

Figure 4. The eigenvalues of the characteristic equation of the exceptional Jordan algebra, for quarks and leptons [21].

Furthermore, these Jordan eigenvalues also represent superposition of square-root mass eigenstates in terms of charge eigenstates, with the down quark family interchanged with the charged lepton family. This is because the electric charge values $1/3$, $2/3$, and 1 for the down quark family, up quark family, and electron family, respectively, are numerically also the same as square-root mass numbers respectively for the electron family, up quark family, and down quark family. Thus, the eigenvalues for the up quark family stay unchanged when charge and square-root mass number are interchanged for the electron family and down quark family. As soon as mass eigenstates are superpositions of charge eigenstates, the eigenvalues in the superposition determine the peculiar observed mass ratios, we now elaborate in some detail below using the related data shown in Figure 5.

The first set of three rows are marked $[1/3]D$, where D stays for 'Dirac neutrino' and $1/3$ is the 'electric-charge-square-root-mass' quantum number prior to symmetry breaking, identical for three generations. The 4th, 5th, 6th rows are marked $[2/3]D$ for the same quantum number taking the value $2/3$. The 7th, 8th and 9th rows are marked $[1]D$ because this quantum number takes the value 1. For a given value of this quantum number, the left-right symmetric lepto-quark can be written as superposition of a LH fermion and a RH fermion, with the two Jordan eigenvalues in any of the nine rows giving the numerical coefficient of the superposition. The successive eigenvalues are labelled as the pairs (A1, B1), (A2, B2), (C1, C2), (D1, D2), (E1, E2), (F1, F2), (G1, G2), (H1, H2), (I1, I2). The first Jordan eigenvalue in any given row labels the LH fermion, and the second eigenvalue is the same row labels the right handed (RH) fermion.

$$[1/3]D \to (1/3 - \sqrt{3/8})\ d1 + (1/3 + \sqrt{3/8})\ e1 \quad \textbf{A1 , A2}$$

$$[1/3]D \to (1/3)\ d2 + (1/3)\ e2 \quad \textbf{B1 , B2}$$

$$[1/3]D \to (1/3 + \sqrt{3/8})\ d3 + (1/3 - \sqrt{3/8})\ e3 \quad \textbf{C1, C2}$$

e=B1, mu/e = (C1/A1) (I1/G1) = tau/mu = (C1/A1) (I1/G1)

$$[2/3]D \to (2/3 - \sqrt{3/8})\ u3 + (2/3 + \sqrt{3/8})\ u1 \quad \textbf{D1, D2}$$

$$[2/3]D \to (2/3)\ u1 + (2/3)\ u2 \quad \textbf{E1, E2}$$

$$[2/3]D \to (2/3 + \sqrt{3/8})\ u2 + (2/3 - \sqrt{3/8})\ u3 \quad \textbf{F1, F2}$$

u=E1, c/u = F1/D1, t/c = E1/D1

$$[1]D \to (1 - \sqrt{3/8})\ e1 + (1 + \sqrt{3/8})\ d1 \quad \textbf{G1, G2}$$

$$[1]D \to (1)\ e2 + (1)\ d2 \quad \textbf{H1, H2}$$

$$[1]D \to (1 + \sqrt{3/8})\ e3 + (1 - \sqrt{3/8})\ d3 \quad \textbf{I1, I2}$$

d=H1, s/d=I1/G1, b/s = (I1/G1) (I1/H1)

Figure 5. From the Jordan eigenvalues, towards an understanding of the mass ratios. $e1, e2, e3$ denote the charged lepton family. $d1, d2, d3$ denote the down quark family. $u1, u2, u3$ denote the up quark family. D stays for the 'Dirac neutrino.' See text for more details.

When the left-right symmetry breaks, the $U(1)$ quantum number prior to symmetry breaking becomes the $U(1)$ number electric charge for all LH fermions and takes the same set of values $(0, 1/3, 2/3, 1)$ across the three generations. Analogously, it takes the same set of values $(0, 1/3, 2/3, 1)$ as the square-root mass number across the three generations of RH fermions, with the relative position of the electron and down quark interchanged. The up quark family positions stay unchanged at 2/3. Since we make all measurements using electric charge eigenstates (and not mass eigenstates), the mass eigenstates manifest in measurements as superpositions of electric charge eigenstates, and the ratios of Jordan eigenvalues reveal the observed mass ratios. Had we been making measurements using mass eigenstates, the electric charge eigenstates would manifest as superpositions of mass eigenstates, and we would have observed strange charge ratios.

Note that, in every set of three rows for a given $[1/n]D$, there is a row where the left quantum number is the same as the right quantum number. These are the rows two, five and eight, with the numbers (B1, B2), (E1, E2) and (H1, H2). These give rise to the mass ratios $(1/3, 2/3, 1)$ for the lightest charged fermions: electron, up quark, down quark. The other two rows with any given $[1/n]D$ flip eigenvalues. These are the rows one and three, four and six, seven and nine. That is, (A1, A2) interchanges eigenvalues with (C1, C2), (D1, D2) with (F1, F2), and (G1, G2) with (I1, I2).

The mass ratios arrived at, from these eigenvalues, as follows, simplest case being the up quark family (up, charm, top) with its quarks, labelled $u1$, $u2$, $u3$. The up quark has the square-root mass ratio u = E1 =2/3. The charm to up square-root mass ratio is $c/u = F1/D1 = (2/3 + \sqrt{3/8})/(2/3 - \sqrt{3/8})$. The top to charm ratio is $t/c = E1/D1 = (2/3)/(2/3 - \sqrt{3/8})$.

The down quark family (down, strange, bottom) and the electron family (electron, muon, tau lepton) are mixed, as can be seen in the first three and last three rows. For the

[1/3]D rows, the first entry in every row is from the down quark family, and the second entry from the electron family, whereas the roles are reversed in the [1]D rows.

As per the last three rows, the down quark has the square-root mass ratio d = H1 = 1. The strange quark has the square-root mass ratio s/d = I1/G1 = $(1+\sqrt{3/8})/(1-\sqrt{3/8})$. The bottom to strange ratio is (I1/G1) (I1/H1). The origin of the peculiar factor I1/H1 remains to be understood.

From the first three rows, we see that the electron has the square-root mass ratio e = B1 = 1/3. The muon has the square-root mass ratio mu/e = (C1/A1)/(I1/G1), and this same ratio also holds for tau/mu.

These ratios above are the same as those, shown earlier in Figure 2. We believe that the octonionic theory provides a reasonably good understanding of the observed mass ratios of charged fermions. The mixing of the down quark family and the electron family is possibly the result of a gauge-gravity duality and of the fact that the three generations are not entirely independent of each other but related by the triality property of $SO(8)$ [18]. Another remarkable feature we observed is that the eigenmatrices corresponding to the Jordan eigenvalues for the charged fermions always have the diagonal entry as 1/3, irrespective of whether the associated quantum number is 1/3 or 2/3 or 1. This seems to suggest that all charged fermions are made of three base states that all have an associated quantum number 1/3. The possible consequences of these observations are currently being investigated.

5.4. Octonions and the Koide Formula

The Koide formula [35] is the following observation for the experimentally measured masses of charged leptons:

$$\frac{m_e+m_\mu+m_\tau}{(\sqrt{m_e}+\sqrt{m_\mu}+\sqrt{m_\tau})^2}=0.666661(7)\approx\frac{2}{3}. \tag{15}$$

That is, this ratio is close to (and a little less than) 2/3, but not exactly 2/3. Remarkably, the octonionic theory explains when the ratio is exactly 2/3, and why it departs from that exact value.

Prior to the left-right symmetry breaking, we can consider that a LH-electron-RH-electron state has an associated electric-charge-square-rot-mass of 1, and the neutrino is a Dirac fermion. In this case, the Jordan eigenvalues, as we mentioned above, are $(1-\sqrt{3/2},1,1+\sqrt{3/2})$. These give the superposition amplitudes when RH mass eigenstates are expressed as superposition of LH mass eigenstates. The Koide ratio is then

$$\frac{(1+\sqrt{3/2})^2+(1)^2+(1-\sqrt{3/2})^2}{3^2}=\frac{2}{3}. \tag{16}$$

Thus, the exact value 2/3 is realised prior to symmetry breaking and prior to when the RH electron and RH down quark switch places. This switch might help understand why the mass ratios for charged leptons know about the Jordan eigenvalues $(1+\sqrt{3/8})$ and $(1-\sqrt{3/8})$, which are otherwise associated with the down quark family.

Using our theoretical mass ratios for the charged leptons as reported in Figure 2, we get the following theoretically predicted value for the Koide ratio,

$$\frac{m_e+m_\mu+m_\tau}{(\sqrt{m_e}+\sqrt{m_\mu}+\sqrt{m_\tau})^2}=0.669163\approx\frac{2}{3}, \tag{17}$$

which is greater than the experimentally measured value of the Koide ratio and also greater than 2/3. The departure from the exact value of 2/3 is a consequence of the left-right symmetry breaking and of the switch between the RH down quark and RH electron. (It remains to be seen if the Karolyhazy correction will predict an exact match between theory and experiment).

Since unification already takes place at low energies (i.e., whenever the system is quantum and not yet measured upon), it follows that, before we make a measurement on the charged leptons to measure their masses, the Koide ratio is exactly 2/3. After the measurement is made, the theoretical prediction for the resulting value is 0.669163, whereas the measured value is smaller than 2/3. The uncertainty in the mass of the tau-lepton 1776.86(12) MeV is such that, by demanding the Koide ratio to be 2/3, one can predict the mass of the tau-lepton to be 1776.969 MeV. At the upper limit 1776.98 MeV of the experimentally measured tau-lepton mass, the ratio is 0.66666728706, i.e., larger than 2/3, but smaller than our predicted theoretical value for Minkowski spacetime (the value realised after measurement).

The above is an important result as we now know when the Koide ratio is exactly 2/3 (it is when the electron is not being observed). In addition, we understand why the measured value of this ratio is not exactly 2/3. Actually, the measured value coud be demanded to be equal to the theoretical value, and thereby fixing the mass of the tau-lepton. It turns out there are no such allowed values for the tau-lepton mass, which is further evidence that the measured value could lie between the spinor spacetime value (2/3) and the Minkowski spacetime value if the mass of the tau lepton is greater than 1776.969 MeV. This is also indirect evidence that sterile neutrinos exist.

5.5. Why Is Matter Electrically Neutral?

When left-right symmetry breaking mechanism in the early universe separated matter from anti-matter, particles were segregated from their anti-particles. In addition, however, the sign of the electric charge was not the criterion for deciding what went where. Matter has the positively charged up quark (2/3) and the negatively charged down quark ($-1/3$) and the electron (-1). Anti-matter has their anti-particles. If sign of electric charge was the deciding criterion for separating matter from anti-matter, all particles in our universe ought to have had the same sign of charge. That is not the case, and moreover, the matter is electrically neutral. How could that have come about? Even the algebraic proof based on the octonions, which shows quantisation of electric charge, naturally clubs positively charged particles together, when their states are made from a Clifford algebra:

Particles Electric charge	Anti–particles Electric charge
anti–neutrino 0	neutrino 0
anti–down quark 1/3	down $-1/3$
up–quark 2/3	anti–up $-2/3$
positron 1	electron -1

What picks the up quark from the left, and down and electron from the right, and clubs them as matter, and yet maintain electrical neutrality? We have proposed that the criterion distinguishing matter from anti-matter is square-root of mass, not electric charge. One can make a new Clifford algebra afresh from the octonions, and show that square-root of mass is quantised, as in the above mass table. Let us now calculate the net electric charge of matter, remembering that there are three down quarks (color) and three up quarks (color): $0 + (-1 \times 1) + (3 \times 2/3) + (3 \times -1/3) = 0$. It seems remarkable that the sum of the electric charges of matter (particles with +ve sqrt mass) comes out to be zero. It need not have been so. This demonstration might help understand how matter-antimatter separation preserved electrical neutrality. Before this separation, the net square-root mass of matter and anti-matter was zero, even though individual sqrt masses were non-zero. In this, we differ from the standard gauge-theoretic picture of EW symmetry breaking and mass acquisition. In EW, particles are massless before symmetry breaking because a mass term in the Lagrangian breaks gauge invariance. However, for us, sqrt mass is not zero before the symmetry breaking—its non-zero value was already set at the Planck scale (and cosmological expansion scaled down actual mass values while preserving mass ratios). Indeed, it is rather peculiar if, prior to the symmetry breaking, particles have electric charge but no mass. For us, QFT (quantum field theory) on a spacetime background (and hence gauge theories) are not valid before the left-right symmetry breaking. In fact, spacetime itself, along with gravitation, emerge after this symmetry breaking, as a result of the quantum to classical transition. Spacetime emerges iff classical matter emerges. Prior to the symmetry breaking, dynamics are described by trace dynamics, there is no spacetime, and we have 'atoms' of space-time-matter. The concepts of electric charge and mass are not defined separately; there is only a charge-square-root mass (a hypercharge can also be defined, as for EW), and this is the source for a unified force in octonionic space.

5.6. Octonions, Scale Invariance, and a CPT Symmetric Universe: A Possible Explanation for the Origin of Matter–Antimatter Asymmetry

In the octonionic theory, prior to the so-called left-right symmetry breaking, the symmetry group is $E_8 \times E_8$ and the Lagrangian of the theory is scale invariant. There is only one parameter, a length scale, which appears as an overall multiplier of the trace Lagrangian. Something dramatic happens after the symmetry breaking. Three new parameters emerge, to characterise the fermions: Electric charge has two signs, sign change operation C is complex conjugation, and ratios are (0, 1/3, 2/3, 1). Chirality/spin has two signs, sign change operation P is octonionic conjugation. The ratios are $(1/2, -1/2)$. Square-root of mass has two signs, sign change operation, T, is time reversal, $t \rightarrow -t$. Ratios are $(0, 1/3, 2/3, 1)$. Thus, there are $2 \times 2 \times 2 = 8$ types of fermions, based on sign of charge, square-root mass and spin. This could possibly offer an attractive explanation for the origin of matter–antimatter asymmetry: a CPT symmetric universe. The four types of fermions which have positive square root mass become matter, our universe, moving forward in time. The other four types of fermions, which have negative square root mass, become anti-matter, a mirror universe is surprisingly found moving backward in time. The forward moving universe and the backward moving universe together restore CPT symmetry. Our universe by itself violates T, and hence also CP. Matter and anti-matter repel each other gravitationally, thus explaining their separation. This also explains why gravitation in our universe is attractive, even though the underlying pre-gravitation theory is a vector interaction. Prior to the symmetry breaking, an octonionic inflation (scale invariant, time-dependent in Connes time) precedes the 'big bang' creation event, which is the symmetry breaking itself. Freeze out happens when radiation to matter-antimatter transition is no longer favorable. Segregation takes place; our matter universe has a one in a billion excess of matter over anti-matter. The backward in time mirror universe has a one in a billion excess of anti-matter over matter. The mathematics of complex octonions naturally account for the C, P, T operations. Scale invariance is transformed into CPT invariance in the emergent universe. We hope to make this idea rigorous in forthcoming investigations. In an elegant proposal, Turok and Boyle [36] have also recently proposed a CPT symmetric universe (mirror universes). They, however, did not use the octonions.

5.7. Prospects of Tests through Particle Physics Experiments and Phenomenology

In Section XV of [5], we have briefly discussed some of the possible experimental predictions of this theory and prospects for their experimental tests. Below, we discuss some particle physics related predictions and their possible relevance for collider experiments and neutrino experiments. These ideas are largely based on a detailed conversation with Ashutosh Kotwal, and their compilation by Vatsalya Vaibhav.

Our theory predicts specific new particles, though much work remains to be done to predict their specific properties such as masses. We predict that the neutrino is Majorana and that there are three right-handed sterile neutrinos. At present, we do not understand neutrino masses, though there is a very real possibility that the neutrino is massless and flavor oscillations are caused by spacetime being higher dimensional and octonionic, and the neutrinos being spacetime triplets [18]. This possibility could be verified if we can calculate the PMNS (Pontecorvo-Maki-Nakagawa-Sakata) matrix from first principles in the octonionic theory; this will be attempted in future work. The Majorana nature of the neutrino also suggests neutrinoless double beta decay, experimental implications for which we investigate further below. The possibility that the sterile neutrino could be massive and light (hence hot dark matter) or heavy (hence cold dark matter) are also investigated, although it is true that our theory favours MOND over cold dark matter.

Our theory also predicts a new charged Higgs boson, and possibly a doubly charged Higgs, although their masses and the fundamental origin of their scalar nature remains to be understood. In addition, of interest is whether the Higgs triplet in this theory can cause the W mass to depart from the standard model prediction. The hierarchy problem, and the exact mechanism of the electroweak symmetry breaking and the left-right symmetry breaking remain to be understood: these could be the same symmetry breaking and could perhaps be mediated by a quantum-to-classical phase transition. It is also of interest to try and predict the (g-2) anomaly, some related physics was discussed in [10], and this issue might also be related to our derivations of mass ratios and the low energy fine structure constant.

A general line of investigation of serious interest is to note that our predictions are made on octonionic twistor space, which is non-classical, whereas all measurements are in Minkowski spacetime, and based on quantum field theory calculations. In transiting from trace dynamics to emergent quantum field theory, there could be important and smoking gun corrections, which could possibly be cast in the language of effective field theory.

As per our recent study on $E_8 \times E_8$ symmetry, we have been able to account for 208 out of the 496 degrees of freedom [18]. While we have commented briefly in that study on the unaccounted degrees of freedom, more work needs to be done to find if they predict new particles which can confirm or rule out this theory.

5.8. *Why Is There No Black Hole Information Loss Paradox in the Octonionic Theory?*

There appears to be an information loss paradox because we are ignoring the physical process of the quantum-to-classical transition, which keeps the black hole (made of enormously many quantum constituents) classical in the first place. Let us consider the following analogy: Consider a box of gas at thermodynamic equilibrium—this is the maximum entropy and minimum information state. Now, a sudden spontaneous fluctuation sends the entire set of gas molecules to one corner of the box. This is a transient low entropy high information and ordered state, far from thermodynamic equilibrium. At the very next instant, the gas molecules will return to equilibrium, spread all over the box, entropy will have been gained, and information lost. If we ignore the spontaneous fluctuation, which sent the gas to a corner, then we have an information loss paradox. Obviously, there is no paradox in reality: information gained during the spontaneous fluctuation is lost during the return to equilibrium. It is exactly the same physics, when we work with a fundamental theory (generalised trace dynamics valid at Planck time resolution), from which quantum theory and classical gravitation are emergent phenomena. Quantum theory (without classical time) is emergent as the thermodynamic equilibrium state description in a statistical thermodynamic approximation of the underlying theory. Classical gravitation, spacetime, and hence the black hole, is far from the equilibrium transient state which arises from a spontaneous fluctuation caused by non-unitary evolution. At equilibrium, the evolution is unitary, except when a large fluctuation kicks in. In Adler's theory of trace dynamics, a collection of quantum states at thermodynamic equilibrium constitute a state of maximum entropy (the most likely state to emerge from coarse-graining of the underlying system). Departure from equilibrium by way of a spontaneous fluctuation produces the black hole as a state of relatively low entropy and high information. By way of Hawking evaporation, the black hole returns to the state of maximum entropy and thermodynamic equilibrium (one has to be careful here while talking of pure states and mixed states; there is no classical spacetime at thermodynamic equilibrium, in this theory.). To this, one could object that not all black holes are formed in the way suggested in the previous paragraph above. That is true; if one has in mind stellar gravitational collapse ending in black hole formation. However, let us trace the star all the way back to the very early universe: inflationary quantum perturbations became classical, spacetime emerged, density perturbations grew, the star formed, the black holes formed, it Hawking radiated, and is going back to the inflationary stage, which was quantum in nature. The formation of classical large scale structure in the universe is like the gas in the box spontaneously fluctuating away from equilibrium. Hawking evaporation is like the gas returning to equilibrium. Interestingly, the above reasoning also throws light on the current accelerating phase of the universe. de Sitter like inflationary expansion is the natural expansion state of the universe—the equilibrium quantum state. Emergence of classicality gives rise to the radiation dominated and matter dominated eras, where matter density is high enough to dictate a power-law expansion of the scale factor, overriding de Sitter like expansion. In recent cosmic history, matter density is falling, and hence de Sitter like scale-invariant expansion (quantum equilibrium) is again dominating. These dynamics are very likely related to the emergent classical Lagrangian of our theory, which exhibits modification of GR by conformal gravity, and will be investigated further for its cosmological implications.

5.9. How Taking the Square-Root of Minkowski Space-Time Paves the Way for Unification

In summary, we believe we have a promising theory of unification under development, as captured in Figure 6 and explained briefly below.

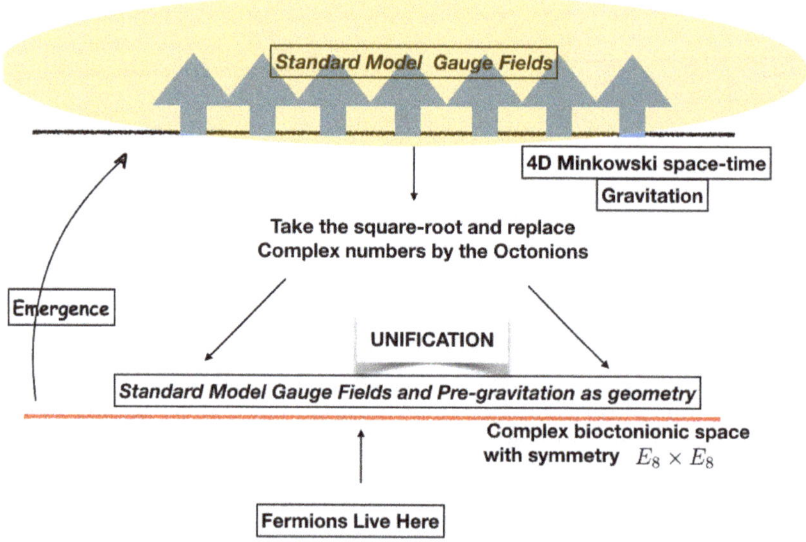

Figure 6. Taking the square-root of 4-dimensional (4D) Minkowski space-time paves the way to unification.

It is like going from the surface of the ocean to the ocean bed. The ocean floor can exist without the surface, but the surface cannot exist without the floor. We live in a 4D Minkowski space-time curved by gravitation, in which standard model gauge fields and fermions reside. However, there is a more precise description. We take the square-root of Minkowski space-time and arrive at Penrose's twistor space, described by complex numbers. In this spinor space-time, replace complex numbers with quaternions, then by octonions. More precisely, complex split bioctonions. We arrive at a space with $E_8 \times E_8$ symmetry whose geometry is a unified description of the standard model and pre-gravitation. The gauge group is $SU(3)_c \times SU(2)_L \times U(1)_Y \times SU(3)_G \times SU(2)_R \times U(1)_g$. Coupling constants are determined by the geometry. In the classical limit, the 4D curved spacetime and the standard model emerge, but with pre-determined values of coupling constants. Fermions span space-time as well as the space of the gauge fields. Taking the square root of Minkowski space-time does not involve change of energy scale. It only gives a more precise mathematical formalism—one that is key to unification, which already takes place at low energies, if we do not restrict ourselves to 4D classical spacetime: only classical systems live in 4D. Quantum systems always live in $E_8 \times E_8$ even at low energies. If we want a comparison with string theory, then this new theory is string theory without compactification. Compactification is effectively achieved by the quantum-to-classical transition; it does not have to be enforced in an ad hoc manner.

Funding: This research received no external funding.

Data Availability Statement: Not applicable.

Acknowledgments: It is a pleasure to thank Priyank Kaushik, Ashutosh Kotwal, Sherry Raj and Vatsalya Vaibhav for very helpful discussions.

Conflicts of Interest: The author declares no conflict of interest.

References

1. Penrose, R. On gravity's role in quantum state reduction. *Gen. Rel. Grav.* **1996**, *28*, 581–600. [CrossRef]
2. Carlip, S. Quantum gravity: A progress report. *Rep. Prog. Phys.* **2001**, *64*, 885. [CrossRef]
3. Adler, S.L. Generalised quantum dynamics. *Nucl. Phys. B* **1994**, *415*, 195–242. [CrossRef]
4. Adler, S.L.; Millard, A.C. Generalised quantum dynamics as pre-quantum mechanics. *Nucl. Phys. B* **1996**, *473*, 199–244. [CrossRef]
5. Singh, T.P. Quantum theory without classical time: A route to quantum gravity and unification. *arXiv* **2021**, arXiv:2110.02062. [CrossRef]
6. Roy, A.K.; Sahu, A.; Singh, T.P. Trace dynamics, and a ground state in spontaneous quantum gravity. *Mod. Phys. Lett. A* **2021**, *36*, 2150019. [CrossRef]
7. Connes, A.; Rovelli, C. Von Neumann algebra automorphisms and time-thermodynamics relation in general covariant quantum theories. *Class. Quantum Grav.* **1994**, *11*, 2899–2918. [CrossRef]
8. Chamseddine, A.H.; Connes, A. The spectral action principle. *Commun. Math. Phys.* **1997**, *186*, 731–750. [CrossRef]
9. Landi, G.; Rovelli, C. General relativity in terms of Dirac eigenvalues. *Phys. Rev. Lett.* **1997**, *78*, 3051–3054. [CrossRef]
10. Meghraj, M.S.; Pandey, A.; Singh, T.P. Why does the Kerr-Newman black hole have the same gyromagnetic ratio as the electron? *arXiv* **2020**, arXiv:2006.05392. [CrossRef]
11. Ramond, P. *Introduction to Exceptional Lie Groups and Algebras*; ERDA Research and Development Report CALT-68-577; California Institute of Technology: Pasadena, CA, USA, 1976. Available online: https://inspirehep.net/literature/111550 (accessed on 10 August 2022).
12. Furey, C. Charge quantisation from a number operator. *Phys. Lett. B* **2015**, *742*, 195–199. [CrossRef]
13. Singh, T.P. Quantum theory without classical time: Octonions, and a theoretical deriva-tion of the fine structure constant 1/137. *Int. J. Mod. Phys. D* **2021**, *30*, 2142010. [CrossRef]
14. Bhatt, V.; Mondal, R.; Vaibhav, V.; Singh, T.P. Majorana neutrinos, exceptional Jordan algebra and mass ratios for charged fermions. *J. Phys. G Nucl. Part. Phys.* **2022**, *49*, 045007. [CrossRef]
15. Todorov, I.; Drenska, S. Octonions, exceptional Jordan algebra and the role of the group F_4 in particle physics. *Adv. Appl. Clifford Algebras* **2018**, *28*, 82. [CrossRef]
16. Dray, T.; Manogue, C. Octonionic Cayley spinors and E_6. *Comment. Math. Univ. Carolin.* **2010**, *51*, 193–207. Available online: https://cmuc.karlin.mff.cuni.cz/cmuc1002/abs/dray.htm (accessed on 10 August 2020).
17. Vaibhav, V.; Singh, T.P. Left-right symmetric fermions and sterile neutrinos from complex split biquaternions and bioctonions. *arXiv* **2021**, arXiv:2108.01858. [CrossRef]
18. Kaushik, P.; Vaibhav, V.; Singh, T.P. An $E_8 \otimes E_8$ unification of the standard model with pre-gravitation, on an octonion-valued twistor space. *arXiv* **2022**, arXiv:2206.06911. [CrossRef]
19. Woit, P. Euclidean twistor quantisation. *arXiv* **2021**, arXiv:2104.05099. [CrossRef]
20. Zyla, P.A.; et al. [Particle Data Group] Review of Particle Properties. *Prog. Theo. Exp. Phys.* **2020**, *8*, 83C01. [CrossRef]
21. Singh, T.P. Quantum gravity effects in the infrared: A theoretical derivation of the low-energy fine structure constant and mass ratios of elementary particles. *Eur. Phys. J. Plus* **2022**, *137*, 664. [CrossRef]
22. Singh, T.P. Quantum gravity, minimum length, and holography. *Pramana* **2021**, *95*, 40. [CrossRef]
23. Baez, J. The octonions. *Bull. Am. Math. Soc.* **2002**, *39*, 145–205. [CrossRef]
24. Gunaydin, M.; Gursey, F. Quark structure and octonions. *J. Math. Phys.* **1973**, *14*, 1651–1667. [CrossRef]
25. Gunaydin, M.; Gursey, F. Quark statistics and octonions. *Phys. Rev. D* **1974**, *9*, 3387–3391. [CrossRef]
26. Gunaydin, M.; Piron, C.; Ruegg, H. Moufang plane and octionic quantum mechanics. *Commun. Math. Phys.* **1978**, *61*, 69–85. [CrossRef]

27. Gursey, F.; Tze, C.-H. *On the Role of Division, Jordan and Related Algebras in Particle Physics*; World Scientific: Singapore, 1996. [CrossRef]
28. Jordan, P.; von Neumann, J.; Wigner, E. On an algebraic generalization of the quantum mechanical formalism. *Ann. Math.* **1934**, *36*, 29–64. [CrossRef]
29. Todorov, I.T.; Drenska, S. Composition algebras, exceptional Jordan algebra and related groups. *J. Geom. Symmetry Phys.* **2017**, *46*, 59–93. [CrossRef]
30. Todorov, I.T.; Dubois-Violette, M. Deducing the symmetry of the standard model from the automorphism and structure groups of the exceptional Jordan algebra. *Int. J. Mod. Phys.* **2018**, *A33*, 1850118. [CrossRef]
31. Dubois-Violette, M. Exceptional quantum geometry and particle physics. *Nucl. Phys. B* **2016**, *B912*, 426–449. [CrossRef]
32. Dubois-Violette, M. Exceptional quantum geometry and particle physics. II. *Nucl. Phys. B* **2019**, *938*, 751–761. [CrossRef]
33. Boyle, L. The standard model,the exceptional Jordan algebra, and triality. *arXiv* **2020**, arXiv:2006.16265. [CrossRef]
34. Singh, T.P. Octonions, trace dynamics and noncommutative geometry—A case for unification in spontaneous quantum gravity. *Z. Naturforsch. A* **2020**, *75*, 1051–1062. [CrossRef]
35. Koide, Y. New view of quark and lepton mass hierarchy. *Phys. Rev. D* **1983**, *28*, 252–254. [CrossRef]
36. Boyle, L.; Finn, K.; Turok, N. CPT-symmetric universe. *Phys. Rev. Lett.* **2018**, *121*, 251301. [CrossRef] [PubMed]

physics

MDPI

Review

The Barbero–Immirzi Parameter: An Enigmatic Parameter of Loop Quantum Gravity

Rakshit P. Vyas * and Mihir J. Joshi

Department of Physics, Saurashtra University, Rajkot 360005, India
* Correspondence: rakshitvyas33@gmail.com

Abstract: The Barbero–Immirzi parameter, (γ), is introduced in loop quantum gravity (LQG), whose physical significance is still the biggest open question because of its profound traits. In some cases, it is real valued, while it is complex valued in other cases. This parameter emerges in the process of denoting a Lorentz connection with a non-compact group $SO(3,1)$ in the form of a complex connection with values in a compact group of rotations, either $SO(3)$ or $SU(2)$. Initially, it appeared in the Ashtekar variables. Fernando Barbero proposed its possibility for inclusion within formalism. Its present value is fixed by counting micro states in loop quantum gravity and matching with the semi-classical black hole entropy computed by Stephen Hawking. This parameter is used to count the size of the quantum of area in Planck units. Until the discovery of the spectrum of the area operator in LQG, its significance remained unknown. However, its complete physical significance is yet to be explored. In the present paper, an introduction to the Barbero–Immirzi parameter in LQG, a timeline of this research area, and various proposals regarding its physical significance are given.

Keywords: loop quantum gravity; Ashtekar variable; Barbero–Immirzi parameter; area operator; black hole entropy

Citation: Vyas, R.P.; Joshi, M.J. The Barbero–Immirzi Parameter: An Enigmatic Parameter of Loop Quantum Gravity. *Physics* **2022**, *4*, 1094–1116. https://doi.org/10.3390/physics4040072

Received: 26 July 2022
Accepted: 29 August 2022
Published: 20 September 2022

1. Introduction

Loop quantum gravity (LQG) is one of the supposed candidates of the theory of quantum gravity. It can unify general relativity (GR) with quantum field theory (QFT). It is a non-perturbative and background independent approach to quantum gravity theory. LQG begins with GR; thereafter, it takes some conceptual basis from QFTs to deliver a quantum theory of gravity. LQG is a theory of constraints, in which various constraints such as Hamiltonian, diffeomorphism and Gauss constraints are converted into operators. In the canonical quantization approach of LQG (ADM formalism), 3 + 1 decomposition of spacetime is necessary to quantize gravity; however, the covariant approach (sum over geometry) follows a different strategy. Here, due to limited space, basics of LQG are not given. There are many classic texts [1–10] and papers [11–23] that explain LQG lucidly.

In 1986, Abhay Ashtekar [24] found new kind of variables (Ashtekar's variable) in classical and quantum gravity. In Ashtekar's formulations, the constraints are simplified by considering a complex valued form for the connection and tetrad variables, and these are known as Ashtekar's variables [1–5].

While dealing with the reality condition of the formalism of Ashtekar's variable, Barbero [25,26] firstly introduced a free parameter in the expression of Ashtekar's variable and then in the expression of constraints. Thereafter, Immirzi [27,28] used various possibilities of this free parameter in the expression of LQG. This free parameter is nowadays known as the Barbero–Immirzi (BI) parameter, γ. Given γ is complex or real, it provides a number of results in LQG. In some cases, the real valued BI parameter is required, while in the other cases, the complex valued BI parameter is necessary [1–5].

The physical significance of the area operator in LQG with the complex BI parameter becomes ambiguous. The LQG kinematics, i.e., kinematical Hilbert space can only be comprehended if the γ is a real number. The $SU(2)$ spin network of LQG can only be created with the real value of the BI parameter [1–5].

With the complex value of the BI parameter ($\gamma = i$), the spatial connection can be seen as spacetime connection, since it transforms under diffeomorphism in the right way. There are also some cases that show that the complex valued BI parameter is also crucial in LQG formalism. For instance, the form of Hamiltonian constraints becomes simpler if $\gamma = i$ is taken [1–5].

In the next Section, various proposals regarding the BI parameter are briefly reviewed in which some proposals advocate the real valued BI parameter, while the other advocates the complex valued BI parameter.

1.1. Ashtekar's Formalism

Before the discovery of Ashtekar's variables, the Palatini action, i.e., the first-order formulation, was incomplete. However, Ashtekar formalism made it complete. In the Palatini action, the tetrad, e^{μ}_{J}, and the spin connection, ω^{JK}_{μ}, are used as independent variables. In GR, the Palatini action is written as [1–5]

$$S_P = \int d^4 x e e^{\mu}_{J} e^{\eta}_{K} \Omega^{JK}_{\mu\eta}[\omega], \tag{1}$$

where $e = \sqrt{-g}$, where g is the determinant of the 4-dimensional metric, $g_{\mu\nu}$, and $\Omega^{JK}_{\mu\eta}$ is the curvature. The capital Latin letters are the internal indices, the Greek and lower-case Latin letters denote the time (0) and space coordinates. The $[\omega]$ denotes the spin connection. Using the Palatini action, the Einstein field equation could be derived, but the form of equation of constraints within this formalism is mathematically complicated [1–5,12,15,17]. The generalization of the Palatini action is equivalent to the Ashtekar formalism, and it is achieved by the Holst action [29].

In the Ashtekar formalism, by converting tetrads into triads, i.e., three-dimensional hypersurfaces Σ_t, one gets $e^{J}_{\mu} \to e^{j}_{c}$, where $\mu \to c \in \{1,2,3\}$, $J \to j \in \{1,2,3\}$ and the spin connection is also transformed as $\Gamma^{j}_{c} = \omega_{ckl}\varepsilon^{klj}$, where ε^{klj} is the Levi-Civita tensor [1–5,12,15,17].

The Hamiltonian constraint is a complicated non-polynomial function in Palatini formulation; thus, canonical quantization is not easy within this formalism. In Palatini formulation, the variables of phase space are $(e^{j}_{c}, \Gamma^{j}_{c})$, where e^{j}_{c} is the intrinsic metric of the spacelike manifold Σ and Γ^{j}_{c} is a function of its extrinsic curvature [1–5,12,15,17].

In the Ashtekar's formalism, complex valued connection Γ^{j}_{c} replaces the real connection ω^{JK}_{μ} with duality (either self (+1) or anti-self (−1)) [1–5,12,15,17],

$$\tilde{E}^{c}_{j} \to \frac{1}{i}\tilde{E}^{c}_{j}, K^{j}_{c} \to A^{j}_{c} = \Gamma^{j}_{c} - iK^{j}_{c}, \tag{2}$$

where $\tilde{E}^{c}_{j}$ is the scalar density or triad electric field, A^{j}_{c} is the Ashtekar–Barbero connection or spatial connection, $K^{j}_{c} = k_{cd}e^{dj}$ with k_{cd} the extrinsic curvature of Σ. Thus, there are two phase space variables, i.e., A^{j}_{c} and $\tilde{E}^{c}_{j}$ [1–5,12,15,17].

Since the Ashtekar's connection formulation variables, i.e., A^{j}_{c} and $\tilde{E}^{c}_{j}$, follow rotation of $SU(2)$ symmetry with respect to the internal indices, the Ashtekar's formalism plays the role of $SU(2)$ gauge theory, and this $SU(2)$ group is a subgroup of $SL(2,\mathbb{C})$ [1–5,12,15,17].

All three constraints are simplified in Ashtekar's variables, and their expressions are [1–5,12,15,17]:

$$\mathcal{G}_j = D_c \tilde{E}^{c}_{j}, \tag{3}$$

$$\mathcal{C}_c = \tilde{E}^{d}_{j} F^{j}_{cd} - A^{j}_{c}\mathcal{G}_j, \tag{4}$$

$$\mathcal{H} = \varepsilon_l^{jk} \tilde{E}_j^c \tilde{E}_k^d F_{cd}^l. \tag{5}$$

Here, D_c is the covariant derivative.

Equations (3), (4), and (5) are Gauss, diffeomorphism, and Hamiltonian constraints, respectively. In Ashtekar's formalism, the Einstein–Hilbert–Ashtekar (EHA) Hamiltonian of GR reads [1–5,12,15,17]:

$$\mathcal{H}_{\text{EHA}} = N^c \mathcal{C}_c + N\mathcal{H} + T^j \mathcal{G}_j = 0 \tag{6}$$

where $\mathcal{C}_c, \mathcal{H}, \mathcal{G}_j, N^c$, and N are the vector constraint, the scalar constraint, the Gauss constraints, the shift, and the lapse, respectively. The T^j is a Lie algebra valued function over spatial surface [1–5,12,15,17].

The unit imaginary, i.e., $i = \sqrt{-1}$, which appears in Equation (2), makes the formalism complex valued. Therefore, some restrictions in terms of reality conditional on the possible solutions of the theory must be applied to achieve tangible physical results relevant to a metric valued in $\mathbb{R}$ instead of in $\mathbb{C}$ [1–5,12,15,17].

For example, if $\dot{Z}$ is used to represent the time derivative of Z, then the reality condition and constraints, i.e., $\mathcal{G}_j = D_c \tilde{E}_j^c$ must be satisfied by solutions. In this case, there are two reality conditions, and the second condition is the time derivative of the first condition; thus [1–5,12,15,17],

$$\tilde{E}_j^c \tilde{E}_k^d \delta^{jk} \in \mathbb{R}, \tag{7}$$

$$\{\tilde{E}_j^c \tilde{E}_k^d \delta^{jk}\}^{\bullet} \in \mathbb{R}. \tag{8}$$

with δ^{jk} being the Kronecker delta.

In a standard form, the Ashtekar variables are given as [1–5,12,15,17]

$$\tilde{E}_j^c \to \frac{1}{\gamma}\tilde{E}_j^c, K_c^j \to A_c^j = \Gamma_c^j - \gamma K_c^j. \tag{9}$$

If $\gamma = i$, then the equation takes the original form.

The Poisson brackets are written as [1–5,12,15,17]

$$\{K_c^j(x), \tilde{E}_k^d(y)\} = \{A_c^j(x), \tilde{E}_k^d(y)\} = k\delta_k^j \delta_c^d \delta(x,y), \tag{10}$$

where $k = 8\pi G\gamma$ with G the gravitational constant.

In a standard form [1–5,12,15,17],

$$\{A_c^j(x), \tilde{E}_k^d(y)\} = 8\pi G\gamma \delta_k^j \delta_c^d \delta^3(x,y). \tag{11}$$

The reality condition is not necessary for a real value, and as a result, new variables and constraints are also real [1–5,12,15,17].

The form of the Hamiltonian constraint becomes complicated with the real value of the γ, i.e.,

$$\mathcal{H} = \varepsilon_l^{jk} \tilde{E}_j^c \tilde{E}_k^d F_{cd}^l - 2(1+\gamma^2)\tilde{E}_j^{[c} \tilde{E}_k^{d]} K_c^j K_d^k \approx 0. \tag{12}$$

If, $\gamma = i$; then the form of Hamitonian constraints simplifies [1–5,12,15,17].

1.2. Why the BI Parameter Was Introduced in LQG?

As mentioned, the complex valued Ashtakar's variables simplified constraints of quantum gravity based on canonical quantization, i.e., LQG. Thereafter, Barbero [25,26] came up with a new strategy to tackle Ashtekar's variable with real value for Lorentzian signature space-times. In [25,26] Barbero wrote down Ashtekar's variable with a free parameter (Equation (9)); namely, γ (denoted β there). Ashtekar used $SU(2)$ and $SL(2,\mathbb{C})$

groups of Yang–Mill theory to deliver complex valued constraints, i.e., Ashtekar's variables. Meanwhile, Barbero showed that one can use $SO(3)$ Yang–Mill phase space to expressthe modified Hamiltonian constraint with Lorentz signatures without complex variables to elaborate space-times without losing the features of Ashtekar's variables [1–5,25].

Barbero [25,26] also showed that for simple forms of a Hamiltonian constraint, complex variable is required, while for a complicated form, this constraint could be written with real variables; for instance, a loop variable of LQG. Barbero derived a Hamiltonian constraint with $\gamma^2 = 1$ (real valued and Euclidean signature). Meanwhile, the Lorentzian signature again yields a complex valued form of the equation. Barbero also derived a Hamiltonian constraint with $\gamma = -1$. The Hamiltonian constraint could also be written with real Ashtekar variables for Lorentzian general relativity with $SO(3)$ ADM formalism [25].

Thereafter, Immirzi [27,28] further clarified on the importance of this parameter. In these papers, Immirzi explained canonical quantization of gravity, i.e., LQG with the Regge calculus. In Immirzi elaborated the basics of LQG with the discussion on the γ. Immirzi discussed various possibilities of the value of the γ and named this arbitrariness of the γ the γ crisis [27,28].

Since Barbero introduced this free parameter and Immirzi used it to explain the canonical quantization method along with Regge calculus, the γ is known as the Barbero–Immirzi parameter. In short, Barbero used one-parameter scale transformation to generalize the Ashtekar canonical transformation to a $U(\gamma)$. Meanwhile, Immirzi observed that such a transformation modifies the spectra of geometrical quantities of LQG [1–5,27,28].

1.3. The Holst Action and the BI Parameter

Today, there are two types of version for the connection variables: $SL(2,\mathbb{C})$ with a self duality of Yang–Mills type of connection i.e., the Ashtekar connection, and the connection with a real $SU(2)$ the Barbero connection. In the latter type of connection, the issue of the reality condition is not present. With the aid of the Holst action, both of the connections can be obtained. The γ is introduced in the Holst action as a multiplicative constant that governs the strength of the dual curvature correction.

The Holst action generalizes the Hilbert–Palatini action using the γ. The Holst action can be derived in the following way using the Einstein–Hilbert action (EH). In GR, the EH action is written as [1–5,15,29]

$$S_{\text{EH}}(g_{\mu\nu}) = \frac{1}{16\pi G}\int d^4x\sqrt{-g}g^{\mu\nu}R_{\mu\nu}, \tag{13}$$

where $R_{\mu\nu}$ is the Ricci tensor.

If $e = \sqrt{-g}$ and $8\pi G = 1$ are taken; then [1–5,15,29],

$$\begin{aligned} S_{\text{EH}}(g_{\mu\nu}(e)) &= \int d^4x e e^\mu_J e^{\nu J} R_{\mu\nu\eta\tau} e^\eta_K e^{\tau K} \\ &= \int d^4x e e^\mu_J e^\eta_K F^{JK}_{\mu\eta}(\omega(e)), \end{aligned}$$

where $F^{JK}_{\mu\eta}(\omega(e)) = e^{J\mu}e^{\eta K}R_{\mu\nu\eta\tau}(e)$,

$$S_{\text{EH}}(g_{\mu\nu}(e)) = \int d^4x \frac{1}{4}\epsilon_{JKLM}\epsilon^{\mu\eta\alpha\beta}e^L_\alpha e^M_\beta F^{JK}_{\mu\eta}(\omega(e)).$$

Hence, as a functional of a densitized triad, the Einstein–Hilbert action takes the form [1–5,15,29],

$$\therefore S\left(e^J_\mu, \omega^{JK}_\mu\right) = \frac{1}{2}\epsilon_{JKLM}\int e^J \wedge e^K \wedge F^{LM}(\omega). \tag{14}$$

By considering the Palatini identity, i.e., $\delta_\omega = F^{LM}(\omega) = d_\omega \delta^{LM}_\omega$ where d_ω denotes derivative with respect to ω, and taking the variation of Equation (14), one gets [1–5,15,29]:

$$\delta_\omega S\left(e^J_\mu, \omega^{JK}_\mu\right) = \frac{1}{2}\epsilon_{JKLM}\int e^J \wedge e^K d_\omega \delta^{LM}_\omega,$$

$$\therefore \delta_\omega S\left(e^J_\mu, \omega^{JK}_\mu\right) = -\frac{1}{2}\epsilon_{JKLM}\int d_\omega\left(e^J \wedge e^K\right) \wedge \delta^{LM}_\omega. \tag{15}$$

If the coupling constant, $1/\gamma$, is added in Equation (14), then one gets the Holst action [1–5,15,29], i.e,

$$S(e,\omega) = \left(\frac{1}{2}\epsilon_{JKLM} + \frac{1}{\gamma}\delta_{JKLM}\right)\int e^J \wedge e^K \wedge F^{LM}(\omega). \tag{16}$$

The Holst action is equivalent to the Ashtekar Hamiltonian if $\gamma = i$ is set in Equation (16) [1–5,15,29].

In general, the Holst action is written as [1–5,15,29]

$$S[e,A] = \frac{1}{8\pi G}\left(\int d^4xee^\mu_J e^\nu_K F^{JK}_{\mu\nu} - \frac{1}{\gamma}\int d^4xee^\mu_J e^\nu_K * F^{JK}_{\mu\nu}\right) \tag{17}$$

where the symbol "$*$" denotes the self-duality in Equation (17) in the presence of the BI parameter.

2. Various Proposals on the Physical Significance of the BI Parameter

In this Section, the historical time line of the research in the BI parameter and various proposals on the physical significance of the BI parameter are briefly discussed. After introducing each of the proposals, pros and cons of a proposal are given and the role of the BI parameter is explained. Here, all proposals are explained only in the context of the BI parameter to an extent relevant to the necessary mathematical treatment.

2.1. Historical Timeline

Table 1 shows the historical timeline of research on the BI parameter in LQG in a chronological order. The table lists the enriched literature of the BI parameter and clarifies on the importance of the study of the BI parameter in LQG. The listing implies that the BI parameter is itself a crucial research area in LQG.

Table 1. Timeline of research on the Barbero–Immirzi parameter.

Year	Research on the BI Parameter and Its Significance
1986	Discovery of the Ashtekar variables
1995	Real Ashtekar variables for Lorentzian signature space–times
1996	Barbero's Hamiltonian derived from a generalized Hilbert–Palatini action
1996	Black hole entropy from loop quantum gravity
1996	From Euclidean to Lorentzian general relativity: the real way
1996	Real and complex connections for canonical gravity
1997	Quantum gravity and Regge calculus
1997	Counting surface states in loop quantum gravity (LQG)
1997	Immirzi parameter in quantum general relativity
1997	On the constant that fixes the area spectrum in canonical quantum gravity
1998	Quantum geometry and black hole entropy
2000	Is Barbero's Hamiltonian formulation a gauge theory of Lorentzian gravity?
2001	Comment on "Immirzi parameter in quantum general relativity"
2003	Quasinormal modes, the area spectrum, and black hole entropy
2004	Black-hole entropy in loop quantum gravity
2004	Black-hole entropy from quantum geometry
2005	Origin of the Immirzi parameter
2005	Physical effects of the Immirzi parameter
2005	On choice of connection in LQG
2007	On a covariant formulation of the Barbero–Immirzi connection
2007	Renormalization and black hole entropy in Loop Quantum Gravity
2008	From the Einstein–Cartan to the Ashtekar–Barbero canonical constraints, passing through the Nieh–Yan functional
2008	The Barbero–Immirzi parameter as a scalar field: K-inflation from LQG?
2008	Topological interpretation of Barbero–Immirzi parameter

Table 1. *Cont.*

Year	Research on the BI Parameter and Its Significance
2009	Peccei–Quinn mechanism in gravity and the nature of the Barbero–Immirzi parameter
2010	A relation between the Barbero–Immirzi parameter and the standard model
2011	Complex Ashtekar variables, the Kodama state and spinfoam gravity
2012	The quantum gravity Immirzi parameter—A general physical and topological interpretation
2012	Complex Ashtekar variables and realitycConditions for Holst's action
2013	Black Hole Entropy from complex Ashtekar variables
2014	Geometric temperature and entropy of quantum isolated horizons
2014	A Correction to the Immirizi Parameter of SU(2) Spin Networks
2014	The Microcanonical Entropy of quantum isolated horizon, "quantum hair" N and the Barbero–Immirzi parameter fixation
2015	The holographic principle and the Immirzi parameter of loop quantum gravity
2017	Immirzi parameter without Immirzi ambiguity: conformal loop quantization of scalar-tensor gravity
2018	Horizon entropy with loop quantum gravity methods
2018	Generalizing the Kodama state. I: construction
2018	Generalizing the Kodama state. II: properties and physical interpretation
2018	Chiral vacuum fluctuations in quantum gravity
2018	Black hole entropy from the SU(2)-invariant formulation of Type I isolated horizons
2018	Black hole entropy and SU(2) Chern–Simons theory
2020	On the value of the Immirzi parameter and the horizon entropy

2.2. The Area Operator and the BI Parameter

In LQG, the loop states as a graph or network Θ with edges e_i denoted by elements of some gauge group. In general, this gauge group can be $SU(2)$ or $SL(2,\mathbb{C})$ [1–5,12,15,17],

$$\psi_\Theta = \psi(g_1, g_2, \ldots, g_k), \tag{18}$$

where $k = 0, 1, 2, \ldots, n$ and g_k is the holonomy (group element) of connection A on the kth edge; see Figure 1. In LQG, the spin network is used to describe these loop states. Penrose [30,31] gave the notion of the spin network. In the spin network, the combinatorial principle of angular momentum is used, and it defines the space–time in discrete way. In LQG, the spin network is essential for representing the loop state [1–5,12,15,17].

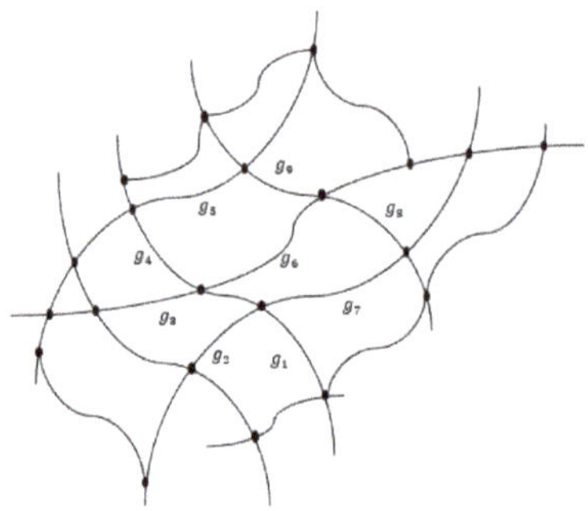

Figure 1. A diagram of spin network. See text for details.

The area of a two-dimensional (2D) surface, S, that is embedded in any manifold, Σ, is defined as [1–5,12,15,17]

$$\mathcal{A}_S = \int d^2x \sqrt{^{(2)}m}, \tag{19}$$

where $^{(2)}m$ is the determinant of the metric $^{(2)}m_{EF}$. The area is 2D; hence, the components of the 2D metric $^{(2)}m_{EF}$ can be denoted as the dyad basis, e^J_E, and $E, F \in \{x, y\}$ here are spatial indices

$$^{2}m_{EF} = e^{J}_{E}e^{K}_{F}\delta_{JK}. \tag{20}$$

The determinant of $^{2}m_{EF}$ can be written as [1–5,12,15,17]

$$\det\left(^{2}m_{EF}\right) = m_{11}m_{22} - m_{12}m_{21} = \vec{e}_z \cdot \vec{e}_z. \tag{21}$$

Hence, Equation (19) becomes [1–5,12,15,17]:

$$A_S = \int d^2x\sqrt{\vec{e}_z \cdot \vec{e}_z}. \tag{22}$$

In LQG, the frame field e^{k}_{E} and the connection A^{E}_{k} are conjugates. For instance, $e^{k}_{E} \to -i\hbar\frac{\delta}{\delta A^{E}_{k}}$, where $\hbar$ is the reduced Planck's constant. Inserting the latter into Equation (22) [1–5,12,15,17],

$$\hat{A}_S = \int d^2x\sqrt{\delta_{JK}\frac{\delta}{\delta A^{z}_{J}}\frac{\delta}{\delta A^{z}_{K}}}. \tag{23}$$

In LQG, $e^{J}_{E}e^{K}_{F} = \frac{\delta}{\delta A^{E}_{J}}\frac{\delta}{\delta A^{E}_{K}} = n_E n_F J^J J^K$. For $SO(3)$ group, the generator is the angular momentum operator, J^J. Here, n_E and n_F are unit tangent vectors. Therefore, the equation of the area operator becomes [1–5,12,15,17]

$$\hat{A}_S = \Sigma_p\sqrt{\delta^{JK}n_E n_F J^J J^K} = \Sigma_p\sqrt{\delta^{JK}\hat{J}_J\hat{J}_K}\Psi_\Theta(\because n^c n_c = 1), \tag{24}$$

$$\therefore \hat{A}_S = \Sigma_p\sqrt{\mathbf{J}^2}. \tag{25}$$

However, the quantum state of J is $\mathbf{J}^2|j\rangle = \hbar^2 j(j+1)|j\rangle$, where j represents the value of the quantum spin; hence, the equation is [1–5,12,15,17]:

$$\hat{A}_S\Psi_\Theta = l_P^2\Sigma_p\sqrt{j_p(j_p+1)}\Psi_\Theta, \tag{26}$$

where $l_P^2 = G\hbar/c^3$ is the Planck area with c the speed of light.

In LQG, lines of the spin network can intersect. Any surface Σ acquires area through the puncture of these lines [1–5,12,15,17]; see Figure 2.

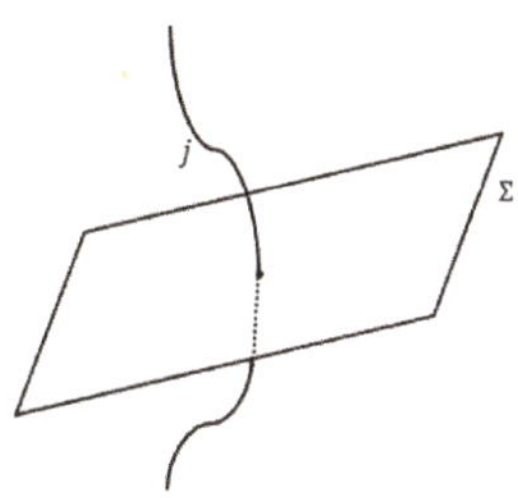

Figure 2. A diagram of surface puncture. See text for details.

In standard form, the area operator with the γ can also be written as [1–5,12,15,17]

$$\therefore \hat{A}_S = \gamma l_P^2\Sigma_p\sqrt{j_p(j_p+1)}. \tag{27}$$

The proportionality coefficient in the formula of area operator in LQG includes the γ [1–5,23]. In 1998, Krasnov [32] found that the multiplicative factor of the area operator is $8\pi\gamma$. Hence, the equation is [1–5,12,15,17]

$$A = 8\pi\gamma l_P^2 \sqrt{j(j+1)}. \tag{28}$$

Similar to the area operator, the BI parameter also appears in the volume operator. The spectrum of volume operator can only be understood; if the γ is real valued [1–5].

Pros and Cons of the Area Operator and the BI Parameter

The spectrum of the area and the volume operator and its eigenvalue can only be understood with the real valued γ. As mentioned, the complex valued γ makes these operators complex valued and the significance of these complex valued operators is ambiguous. Is there any valid significance of the area operator and the volume operator with the complex valued γ? This question is still unresolved.

In short, there are two difficulties for the complex valued γ for these geometrical operators. (1) The physical significance of the complex valued geometrical operators such as the area and the volume operator is yet to be found. (2) The mathematical structure of the complex valued geometrical operators is not yet clear and complete.

2.3. The BI Parameter and Black Hole Entropy Calculation in LQG

The expression of entropy of a black hole in Planck units calculated semi-classically by Hawking is written as [1–5,23]

$$S = A/4. \tag{29}$$

In 1996, Rovelli [33] calculated black hole entropy within LQG using the statistical framework.

In LQG, any surface obtains area when the link of the spin network punctures that surface. One can allot micro states to each surface puncture. Thus, the micro states are associated with the discrete pieces of the surface, which provide the value of area spectrum by puncturing. So, the entropy, S, is proportional to the log of the number of ways in which the sphere can be punctured that provides an area within each macroscopic interval (see Figure 3) [33].

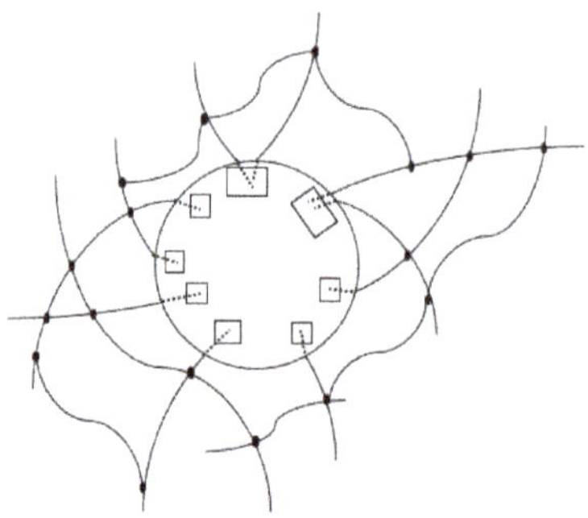

Figure 3. The black hole entropy through puncture in the surface of the event horizon. The rectangles show micro states.

If the eigenvalue, A_p, of the area operator is expressed via the j_p of the form $m_p/2$ ($m_p \in Z$), then [33]

$$A_p = 4\pi\gamma l_P^2 \sqrt{m_p(m_p+2)}. \tag{30}$$

For an interval $[A+\delta A, A-\delta A]$, where δA is some small interval ($\delta A/A \ll 1$), A is a macroscopic value of area. The allowed number $N(M)$ of sequences of integers $\{m_p, \ldots, m_N\}$, $p = 1,2,3,\ldots,N$, is determined by the number N of edges that puncture the surface, so that the determined value for the total area lies within the given interval, $M = A/(4\pi\gamma l_P^2)$ [33].

The number of sequences $N(M)$, in which each sequence is $\{m_p\}$, can be given as [33]

$$M = \frac{A}{4\pi\gamma l_P^2} = \Sigma_p \sqrt{m_p(m_p+2)}. \tag{31}$$

The number of sequences is indicated by $N_+(M)$, such that $\Sigma_p m_p = M$ and the number of sequences is indicated by $N_-(M)$, such that $\Sigma_p(m_p+1) = M$. Hence, the given set of inequalities implies that [33]

$$N_-(M) < N(M) < N_+(M). \tag{32}$$

From calculation, $\ln N_+(M) = \ln 2$, $\ln N_-(M) = \ln \frac{1+\sqrt{5}}{2}$ and $\ln N(M) = dM$. From Equation (32), the inequalities are now given as [33]

$$\begin{aligned} \ln \tfrac{1+\sqrt{5}}{2} < d < \ln 2, \\ 0.48 < d < 0.69. \end{aligned} \tag{33}$$

By taking $M = A/(8\pi\hbar G)$, one gets

$$\begin{aligned} \ln N(A) &= d\tfrac{A}{8\pi\hbar G}, \\ S(A) &= k \ln N(A), \end{aligned} \tag{34}$$

$$\therefore S(A) = c\frac{k}{\hbar G}A, \tag{35}$$

where $c = d/(8\pi) = 1/4$ is constant.

In 1997, Ashtekar et al. [34] showed that spin networks explains spacetime geometry outside a black hole. Some edges of this spin network puncture the event horizon and provide the value of area through this contribution. The $U(1)$ Chern–Simons theory explains the quantum geometry of the horizon. In this formalism, the rotation of $SO(2)$ describes 2D geometry, which is isomorphic to $U(1)$. The entropy of a black hole is calculated by counting the spin network states relevant to an event horizon. Thus, the expression of black hole entropy in LQG is [1–5,34],

$$S = \frac{\gamma_0 A}{4\gamma}. \tag{36}$$

There are two possibilities for the value of γ_0 [1–5,34], i.e.,

$$\gamma_0 = \frac{\ln 2}{\sqrt{3}\pi} \tag{37}$$

or

$$\gamma_0 = \frac{\ln 3}{\sqrt{8}\pi}. \tag{38}$$

The value of γ_0 relies on the choice of the gauge group. By taking $\gamma_0 = \gamma$, one gets actual black hole entropy formula, calculated by Hawking [1–5,34], i.e.,

$$S = \frac{\gamma_0 A}{4\gamma_0} = \frac{A}{4}. \tag{39}$$

This calculation is true for each sort of black hole. The black hole entropy calculation in LQG is a quite enriched research area. The value of the γ and black hole entropy formula in LQG is a topic whose implications are far-reaching [1–5,34].

In 2002, Dreyer [35] fixed the value of the γ using classical quasinormal mode spectrum of a black hole and gave black hole entropy formula in LQG with $SO(3)$ group instead of

$SU(2)$. Instead of $j_{\min} = 1/2$, the value of $j_{\min} = 1$ is taken in the expression of the area operator, i.e.,

$$\Delta A = A(j_{\min}) = 8\pi\gamma l_P^2 \sqrt{j_{\min}(j_{\min}+1)}. \tag{40}$$

In this case, the change in the mass due to the frequency of the quasinormal modes (QNM) is [35]

$$\Delta M = \hbar\omega_{\text{QNM}} = \frac{\hbar \ln 3}{8\pi M}. \tag{41}$$

For a Schwarzschild black hole, the area and the mass are related to each other by the relation $A = 16\pi M^2$. A change in the area corresponding to the mass change is given as [35]

$$\Delta A = 4\ln 3\, l_P^2. \tag{42}$$

The expression for the γ is obtained through comparison between Equations (40) and (42) [35],

$$\gamma = \frac{\ln 3}{2\pi\sqrt{j_{\min}(j_{\min}+1)}}. \tag{43}$$

If $j_{\min} = 1$ is taken in Equation (43), then the fixed value of the BI parameter is [35]

$$\gamma = \frac{\ln 3}{2\pi\sqrt{2}}. \tag{44}$$

Thereafter, Meissner [36] fixed the value of the area in LQG by fixing γ_M and γ by comparing the Bekenstein–Hawking entropy formula with the derived formula. The Bekenstein–Hawking entropy formula reads:

$$S = \frac{1}{4}\frac{A}{l_P^2}. \tag{45}$$

In [36], the derived expression of the black hole entropy formula is:

$$S = \ln N(a) = \frac{\gamma_M}{4\gamma}\frac{A}{l_P^2} + \mathcal{O}(\ln A). \tag{46}$$

Here, by comparing the derived black hole entropy Formula (46) with the Bekenstein–Hawking Formula (45), one gets [1–5,36].

$$\gamma = \gamma_M. \tag{47}$$

The calculated value of γ_M is [36]

$$\gamma_M = 0.2375\ldots \tag{48}$$

or

$$\gamma_M = 0.2739\ldots \tag{49}$$

In 2004, Domagala and Lewandowski [37] defined microscopic degrees of freedom to count the black-hole entropy. On the basis of a ratio, i.e., $\ln(N(a))/a$, for large a, the value of entropy (the eigen value of area operator is equal to or less than a that is a number; for quantum states of black holes) is

$$\frac{\ln 2}{4\pi\gamma l_P^2} a \le \ln N(a) \le \frac{\ln 3}{4\pi\gamma l_P^2} a. \tag{50}$$

Hence, the upper and lower bounds value for γ are [37]

$$\frac{\ln 2}{\pi} \leq \gamma \leq \frac{\ln 3}{\pi}. \tag{51}$$

Since the spin greater than 1/2 also contributes to the entropy, this contribution is also considered here [37].

In 2007, Jacobson [38] studied the renormalization and the black hole entropy in LQG. For black holes, he found that the microscopic state counting is related to Newton's universal constant

G:

$$S_{\text{LQG}} = \frac{b}{\gamma}\frac{A}{\hbar G}, \tag{52}$$

where b is a numerical constant. In LQG, from Equation (52), one can see that the entropy is related to the area of the horizon A ($S_{\text{LQG}} \propto A$) and the gravitational constant G ($S_{\text{LQG}} \propto 1/G$) [38].

Jacobson found that one should compare this formula with the actual Bekenstein–Hawking entropy formula after accounting the scale dependence of Newton's constant and area. For any value of the γ, if some property of the renormalization is followed, then one can compare both entropy formulas. The BI parameter to be $\gamma = 4b$ to match the black hole entropy in LQG with the Bekenstein–Hawking Formula (45) [38].

In 2013, Frodden et al. [39] found that by taking complex valued Ashtekar variables, the black hole entropy formula is achieved in certain conditions. In this case, the BI parameter can be complex valued ($\gamma = \pm i$). Ref. [39] shows that the number of micro states $N_\Gamma(A, \gamma \to \pm i)$ acts as $\exp\big(A/(4l_P^2)\big)$ for certain case, i.e., for large area A in the large spin semi-classical limit. With respect to the complex self-dual Ashtekar connections, $N_\Gamma(A, \pm i)$ is the number of states for a theory to be defined.

The $SU(2)$ Chern–Simons theory is related to the study of the black hole entropy in LQG. From $SU(2)$ to $SL(2,\mathbb{C})$ representation, the formula for the Hilbert space of $SU(2)$ Chern–Simons theory follows a specific analytic continuation with constraints of self-duality. The complex formulation (with the Ashtekar variables) within this proposal gives the derivation of the black hole entropy within LQG formalism for a large spin asymptotic domain which is semi classical in nature [39]. Hence,

$$\log(N_\Gamma(A, \pm i)) \sim \frac{A}{4\, l_P^2}. \tag{53}$$

One can list more papers on black hole entropy in LQG [40–48].

Pros and Cons of the BI Parameter and Black Hole Entropy Calculation in LQG

On the basis of the black hole entropy calculation in LQG, one gets various expressions for the real valued γ, such as $\gamma_0 = \frac{\ln 2}{\sqrt{3}\pi}$, $\gamma_0 = \frac{\ln 3}{\sqrt{8}\pi}$ and $\gamma = \frac{\ln 3}{2\pi\sqrt{2}}$. The numerical value for the γ is either 0.2375… or 0.2739… based on the calculation. The BI parameter is also expressed in terms of a numerical constant b, i.e., $\gamma = 4b$. With the complex valued γ, the black hole entropy can also be calculated using $SU(2)$ Chern–Simons theory. Whether it is the real valued or the complex valued γ; the black hole entropy can be calculated in LQG. However, the interpretation of the complex valued BI parameter within the black hole entropy formula in LQG is difficult to comprehend.

One criticism for the black hole entropy calculation in LQG is regarding the different value of the γ. However, this criticism is easy to address, because a value of the real valued γ can be applied to all kinds of black holes. The γ is a free parameter; its same value is applied to all kinds of black hole.

2.4. The BI Parameter as Immirzi Ambiguity

In LQG, the geometrical observables, such as the area and the volume, are quantized and exhibit a discrete spectrum. In 1996, Immirzi noticed that LQG does not determine the complete scale of these spectra [27]. Immrizi also observed that one can have different spectra for the same geometrical quantities, if starting with the scaled elementary variables. The algebra of holonomy relies on a free parameter that gives the family of one parameter of quantum theories with inequivalence. The γ represents this family of one parameter [49].

There is a certain symmetry under study, according to which classical theory is identified as a canonical transformation; however, one cannot identify it as a unitary transformation of quantum theory. Since the holonomy is the operator of LQG and because of weird sort of representation of LQG, one has to consider the γ as an ambiguity [49].

In LQG, there are two connections, i.e., A and Γ. Therefore, one has to create the γ-scaled connection, namely, A_γ, via interpolation between different connections. Thus, the elementary excitation of LQG—namely the Wilson loop of A_γ—has different results for various value of the γ. Therefore, some physical spectrum of quantity of LQG relies on the γ [49].

Additionally, the metric information resides in the E (conjugate variable). Since E is a conjugate for connection, in quantum formalism, it is written as a derivative operator that acts on functions over the group. Over the group manifold, any geometrical quantity that is a function of E behaves as an elliptic operator that results in the discrete spectrum. Such elliptic operators possess non-vanishing scalar dimension relative to the affine scaling of the connection. Hence, in the elliptic geometric operators spectrum, ambiguity is introduced, i.e., the Immirzi ambiguity. This ambiguity influences the discreteness of the space in LQG. In [49], the The authors also described that various interpretations regarding the γ are incorrect in context with the Immirzi ambiguity. The authors also gave various models such as harmonic oscillator with no Immirzi ambiguity, a particle on a circle with no Immirzi ambiguity, and a simple model with γ as a free parameter. Due to lack of space, here, only the cause of the Immirzi ambiguity and its effect in LQG are given [49].

In the year 2001, Samual [50] commented that interpretations of the Immirzi ambiguity are unclear and do not give any agreement on its origin and significance. All interpretations of the BI parameter as Immirzi ambiguity seem unclear, and do not give any satisfactory explanation about the origin and significance of it. Moreover, the examples of the Immirzi ambiguity are not real, but are artificially generated through the compactification of the configuration space.

In 2017, Veraguth and Wang [51] gave a proposal in which they explained LQG without Immirzi ambiguity using conformal LQG. The conformal LQG provides a way to achieve loop quantization through a conformally equivalent class of metrics. The conformal geometry gives an extended symmetry to permit a reformulated BI parameter. In scalar–tensor gravity, this can be achieved via conformal frame transformations. In this proposal, the authors showed that the LQG, along with a conformally transformed Einstein metric which has dissimilar values of the relevant BI parameter, are connected by a conformal frame with global change. The conformal LQG is free from the Immirzi ambiguity. They defined the Ashtekar variables in the following way [51]:

$$A_c^{'j} = \Gamma_c^j + \gamma\kappa K_c^j, \frac{1}{\gamma\kappa} E_c^j. \tag{54}$$

Pros and Cons of the BI Parameter as Immirzi Ambiguity

Research on the Immirzi ambiguity is still incomplete. As mentioned, there is a specific symmetry under study, according to which classical theory is identified as a canonical transformation, but one cannot identify it as a unitary transformation of quantum theory. This is the reason behind the Immirzi ambiguity. One-parameter family of the BI parameter is another reason. The γ-scaled connection, namely A_γ for various values of the BI parameter, is also different. The elliptic nature of the geometric operator due to the frame field E that is a conjugate to the connection A is also responsible for the Immirzi ambiguity. However, the conformal formalism of LQG may remove the Immirzi ambiguity.

The BI parameter was added into the LQG framework to remove mathematical complexities (in the connection formalism, constraints equations, geometrical operators equations and many other important equations); however, it emerged as ambiguity because of the above-mentioned causes.

2.5. Origin of the BI Parameter

In 2005, Chou et al. [52] found a technique, through which a ratio which equals the γ is obtained. They used quadratic spinor techniques, in which the physical significance and effect of the γ become obvious in GR. The authors also inferred that without other matter fields in GR, the γ as a observable is a physical property of the sector of gravity.

Firstly, the Holst action is defined in a novel way [52], i.e.,

$$S[\mathbf{e},\omega] = \alpha \int *\left(\mathbf{e}^c \wedge \mathbf{e}^d\right) \wedge R_{cd}(\omega) + \gamma\left(\mathbf{e}^c \wedge \mathbf{e}^d\right) \wedge R_{cd}(\omega), \tag{55}$$

where the γ is a ratio, i.e., $\gamma = \alpha/\beta$, $c, d, \ldots = 0,1,2,3$, and "$*$" denotes duality. Thereafter, the authors compared the Equation (55) with the quadratic spinor Lagrangian [52], i.e.,

$$\mathcal{L}_\psi = 2D(\bar{\psi}\mathbf{e})\gamma_5 D(\psi\mathbf{e}), \tag{56}$$

where ψ is auxiliary spinor field, $\mathbf{e}$ is $\mathbf{e} = \mathbf{e}^c\gamma_c$, D is the covariant derivative, and γ_c is the Dirac gamma matrix.

By defining the spinor curvature identity [52],

$$\begin{aligned} 2D(\bar{\psi}\mathbf{e})\gamma_5 D(\mathbf{e}\psi) &= \bar{\psi}\psi R_{cd} \wedge *\left(\mathbf{e}^c \wedge \mathbf{e}^d\right) \\ &\quad + \bar{\psi}\gamma_5\psi R_{cd} \wedge \mathbf{e}^c \wedge \mathbf{e}^d + d[D(\bar{\psi}\mathbf{e})\gamma_5\mathbf{e}\psi + \bar{\psi}\mathbf{e}\gamma_5 D(\mathbf{e}\psi)]. \end{aligned} \tag{57}$$

For $\omega[e]$, the equation of motion is [52]:

$$D[\bar{\psi}\psi * \left(\mathbf{e}^c \wedge \mathbf{e}^d\right) + \bar{\psi}\gamma_5\psi\left(\mathbf{e}^c \wedge \mathbf{e}^d\right)] = 0. \tag{58}$$

It was found that the γ can be written as a ratio of scalar and pseudoscalar contributions in the theory [52], i.e.,

$$\gamma = \frac{\langle\bar{\psi}\psi\rangle}{\langle\bar{\psi}\gamma_5\psi\rangle}. \tag{59}$$

If $\bar{\psi}\psi = 1$ and $\bar{\psi}\gamma_5\psi = 0$, then $\gamma = \infty$. The BI parameter $\gamma = i$ corresponds to Ashtekar formalism with self duality; meanwhile, $\gamma = 1$ satisfies the action of the Hamiltonian given by Barbero. Therefore, the γ implies that Einstein gravity can be distinguished from the other gravitation theories via general covariance. In other words, this ratio the ratio (59) can be seen as a measure of how gravity differs from covariant gravity. Such a technique permits the renormalization scale, μ, regarding the γ via spinor's expectation value in quantization process ($\therefore$ $\langle\bar{\psi}\psi\rangle_\mu, \langle\bar{\psi}\gamma_5\psi\rangle_\mu$). Here, $\bar{\psi}\gamma_5\psi$ is not a real function. To get $\bar{\psi}\gamma_5\psi$ to be real, one has to use an anti-commuting spinor to achieve the real Ashtekar variables [52].

Pros and Cons of Origin of the BI Parameter

This proposal gives the origin of the γ using quadratic spinor techniques in which a ratio of scalar and pseudoscalar contributions is defined as the γ. In essence, in this proposal, the γ can be real as well as complex valued under different condition. The anti-commuting spinor is necessary to get real valued $\bar{\psi}\gamma_5\psi$ and the Ashtekar variables.

2.6. On a Covariant Formulation of the BI Connection

In 2007, Fatibene et al. [53] gave a proposal on covariant formulation of the BI connection in which they defined a global covariant $SU(2)$-connection over whole spacetime that limits generalizations of the Barbero–Immirzi connection on a given slice of space. The BI connection is a collective $SU(2)$ gauge connection on a 3D surface $S \subset \mathcal{M}$ in 4D spacetime $\mathcal{M}$. On the basis of groups and spacetime involved in the theory, the BI connection is global. However, the $SU(2)$ principal bundles $^{+}\Sigma$ over one 3D base S should be trivial. In this paper, the global aspects of the BI connection, the covariant formulation of the BI connection with its spacetime interpretation and the Lorentzian case are investigated. Here, the point of study is the BI connection. The γ is less emphasized in this study. Thus, this investigation is covered with the necessary details.

Pros and Cons of a Covariant Formulation of the BI Connection

This proposal advocates the usual interpretation of the γ (real valued) on the basis of the black hole entropy calculation in LQG. In this proposal, the complex valued γ is also less emphasized due to its obscured significance.

2.7. The BI Parameter as a Scalar Field

In 2008, Taveras and Yunes [54] (see also [55]) gave a proposal on the γ as a scalar field. They studied the LQG-based generalization of GR, in which they modified the Holst action was modifeid.

The authors scalarized the γ in the Holst action. This meant that the γ was promoted as a field under the integral of the dual curvature term. In this formalism, the γ acts as a dynamical scalar field. This formalism gives a non-zero torsion tensor which modifies the field equations through quadratic first derivatives of the BI field. Such a modification is similar to the general theory of relativity with non-trivial kinetic energy in the presence of a scalar field [54].

Before promoting the the authors firstly modified the Holst action [54], i.e.,

$$S = \frac{1}{4\kappa}\int \epsilon_{JKLM}e^{J}\wedge e^{K}\wedge F^{LM} + \frac{1}{2\kappa}\int \bar{\gamma}e^{J}\wedge e^{K}\wedge F_{LM} + S_{\text{mat}}, \tag{60}$$

where $\kappa = 8\pi G$, the coupling field $\bar{\gamma}$ is $\bar{\gamma} = 1/\gamma$, andS_{mat} is the action for additional matter degrees of freedom. To introduce torsion and contorsion in Equation (60) one can simplify the Holst action as follows [54]:

$$\begin{aligned} S &= \tfrac{1}{4\kappa}\int \epsilon_{JKLM}e^{J}\wedge e^{K}\wedge e^{Q}\wedge e^{R}\tfrac{1}{2}F^{LM}_{QR} + \tfrac{1}{2\kappa}\int \bar{\gamma}e^{J}\wedge e^{K}\wedge e^{L}\wedge e^{M}\tfrac{1}{2}F_{JKLM} + S_{\text{mat}} \\ &= \tfrac{1}{8\kappa}\int \epsilon_{JKLM}(-\tilde{\sigma})\epsilon^{JKQR}F^{LM}_{QR} + \tfrac{1}{4\kappa}\int \bar{\gamma}(-\tilde{\sigma})\epsilon^{JKLM}F_{JKLM} + S_{\text{mat}}. \end{aligned} \tag{61}$$

Through simplification, one gets [54]

$$\therefore S = \frac{1}{2\kappa}\int \tilde{\sigma}\left[\delta^{[QR]}_{LM}F^{LM}_{QR} - \frac{\bar{\gamma}}{2}\epsilon^{JKLM}F_{JKLM}\right] + S_{\text{mat}}, \tag{62}$$

where $\tilde{\sigma} = d^{4}xe = d^{4}x\sqrt{-g}$, and $e^{J}\wedge e^{K}\wedge e^{L}\wedge e^{M} = -\tilde{\sigma}\epsilon^{JKLM}$.

In the simpler form, the modified form of the Holst action is [54]

$$S = \frac{1}{2\kappa}\int d^{4}xep^{JK}_{LM}e^{\mu}_{J}e^{\nu}_{K}F^{LM}_{\mu\nu} \tag{63}$$

where $p^{JK}_{LM} = \delta^{[L}_{J}\delta^{M]}_{K} - \frac{\bar{\gamma}}{2}\epsilon^{LM}_{JK}$.

Thereafter, the authors gave field equations with the modified Holst action and its solutions. They also gave effective action and the inflation with the γ as a dynamical scalar field [54].

In 2009, Calcagni and Mercuri [54] also promoted the γ as a field in the canonical formalism of pure gravity. In this paper, the authors investigated the parity properties of the field of the γ by performing the decomposition of torsion into irreducible components. Under a local Lorentz group, they suggested that the γ ought to be pseudoscalar to conserve the transformation properties of these components.

To understand the Riemann–Cartan space–time, one has to generalize the Holst formalism. This can be achieved by adding a torsion part in the Holst action. It gives net coupling with the γ, which gives rise to Nieh–Yan density [54].

The field of the γ is a real canonical pseudoscalar field for $\gamma = \gamma(x)$ coupled with the Nieh–Yan invariant. The γ is pseudoscalar in nature because of the axial component of torsion, which is proportional to the partial derivative of the In the absence of matter, the field of the was studied in the first-order Hamiltonian formalism. Here, the derivation of the action in the Lagrangian formalism is avoided, since the subject of study is the field of the BI parameter in the Hamiltonian formalism (canonical). The authors also compared the Holst case with the Nieh–Yan case. The total Hamiltonian in the form of the action in the Holst case is expressed as [54]

$$H_D = \int d^3x \left(\Lambda^j \mathcal{R}_j + N^\beta \mathcal{H}_\beta + N\mathcal{H} \right), \tag{64}$$

where Λ^j, N^β and N are Lagrange-undetermined multipliers. $\mathcal{R}_j$, H_β and $\mathcal{H}$ are the rotation, super momentum and super Hamiltonian constraints, respectively.

The expression of all constraints is written as [54],

$$\mathcal{R}_j \equiv \epsilon^l_{jk} K^k_\beta E^\beta_l \,, \tag{65}$$

$$\mathcal{H}_\beta \equiv E^\eta_j D_{[\beta} K^j_{\eta]} + \Pi \partial_\beta \gamma \approx 0 \,, \tag{66}$$

$$\mathcal{H} \equiv -\frac{1}{2e} E^\beta_j E^\eta_k \left(\epsilon^{jk}_l R^l_{\beta\eta} + 2K^j_{[\beta} K^k_{\eta]} \right) + \frac{1+\gamma^2}{3e} \Pi^2 - \frac{3}{4} \frac{e}{1+\gamma^2} \partial_\beta \gamma \partial^\beta \gamma \approx 0 \,. \tag{67}$$

Here, ∂_β denotes the coordinate derivative.

For the Nieh–Yan case, the super Hamiltonian is denoted as [54] $\mathcal{H}$, i.e.,

$$\mathcal{H} \equiv -\frac{1}{2e} E^\beta_j E^\eta_k \left(\epsilon^{jk}_l R^l_{\beta\eta} + 2K^j_{[\beta} K^k_{\eta]} \right) + \frac{1}{3e} \Pi^2 - \frac{3}{4} e \partial_\beta \gamma \partial^\beta \gamma \approx 0 \,. \tag{68}$$

Here, in the canonical formalism, the factor $(1+\gamma^2)$ disappears in the contribution of the pseudo-scalar field for the Nieh–Yan term. The Nieh–Yan term exhibits a shift symmetry, i.e., $\gamma \rightarrow \gamma + \gamma_0$ [54].

Pros and Cons of the BI Parameter as a Scalar Field

Taveras and Yunes defined the γ as a dynamical scale field in the Holst action with non zero torsion tensor, while Calcagni and Mercuri defined the γ as a field in the canonical formalism. These proposals provide new significance of the BI parameter; however, the γ, which is sometimes complex valued, is still unclear.

2.8. *Topological Interpretation of the BI Parameter*

In 2008, Date et al. [56] gave a proposal on the topological interpretation of the γ.

In terms of the Holst formalism, the Hilbert–Palatini Lagrangian as the Lagrangian density can be written as [56]

$$\mathcal{L} = \frac{1}{2} e \Sigma^{\mu\nu}_{JK} R^{JK}_{\mu\nu}(\omega) + \frac{\gamma}{2} e \Sigma^{\mu\nu}_{JK} \tilde{R}^{JK}_{\mu\nu}(\omega) \tag{69}$$

where $\Sigma^{\mu\nu}_{JK} := \frac{1}{2}\left(e^\mu_J e^\nu_K - e^\mu_K e^\nu_J \right)$, $R^{JK}_{\mu\nu}(\omega) := \partial_{[\mu} \omega^{JK}_{\nu]} + \omega^{JK}_{[\mu} \omega^J_{\nu]L}$ and $\tilde{R}^{JK}_{\mu\nu}(\omega) := \frac{1}{2} \epsilon_{JKLM} R_{\mu\nu LM}(\omega)$.

Here, with γ^{-1}, the second term is the Holst term, while with $\gamma = -i$, this Lagrangian density gives the complex value $SU(2)$ Ashtekar connection. For $\gamma = 1$, one gets the real valued $SU(2)$ Barbero connection [56]. This has been already discussed in the introduction.

The expression of the Nieh–Yan (NY) density is given as [56]

$$I_{\text{NY}} = \epsilon^{\mu\nu\beta\eta} \left[D_\mu(\omega) e^J_\nu D_\beta(\omega) e_{J\beta} - \frac{1}{2} \Sigma^{JK}_{\mu\nu} R_{\beta\eta JK}(\omega) \right], \tag{70}$$

where $D_\mu(\omega)e^J_\nu = \partial_\mu e^J_\nu + \omega^J_{\mu K}e^K_\nu$. For a torsion-free connection, the Nieh–Yan density disappears.

In this proposal, in the time gauge, Nieh–Yan topological density of a theory of gravity permits us to explain gravity as a real $SU(2)$ connection. For $\gamma = 1$, the set of constraints for both the Hamiltonian and the Barbero formalism are the same. For the rest of the values of the γ, the Immirzi formulation exhibits $\frac{1}{\gamma}$. The parameter γ is analogous to the θ parameter of the quantum chromodynamics. This parameter (γ) implies an enriched vacuum structure of gravity. The Nieh–Yan density is fully constructed from geometric quantities, while the modified Holst terms exhibit fields of matter. With the aid of connection equation of motion, both are connected [56].

Pros and Cons of Topological Interpretation of the BI Parameter

In this proposal, the authors gave the interpretation of the γ topologically using the Nieh–Yan density with the real valued $SU(2)$ connection. From this proposal, the γ can be compared to the θ parameter of the quantum chromodynamics. The complex valued γ is also less emphasized in this proposal.

2.9. The Peccei–Quinn Mechanism in Gravity and the Nature of the BI Parameter

Promoting the γ as a field was an active research area between 2007 and 2011; that added new physical significance for the γ with the topological perspective in LQG.

In 2009, Mercuri [57] gave a proposal on the nature of the γ using the Peccei–Quinn mechanism in gravity, in which the γ is taken as a field. Using the Holst formalism, the modified Hilbert–Palatini (HP) action is obtained. This modified (total) action with the Nieh–Yan invariant (spacetime with the torsion) and the matter coupling is given as $S_{\text{tot}} = S_{\text{HP}}[e,\omega] + S_{\text{NY}}[e,\omega] + S_{\text{mat}}$. Hence,

$$\begin{aligned} S_{\text{tot}} = &-\frac{1}{16\pi G}\int e_c \wedge e_d \wedge \star R^{cd} + \frac{\gamma}{16\pi G}\int \left(T^c \wedge T_c - e_c \wedge e_d \wedge R^{cd}\right) \\ &+\frac{i}{2}\int \star e_c \wedge \left(\bar{\psi}\gamma^c D\psi - \overline{D\psi}\gamma^c\psi + \frac{i}{2}me^c\bar{\psi}\psi\right), \end{aligned} \tag{71}$$

where T_c denotes the torsion two-form. The author suggested promoting γ as a field, the contribution from divergence to the chiral anomaly must be reabsorbed [57].

The author also implied that the Peccei–Quinn mechanism (this mechanism is used for charge-parity (CP) conservation for a strong force, in which pseudo-particle effects are considered with a scalar field) permits one to connect the constant value of the γ to the other certain topological ambiguities. This connection creates an interaction between gravity and the field of the BI parameter [57].

From the spontaneous symmetry breaking ($SU(2) \times U(1)$), the obtained quark mass matrices M is not diagonal and Hermitian. A chiral rotation is needed to diagonalize it. Similar to this, the chiral rotation of the fermionic measure in the Euclidean path integral produces a NY term with Pontryagin class in space–time with torsion that is diverged as the square of the regulator [57].

The following equation is a part of regularization procedure when γ is considered as a field:

$$\delta\psi\delta\bar{\psi} \to \delta\psi\delta\bar{\psi}\exp\left\{\frac{i}{8\pi^2}\int \beta\left[R_{cd}\wedge R^{cd} + 2\Lambda^2\left(T_c \wedge T^c - e_c \wedge e_d \wedge R^{cd}\right)\right]\right\}, \tag{72}$$

where β is the transformation parameter and Λ is a regulator. In short, Equation (72) shows the regularization. Hence, the effective action after the regularization procedure in Equation (72) reads:

$$S_{\text{eff}} = S_{\text{HP}}[e,\omega] + S_D[e,\omega,\psi,\bar{\psi}] + \frac{1}{8\pi^2}\beta \int R_{cd} \wedge R^{cd} + \frac{1}{16\pi G}\left(\gamma + \frac{4G}{\pi}\beta\Lambda^2\right) \times \int \left(T^c \wedge T_c - e_c \wedge e_d \wedge R^{cd}\right). \tag{73}$$

Any attempt at removal of the regulator results in the divergence of β. By promoting γ as a field, this divergence is reabsorbed [57]. Hence,

$$S_{\text{eff}} = S_{\text{HP}}[e,\omega] + S_D[e,\omega,\psi,\bar{\psi}] + \frac{1}{16\pi G} \times \int \gamma'(x)\left(T^c \wedge T_c - e_c \wedge e_d \wedge R^{cd}\right), \tag{74}$$

where $\gamma'(x) = \gamma(x) + \frac{4G}{\pi}\beta\Lambda^2$.

Pros and Cons of the Peccei–Quinn Mechanism in Gravity and the Nature of the BI Parameter

In this proposal, the γ is promoted as a field using the Peccei–Quinn mechanism in LQG. This notion is essential, since it removes divergence from the total effective action. However, the origin of the complex valued γ is still open for exploration.

2.10. The Kodama State and the BI Parameter

In year 2006, suggested Randono [58,59] the generalization of Kodama states, in which the real valued γ was used to generalize these states and derived physical interpretation of these states. The Kodama state is special, providing an exact solution to all the normal constraints of canonical quantum gravity.

The Kodama state has an unambiguous semi-classical interpretation as a quantum sort of classical spacetime (anti-de Sitter space). However, the structure of the phase space is complex. Therefore, a generalization of the real valued γ state is needed [58].

The state of Lorentzian Kodama is a solution to the quantum constraints in the Ashtekar formalism, in which the connection is complex valued. However, to get the classical GR, one has to execute the reality conditions ensuring the reality of the metric [58].

In the Euclidean framework formalism, the $SO(4)$ group is divided into two left and right parts, as in the complex framework. The Ashtekar variables exhibit a real valued $SO(3)$ connection, and its real valued momentum conjugate for the left-handed part of the group. The corresponding state in the Euclidean framework is a pure phase, because the connection is real. Thus, the state is written as [58],

$$\Psi[A] = \mathcal{N} e^{-\frac{3}{4\kappa\lambda}} \int Y_{\text{CS}}[A], \tag{75}$$

where $\int Y_{\text{CS}}[A]$ is the Chern–Simon term, $\lambda = G\hbar\Lambda$ with Λ the cosmological constant, and $\mathcal{N}$ being topology dependent normalization.

For this result, no reality condition is required, since the structure of the phase space of the Euclidean framework is simple. Hence, the complexification of the phase space (in which the γ is the complex valued $\gamma = i$) is the main reason for the requirement of the reality conditions [58].

The Kodama state beginning with the Holst action with the cosmological constant is given as [58],

$$S_{\text{H}} = \frac{1}{\kappa}\int_{\mathcal{M}} \star e \wedge e \wedge R + \frac{1}{\gamma} e \wedge e \wedge R - \frac{\lambda}{3} \star e \wedge e \wedge e \wedge e, \tag{76}$$

where $e = \frac{1}{2}\gamma_J e^J$, $R = \frac{1}{4}\gamma_{[J}\gamma_{K]}\omega^{JK}$, $\star = -i\gamma^5 = \gamma^0\gamma^1\gamma^2\gamma^3$, and $\mathcal{M}$ is the manifold.

The chiral symmetric Holst action is written as [58]

$$S = \frac{1}{\kappa}\int_{\mathcal{M}} 2(\alpha_L P_L + \alpha_R P_R) \star \Sigma \wedge \left(R - \frac{\lambda}{6}\right)$$
$$= \frac{2}{\kappa}\int_{\mathcal{M}} \alpha_L \star \Sigma_L \wedge \left(R_L - \frac{\lambda}{6}\Sigma_L\right) + \alpha_R \star \Sigma_R \wedge \left(R_R - \frac{\lambda}{6}\Sigma_R\right).$$

Here, $\Sigma_R = (e \wedge e)_R$, P_L and P_R are left-handed and right-handed chiral projection operators, R_L and R_R are the left-handed and right-handed chiral curvatures for $spin(3,1)$ connection.

If $\alpha_L + \alpha_R = 1$ and $\gamma = \frac{-i}{\alpha_L - \alpha_R}$, the equation becomes [58]

$$S = \frac{1}{\kappa}\int_{\mathcal{M}} (\alpha_L + \alpha_R) \star \Sigma \wedge \left(R - \frac{\lambda}{6}\Sigma\right) + i(\alpha_L - \alpha_R)\Sigma \wedge R. \tag{77}$$

In year 2011, Wieland [60] gave a proposal, namely complex Ashtekar variables, the Kodama state and spinfoam gravity, in which the complex valued Ashtekar variable and the real valued γ were used. In Ref. [60], the author used $SL(2,\mathbb{C})$ Kodama state and proposed a spinfoam vertex amplitude.

As per the usual method, it also started with the Holst action with the cosmological constant, Λ [60], i.e.,

$$S[e,\omega] = \frac{\hbar}{4l_P^2}\int_{\mathcal{M}} e^J \wedge e^K \wedge \left(\epsilon_{JKLM}R^{LM}[\omega] - \frac{2}{\gamma}R_{JK}[\omega] - \frac{\Lambda}{6}\epsilon_{JKLM}e^L \wedge e^M\right). \tag{78}$$

where $\gamma \in \mathbb{R}$.

As mentioned, in this proposal, the real valued γ and the complex valued Ashtekar variable are considered [60].

Pros and Cons of the Kodama State and the BI Parameter

The generalization of the Kodama state can only be achieved with the real valued γ, since, the γ with the complex value makes the state complex and ambiguous. Here, the significance of the complex valued BI parameter is also unclear.

2.11. The Quantum Gravity BI Parameter—A General Physical and Topological Interpretation

In the year 2013, El Naschie [61] gave a proposal on general physical and topological interpretation of the γ. This proposal is not directly related to LQG. In this paper, the γ of LQG is considered as a definite quantum entanglement correction.

According to this proposal, the γ is not only a free basic parameter of LQG; it is also an exact sort of a basic constant of the micro-spacetime topology [61].

As mentioned in Section 2.3, one of the fixed values of the γ from black hole entropy calculation in LQG is given as [61]

$$\gamma = \frac{\log 2}{\pi\sqrt{3}} = 0.055322. \tag{79}$$

From Hardy's quantum entanglement, the author proposed that the value of the γ is same as that obtained from quantum entanglement correction [61], i.e.,

$$\gamma = \phi^6 = \left(\frac{\sqrt{5}-1}{2}\right)^6 = 0.055728. \tag{80}$$

Pros and Cons of the Quantum Gravity BI Parameter—A General Physical and Topological Interpretation

This proposal advocates the real valued γ. It does not explain the physical significance of the γ in LQG.

2.12. A Correction to the BI Parameter of SU(2) Spin Networks

In year 2014, Sadiq [62] gave a correction to the γ of $SU(2)$ spin networks. In this paper, by taking $j = 1$ and to preserve the $SU(2)$ symmetry of theory, twice the value of the γ is proposed. Previously in LQG, the γ was fixed by $j = 1$ transitions of spin networks as the dominant processes instead of $j = 1/2$ transitions. This means $SO(3)$ should be a gauge group instead of $SU(2)$.

This proposal begins with [35] (see Section 2.3) and gave a correction to the γ. The author investigated that if $SU(2)$ is the compatible gauge group and $j_{\text{min}} = 1$ process governs, then the change in the mass of the black hole during the transition is [62]

$$\Delta M = 2\hbar\omega_{\text{QNM}}. \tag{81}$$

Since the value of ω_{QNM} is $\omega_{\text{QNM}} = \ln 3/(8\pi M)$, the change in the mass is [62]

$$\Delta M = \frac{2\hbar \ln 3}{8\pi M}. \tag{82}$$

Therefore, the fixed value of γ for $j_{\text{min}} = 1$ is [62]

$$\gamma = \frac{\ln 3}{\pi\sqrt{2}}. \tag{83}$$

Pros and Cons of a Correction to the BI Parameter of $SU(2)$ Spin Networks

In this proposal, the author modified the fixed value of the γ; that was proposed in [35]. This proposal advocates the real valued γ. The physical significance is based on the black hole entropy calculation in LQG.

2.13. Physical Effect of the Immirzi Parameter in LQG

In this proposal, Perez and Rovelli [63] proposed that the BI term in the (Holst) action does not disappear on the shell when fermions are there. The BI term is also present in the equations of motion. The γ governs the coupling constant of a four-fermion interaction (it is mediated by a torsion). In other words, the γ is a coupling constant that governs the strength of a four-fermion interaction. Thus, the γ may show physical effects that can be observed independently from LQG.

The Holst action with the fermionic field is expressed as [63]

$$S[e, A, \psi] = S[e, A] + \frac{i}{2}\int d^4xe\left(\bar{\psi}\gamma^J e^c_J D_c\psi - \overline{D_c\psi}\gamma^J e^c_J\psi\right), \tag{84}$$

where $S[e, A] = \frac{1}{16\pi G}\left(\int d^4xee^c_J e^d_K F^{JK}_{cd} - \frac{1}{\gamma}\int d^4xee^c_J e^d_K * F^{JK}_{cd}\right)$, D_c is a covariant derivative, and γ^J is the Dirac matrix. In this proposal, $D_{[c}e^J_{d]}$ is the fermionic current. In the connection, the fermion current behaves as a source for a torsion component [63].

In the fermionic current, the linear terms are total derivative; hence, the resulting action is

$$S[e, \psi] = S[e] + S_f[e, \psi] + S_{int}[e, \psi],$$

$$\begin{aligned} \therefore S[e, \psi] = &\frac{1}{16\pi G}\int d^4xee^c_J e^d_K F^{JK}_{cd}[\omega[e]] + i\int d^4xe\bar{\psi}\gamma^J e^c_J D_c[\omega[e]]\psi \\ &- \frac{3}{2}\pi G\frac{\gamma^2}{\gamma^2+1}\int d^4xe(\bar{\psi}\gamma_5\gamma_A\psi)\left(\bar{\psi}\gamma_5\gamma^A\psi\right). \end{aligned} \tag{85}$$

The standard coupling of the Einstein–Cartan theory is recovered in the third term with the limit $\gamma \to \infty$ [63].

Pros and Cons of Physical Effect of the Immirzi Parameter in LQG

In this proposal, the γ is a coupling constant that governs the strength of a four-fermion interaction. Here, the γ is indeed free parameter. $\gamma = i$ gives the self-dual Ashtekar canonical formalism and $\gamma = 1$ (real valued) gives the $SU(2)$ Barbero connection.

2.14. *A Relation between the BI Parameter and the Standard Model*

In the year 2010, Broda and Szanecki [64] established the relationship between the γ and the standard model.

In this proposal, Sakharov's method was used with the Nieh–Yan term (in the Holst action) that fixed the γ by considering the Lagrangian density [64], i.e.,

$$\mathcal{L} = \alpha \star \left(e^c \wedge e^d\right) \wedge R_{cd} - \beta\left(T^c \wedge T_d - e^c \wedge e^d \wedge R_{cd}\right), \tag{86}$$

where $\gamma = \alpha/\beta$.

The Einstein–Hilbert Lagrangian is written as [64]

$$\mathcal{L}_{\text{EH}} = -\frac{1}{12}\left(\frac{M}{4\pi}\right)^2 \left(N_0 + N_{\frac{1}{2}} - 4N_1\right) \star \left(e^c \wedge e^d\right) \wedge R_{cd}, \tag{87}$$

where N_1 is the gauge fields number, $N_{\frac{1}{2}}$ the two-component fermion fields number, and N_0 is the minimal scalar degrees of freedom number. Here, the γ is defined as [64]

$$\gamma = \frac{-\frac{1}{12}\left(N_0 + N_{\frac{1}{2}} - 4N_1\right)}{-\frac{1}{4}N_L} = \frac{1}{9} \approx 0.11. \tag{88}$$

Here, N_L is the number of the chiral left-handed modes. By taking $N_0 = 4$ (for Higgs), $N_1 = 12$, $N_{\frac{1}{2}} = 45$ and $N_L = 3$ (3 neutrinos) [64].

This is approximately equal to one of the values of the BI parameter in black hole entropy calculation in LQG [64],

$$\gamma = \frac{\ln 2}{\pi\sqrt{3}} \approx 0.13. \tag{89}$$

Pros and Cons of a Relationship between the BIParameter and the Standard Model

This proposal established a relationship between the γ and the standard model. It advocates the real valued γ by comparing the results with the black hole entropy calculation in LQG. Similar to other proposals, the significance of the complex valued γ is less emphasized.

2.15. *The Holographic Principle and the BI Parameter of LQG*

In the year 2015, Sadiq [65] gave a proposal that correlates the γ in LQG and the holographic principle. In this proposal, the γ is fixed using the equipartition theorem based on LQG at holographic boundary in such a way that the Unruh–Hawking law of temperature holds and follows. Such derived value of the γ exhibits validity universally. In this way, this approach correlates the value of the γ in LQG and the holographic principle. In this proposal, the real valued BI parameter is considered. Since the relation between the holographic principle and LQG demands more research, this proposal is given in brief.

Pros and Cons of the Holographic Principle and the BI Parameter of LQG

Similar to majority of the proposals, this proposal also advocates the real valued BI parameter.

2.16. Discussion

The γ is a free parameter as well as an enigmatic parameter of LQG. It is a free parameter because it can be real valued as well as complex valued. It is enigmatic parameter because its significance for either the real value or with the complex value is still obscure.

As mentioned, there are two kinds of version for the connection variables: $SL(2,\mathbb{C})$, with a self-duality of Ashtekar formalism, $\gamma = i$, and the connection with a real $SU(2)$ Barbero connection, $\gamma = 1$.

If LQG is compared with the other quantum gravity theory, it has only one free parameter in its formalism. Here, several proposals on the physical significance of the γ are discussed. Some argue in favor of the real valued BI parameter, while the others argue in favor of the complex valued BI parameter. The real valued proposals on BI parameter are more tangible. The value of the BI parameter found from the Black hole entropy calculation in LQG has more consent than other proposals because the same value is applied for all sorts of black holes.

Still, the exact physical significance of the γ is yet to be found. There are several questions regarding the choice of the BI parameter, which are discussed via various proposals. The most important question is regarding the physical significance of the area operator and the volume operator spectrum with the complex valued BI parameter. The complex valued BI parameter is also important because it removes the mathematical complexities from the equations of the constraints, especially from the Hamiltonian constraint. Research on the physical significance of the complex valued BI parameter $\gamma = i$ will open up a new direction in the field of quantum gravity. Time will shed light on these mysteries.

3. Concluding Remarks

- In this paper, initially, a short introduction of the Barbero–Immirzi (BI) parameter, γ, along with the introduction to the Ashtekar formalism, the origin of the BI parameter, the Holst action and a historical timeline of research on the physical significance of the γ in LQG are given.
- The value of the γ and its implication are very important, especially in the area operator spectrum and the black hole entropy calculation in LQG; afterwards, these are elaborated on.
- Thereafter, various proposals on the physical significance of the γ in LQG are given in brief with their pros and cons.
- Most of the proposals advocate the real valued BI parameter γ, since the significance of the complex valued BI parameter γ is not yet clear. However, the complex valued γ is also important, as it removes mathematical complexities from the LQG framework. Research on the complex valued BI parameter will shed light on its physical significance in future.
- Hence, the γ, whether it is complex valued or the real valued, is a crucial free parameter of LQG.

Author Contributions: The authors (R.P.V. and M.J.J.) contributed equally to the work. All authors have read and agreed to the published version of the manuscript.

Funding: This research received no external funding.

Data Availability Statement: Not applicable.

Acknowledgments: The authors are thankful to the Physics Department, Saurashtra University, Rajkot, India.

Conflicts of Interest: The authors declare no conflict of interest.

References

1. Thiemann, T. *Modern Canonical Quantum General Relativity*; Cambridge University Press: Cambridge, UK, 2007. [CrossRef]
2. Rovelli, C. *Quantum Gravity*; Cambridge University Press: Cambridge, UK, 2004. [CrossRef]
3. Ashtekar, A.; Pullin, J. *Loop Quantum Gravity. The First 30 Years*; World Scientific Publishing: Singapore, 2017. [CrossRef]

4. Rovelli, C.; Vidotto, F. *Covariant Loop Quantum Gravity*; Cambridge University Press: Cambridge, Cambridge, UK, 2014. [CrossRef]
5. Rovelli, C. Loop quantum gravity. *Living Rev. Relat.* **2008**, *11*, 5. [CrossRef] [PubMed]
6. Gambini, R.; Pullin, J. *Loops, Knots, Gauge Theories and Quantum Gravity*; Cambridge University Press: Cambridge, UK, 1996. [CrossRef]
7. Ashtekar, A. *Lectures on Non-Perturbative Canonical Gravity*; World Scientific: Singapore, 1991. [CrossRef]
8. Bojowald, M. *Quantum Cosmology*; Springer Science+Business Media, LLC: New York, NY, USA, 2011. [CrossRef]
9. Bojowald, M. *Canonical Gravity and Applications*; Cambridge University Press: New York, NY, USA, 2011. [CrossRef]
10. Gambini, R.; Pullin, J. *A First Course in Loop Quantum Gravity*; Oxford University Press: Oxford, UK, 2011. [CrossRef]
11. Rovelli, C.; Gaul, M. Loop Quantum Gravity and the Meaning of Diffeomorphism Invariance. In *Towards Quantum Gravity: Proceed. XXXV Intern. Winter School on Theoretical Physics, Polanica, Poland, 2–11 February 1999*; Kowalski-Glikman, J., Ed.; Springer: Berlin/Heidelberg, Germany, 2000; pp. 277–324. Available online: https://link.springer.com/chapter/10.1007/3-540-46634-7_11 (accessed on 30 August 2022).
12. Ashtekar, A.; Lewandowski, J. Background independent quantum gravity: A status report. *Class. Quant. Grav.* **2004**, *21*, R53. [CrossRef]
13. Alexandrov, S.; Roche, P. Critical overview of loops and foams. *Phys. Rep.* **2011**, *506*, 41–86. [CrossRef]
14. Mercuri, S. Introduction to loop quantum gravity. *PoS ISFTG* **2009**, *81*, 16. [CrossRef]
15. Doná, P.; Speziale, S. Introductory lectures to loop quantum gravity. *arXiv* **2010**, arXiv:1007.0402. [CrossRef]
16. Esposito, G. An introduction to quantum gravity. *arXiv* **2011**, arXiv:1108.3269. [CrossRef]
17. Rovelli, C. Zakopane Lectures on loop gravity. *arXiv* **2011**, arXiv:1102.3660. [CrossRef]
18. Perez, A. The new spin foam models and quantum gravity. *Papers Phys.* **2012**, *4*, 040004. [CrossRef]
19. Rovelli, C. Notes for a brief history of quantum gravity. *arXiv* **2001**, arXiv:gr-qc/0006061. [CrossRef]
20. Rovelli, C. Loop quantum gravity: The first 25 years. *Class. Quant. Grav.* **2011**, *28*, 153002. [CrossRef]
21. Ashtekar, A.; Bianchi, E. A short review of loop quantum gravity. *Rep. Prog. Phys.* **2021**, *84*, 042001. [CrossRef]
22. Corichi, A.; Hauser, A. Bibliography of publications related to classical self-dual variables and loop quantum gravity. *arXiv* **2005**, arXiv:gr-qc/0509039. [CrossRef]
23. Vyas, R.P.; Joshi, M.J. Loop quantum gravity: A demystified view. *Gravit. Cosmol.* **2022**, *28*, 228–262. [CrossRef]
24. Ashtekar, A. New variables for classical and quantum gravity. *Phys. Rev. Lett.* **1986**, *57*, 2244–2247. [CrossRef] [PubMed]
25. Barbero G., J.F. Real Ashtekar variables for Lorentzian signature space-times. *Phys. Rev. D* **1995**, *51*, 5507–5510. [CrossRef]
26. Barbero G., J.F. From Euclidean to Lorentzian general relativity: The real way. *Phys. Rev. D* **1996**, *54*, 1492–1499. [CrossRef]
27. Immirzi, G. Real and complex connections for canonical gravity. *Class. Quant. Grav.* **1997**, *14*, L177–L181. [CrossRef]
28. Immirzi, G. Quantum gravity and Regge calculus. *Nucl. Phys. B Proc. Suppl.* **1997**, *57*, 65–72. [CrossRef]
29. Holst, S. Barbero's Hamiltonian derived from a generalized Hilbert–Palatini action. *Phys. Rev. D* **1996**, *53*, 5966–5969. [CrossRef]
30. Penrose, R. On the Nature of Quantum Geometry. In *Magic Without Magic*; Klauder, J., Ed.; Freeman: San Francisco, CA, USA, 1972; pp. 333–354. Available online: https://math.ucr.edu/home/baez/penrose/Penrose-OnTheNatureOfQuantumGeometry.pdf (accessed on 30 August 2022).
31. Penrose, R. Angular Momentum: An Approach to Combinatorial Space-Time. In *Quantum Theory and Beyond*; Bastin, T., Ed.; Cambridge University Press: Cambridge, MA, USA, 1971; pp. 151–180. Available online: https://math.ucr.edu/home/baez/penrose/Penrose-AngularMomentum.pdf (accessed on 30 August 2022).
32. Krasnov, K. On the constant that fixes the area spectrum in canonical quantum gravity. *Class. Quant. Grav.* **1998**, *15*, L1–L4. [CrossRef]
33. Rovelli, C. Black hole entropy from loop quantum gravity. *Phys. Rev. Lett.* **1996**, *77*, 3288–3291. [CrossRef] [PubMed]
34. Ashtekar, A.; Baez, J.; Corichi, A.; Krasnov, K. Quantum geometry and black hole entropy. *Phys. Rev. Lett.* **1998**, *80*, 904–907. [CrossRef]
35. Dreyer, O. Quasinormal modes, the area spectrum, and black hole entropy. *Phys. Rev. Lett.* **2003**, *90*, 081301. [CrossRef] [PubMed]
36. Meissner, K. Black-hole entropy in loop quantum gravity. *Class. Quant. Grav.* **2004**, **21**, 5245–5252. [CrossRef]
37. Domagala, M.; Lewandowski, J. Black hole entropy from quantum geometry. *Class. Quant. Grav.* **2004**, *21*, 5233–5244. [CrossRef]
38. Jacobson, T. Renormalization and black hole entropy in loop quantum gravity. *Class. Quant. Grav.* **2007**, *24*, 4875–4879. [CrossRef]
39. Frodden, E.; Geiller, M.; Noui, K.; Perez, A. Black-hole entropy from complex Ashtekar variables. *Europhys. Lett.* **2014**, *107*, 10005. [CrossRef]
40. Krasnov, K.V. Counting surface states in the loop quantum gravity. *Phys. Rev. D* **1997**, *55*, 3505–3513. [CrossRef]
41. Pranzetti, D. Geometric temperature and entropy of quantum isolated horizons. *Phys. Rev. D* **2014**, *89*, 104046. [CrossRef]
42. Majhi, A. The microcanonical entropy of quantum isolated horizon, 'quantum hair' N and the Barbero–Immirzi parameter fixation. *Class. Quant. Grav.* **2014**, *31*, 095002. [CrossRef]
43. Magueijo, J.; Benincasa, D.M.T. Chiral vacuum fluctuations in quantum gravity. *Phys. Rev. Lett.* **2011**, *106*, 121302. [CrossRef] [PubMed]
44. Engle, J.; Noui, K.; Perez, A.; Pranzetti, D. Black hole entropy from an $SU(2)$-invariant formulation of Type I isolated horizons. *Phys. Rev. D* **2010**, *82*, 044050. [CrossRef]
45. Engle, J.; Noui, K.; Perez, A. Black hole entropy and $SU(2)$ Chern–Simons theory. *Phys. Rev. Lett.* **2010**, *105*, 031302. [CrossRef] [PubMed]

46. Majumdar, P. Quantum black hole entropy. *arXiv* **1998**, arXiv:gr-qc/9807045. [PubMed]
47. Kaul, R.K.; Majumdar, P. Logarithmic correction to the Bekenstein-Hawking entropy. *Phys. Lett. B* **1998**, *439*, 267–270. [CrossRef]
48. Pranzetti, D.; Sahlmann, H. Horizon entropy with loop quantum gravity methods. *Phys. Lett. B* **2015**, *746*, 209–216. [CrossRef]
49. Rovelli, C.; Thiemann, T. Immirzi parameter in quantum general relativity. *Phys. Rev. D* **1997**, *57*, 1009–1014. [CrossRef]
50. Samuel, J. Comment on "Immirzi parameter in quantum general relativity". *Phys. Rev. D* **2001**, *64*, 048501. [CrossRef]
51. Veraguth, O.J.; Wang, C.H.-T. Immirzi parameter without Immirzi ambiguity: Conformal loop quantization of scalar-tensor gravity. *Phys. Rev. D* **2017**, *96*, 084011. [CrossRef]
52. Chou, C.-H.; Tung, R.-S.; Yu, H.-L. Origin of the Immirzi parameter. *Phys. Rev. D* **2005**, *72*, 064016. [CrossRef]
53. Fatibene, L.; Francaviglia, M.; Rovelli, C. On a covariant formulation of the Barbero–Immirzi connection. *Class. Quant. Grav.* **2007**, *24*, 3055–3066. [CrossRef]
54. Taveras, V.; Yunes, N. Barbero–Immirzi parameter as a scalar field: *K*-inflation from loop quantum gravity? *Phys. Rev. D* **2008**, *78*, 064070. [CrossRef]
55. Calcagni, G.; Mercuri, S. Barbero–Immirzi parameter field in canonical formalism of pure gravity. *Phys. Rev. D* **2009**, *79*, 084004. [CrossRef]
56. Date, G.; Kaul, R.K.; Sengupta, S. Topological interpretation of Barbero–Immirzi parameter. *Phys. Rev. D* **2008**, *79*, 044008. [CrossRef]
57. Mercuri, S. Peccei-Quinn mechanism in gravity and the nature of the Barbero–Immirzi parameter. *Phys. Rev. Lett.* **2009**, *103*, 081302. [CrossRef] [PubMed]
58. Randono, A. Generalizing the Kodama State. I: Construction. *arXiv* **2006**, arXiv:gr-qc/0611073. [CrossRef]
59. Randono, A. Generalizing the Kodama state. II: Properties and physical interpretation. *arXiv* **2006**, arXiv:gr-qc/0611074. [CrossRef]
60. Wieland, W. Complex Ashtekar variables, the Kodama state and spinfoam gravity. *arXiv* **2011**, arXiv:1105.2330.
61. El Naschie, M.S. The quantum gravity Immirzi parameter—A general physical and topological interpretation. *Gravit. Cosmol.* **2013**, *19*, 151–155. [CrossRef]
62. Sadiq, M. A correction to the Immirizi parameter of $SU(2)$ spin networks. *Phys. Lett. B* **2015**, *741*, 280–283. [CrossRef]
63. Perez, A.; Rovelli, C. Physical effects of the Immirzi parameter in loop quantum gravity. *Phys. Rev. D* **2006**, *73*, 044013. [CrossRef]
64. Broda, B.; Szanecki, M. A relation between the Barbero–Immirzi parameter and the standard model. *Phys. Lett. B* **2010**, *690*, 87–89. [CrossRef]
65. Sadiq, M. The holographic principle and the Immirzi parameter of loop quantum gravity. *arXiv* **2015**, arXiv:1510.04243. [CrossRef]

physics

MDPI

Article

Cosmology of a Polynomial Model for de Sitter Gauge Theory Sourced by a Fluid

Jia-An Lu [1,2]

[1] School of Data and Computer Science, Guangdong Peizheng College, Guangzhou 510830, China; ljagdgz@163.com
[2] School of Physics, Sun Yat-sen University, Guangzhou 510275, China

Abstract: In the de Sitter gauge theory (DGT), the fundamental variables are the de Sitter (dS) connection and the gravitational Higgs/Goldstone field ξ^A, where A is a 5 dimensional index. Previously, a model for DGT was analyzed, which generalizes the MacDowell–Mansouri gravity to have a variable cosmological constant, $\Lambda = 3/l^2$, where l is related to ξ^A by $\xi^A\xi_A = l^2$. It was shown that the model sourced by a perfect fluid does not support a radiation epoch and the accelerated expansion of the parity invariant universe. In this paper, I consider a similar model, namely, the Stelle–West gravity, and couple it to a modified perfect fluid, such that the total Lagrangian 4-form is polynomial in the gravitational variables. The Lagrangian of the modified fluid has a nontrivial variational derivative with respect to l, and as a result, the problems encountered in the previous study no longer appear. Moreover, to explore the elegance of the general theory, as well as to write down the basic framework, I perform the Lagrange–Noether analysis for DGT sourced by a matter field, yielding the field equations and the identities with respect to the symmetries of the system. The resulted formula are dS covariant and do not rely on the existence of the metric field.

Keywords: Stelle–West gravity; gauge theory of gravity; cosmic acceleration

Citation: Lu, J.-A. Cosmology of a Polynomial Model for de Sitter Gauge Theory Sourced by a Fluid. *Physics* **2022**, *4*, 1168–1179. https://doi.org/10.3390/physics4040076

Received: 9 July 2022
Accepted: 15 September 2022
Published: 2 October 2022

1. Introduction

The gauge theories of gravity (GTG) aims at treating gravity as a gauge field, in particular, constructing a Yang–Mills-type Lagrangian, which reduces to general relativity (GR) in some limiting cases, while providing some novel falsifiable predictions. A well-founded subclass of GTG is the Poincaré gauge theory (PGT) [1–5], in which the gravitational field consists of the Lorentz connection and the co-tetrad field. Moreover, the PGT can be reformulated as de Sitter gauge theory (DGT), in which the Lorentz connection and the co-tetrad field are united into a de Sitter (dS) connection [6,7]. In fact, before the idea of DGT is realized, a related Yang–Mills-type Lagrangian for gravity was proposed by MacDowell and Mansouri (MM) [8], and reformulated into a dS-invariant form by West [9], which reads:

$$\begin{aligned}\mathcal{L}^{\rm MM} &= \epsilon_{ABCDE}\,\xi^E \mathcal{F}^{AB}\wedge\mathcal{F}^{CD}\\ &= \epsilon_{\alpha\beta\gamma\delta}\,(lR^{\alpha\beta}\wedge R^{\gamma\delta} - 2l^{-1}R^{\alpha\beta}\wedge e^\gamma\wedge e^\delta + l^{-3}e^\alpha\wedge e^\beta\wedge e^\gamma\wedge e^\delta),\end{aligned} \quad (1)$$

where ϵ_{ABCDE} and $\epsilon_{\alpha\beta\gamma\delta}$ are the 5-dimensional (5d) and 4d Levi–Civita symbols, ξ^A is a dS vector constrained by $\xi^A\xi_A = l^2$, l is a positive constant, $\mathcal{F}^{AB}$ is the dS curvature, $R^{\alpha\beta}$ is the Lorentz curvature, and e^α is the orthonormal co-tetrad field. The 5d indexes are denoted by capital Latin letters and take on the values 0, 1, 2, 3, 4, and the 4d indexes are denoted by Greek letters and take on teh values 0 (time), 1, 2, 3 (space). This theory is equivalent to the Einstein–Cartan (EC) theory with a cosmological constant $\Lambda = 3/l^2$ and a Gauss–Bonnet (GB) topological term, as seen in Equation (1).

Note that some special gauges with the residual Lorentz symmetry can be defined by $\xi^A = \delta^A{}_4 l$, where $\delta^A{}_B$ is the Kronecker delta. Henceforth, ξ^A is akin to an unphysical

Goldstone field. To make ξ^A physical, and become the gravitational Higgs field, one may replace the constant l by a dynamical l, resulting in the Stelle–West (SW) theory [7]. The theory is further explored in Refs. [10,11] (see also the review [12]), in which the constraint $\xi^A\xi_A = l^2$ is completely removed, in other words, $\xi^A\xi_A$ needs not to be positive. Suppose that $\xi^A\xi_A = \sigma l^2$, where $\sigma = \pm 1$. When $l \neq 0$, the metric field can be defined by $g_{\mu\nu} = (\widetilde{D}_\mu \xi^A)(\widetilde{D}_\nu \xi_A)$, where $\widetilde{D}_\mu \xi^A = \widetilde{\delta}^A{}_B D_\mu \xi^B$, $\widetilde{\delta}^A{}_B = \delta^A{}_B - \xi^A\xi_B/\sigma l^2$, $D_\mu \xi^A = d_\mu \xi^A + \Omega^A{}_{B\mu}\xi^B$, and $\Omega^A{}_{B\mu}$ is the dS connection. It was shown [11] that $\sigma = \pm 1$ corresponds to the Lorentz/Euclidean signature of the metric field, and the signature changes when $\xi^A\xi_A$ changes its sign.

On the other hand, it remains to check whether the SW gravity is viable. Although the SW Lagrangian reduces to the MM Lagrangian when l is a constant, the field equations do not. In the SW theory, there is an additional field equation coming from the variation with respect to l, which is nontrivial even when l is a constant. Actually, a recent study [13] presents some negative results for a related model, whose Lagrangian is equal to the SW one times $(-l/2)$. For a homogeneous and isotropic universe with parity-invariant torsion, it is found that l being a constant implies the energy density of the material fluid being a constant, and so l should not be a constant in the general case. Moreover, in the radiation epoch, the l equation forces the energy density to be equal to zero; while in the matter epoch, a dynamical l only works to renormalize the gravitational constant by some constant factor, and hence, the cosmic expansion decelerates as in GR.

In this paper, it is shown that the SW gravity suffers from similar problems encountered in the model considered in Ref. [13]. Furthermore, I solve these problems by using a new fluid with the Lagrangian being a polynomial in the gravitational variables. The merits of a Lagrangian polynomial in some variables are that it is simple and nonsingular with respect to those variables. In Refs. [14,15], the polynomial Lagrangian for gravitation and other fundamental fields were proposed, while in this paper, the polynomial Lagrangian for a perfect fluid is proposed, which reduces to the Lagrangian of a usual perfect fluid when l is a constant. It turns out that, in contrast to the case with an ordinary fluid, the SW gravity coupled with the new fluid supports the radiation epoch and naturally drives the cosmic acceleration. In addition, when writing down the basic framework of DGT, a Lagrangian–Noether analysis is performed, which generalizes the results of Ref. [16] to the cases with arbitrary matter field and arbitrary ξ^A.

The article is organized as follows. In Section 2.1, a Lagrangian–Noether analysis is conducted for the general DGT sourced by a matter field. In Section 2.2, I reduce the analysis of Section 2.1 in the Lorentz gauges, and show how the two Noether identities in PGT can be elegantly unified into one identity in DGT. In Section 3.1, the SW model of DGT is introduced, with the field equations derived both in the general gauge and the Lorentz gauges. Further, the matter source is discussed in Section 3.2, where a modified perfect fluid with the Lagrangian polynomial in the gravitational variables is constructed, and a general class of perfect fluids is defined, which contains both the usual and modified perfect fluids. Then, I couple the SW gravity with the class of fluids and study the coupling system in the homogeneous, isotropic, and parity-invariant universe. The field equations are deduced in Section 4.1, solved in Section 4.2 for the vacuum case, and, in Section 4.3, for the material case. In Section 4.4, the above results are compared with observations, which determines the value of the coupling constant. In the last section, I give some conclusions, and discuss the remaining problems, possible solutions, and extensions.

2. De Sitter Gauge Theory

2.1. Lagrangian–Noether Machinery

The DGT sourced by a matter field is described by the Lagrangian 4-form:

$$\mathcal{L} = \mathcal{L}(\psi, D\psi, \xi^A, D\xi^A, \mathcal{F}^{AB}), \tag{2}$$

where ψ is a p-form valued at some representation space of the dS group $SO(1,4)$, $D\psi = d\psi + \Omega^{AB}T_{AB} \wedge \psi$ is the covariant exterior derivative, $T_A{}^B$ are representations of the dS generators, ξ^A is a dS vector, $D\xi^A = d\xi^A + \Omega^A{}_B\xi^B$, $\Omega^A{}_B$ is the dS connection 1-form, and $\mathcal{F}^A{}_B = d\Omega^A{}_B + \Omega^A{}_C \wedge \Omega^C{}_B$ is the dS curvature 2-form. The variation of $\mathcal{L}$ resulted from the variations of the explicit variables reads:

$$\begin{aligned} \delta\mathcal{L} \;=\;& \delta\psi \wedge \partial\mathcal{L}/\partial\psi + \delta D\psi \wedge \partial\mathcal{L}/\partial D\psi + \delta\xi^A \times \partial\mathcal{L}/\partial\xi^A + \delta D\xi^A \wedge \partial\mathcal{L}/\partial D\xi^A \\ & + \delta\mathcal{F}^{AB} \wedge \partial\mathcal{L}/\partial\mathcal{F}^{AB}, \end{aligned} \tag{3}$$

where $(\partial\mathcal{L}/\partial\psi)_{\mu_{p+1}\cdots\mu_4} \equiv \partial\mathcal{L}_{\mu_1\cdots\mu_p\mu_{p+1}\cdots\mu_4}/\partial\psi_{\mu_1\cdots\mu_p}$, and the other partial derivatives are similarly defined. The variations of $D\psi$, $D\xi^A$, and $\mathcal{F}^{AB}$ can be transformed into the variations of the fundamental variables ψ, ξ^A, and Ω^{AB}, leading to:

$$\begin{aligned} \delta\mathcal{L} \;=\;& \delta\psi \wedge V_\psi + \delta\xi^A \times V_A + \delta\Omega^{AB} \wedge V_{AB} \\ & + d(\delta\psi \wedge \partial\mathcal{L}/\partial D\psi + \delta\xi^A \times \partial\mathcal{L}/\partial D\xi^A + \delta\Omega^{AB} \wedge \partial\mathcal{L}/\partial\mathcal{F}^{AB}), \end{aligned} \tag{4}$$

where,

$$V_\psi \equiv \delta\mathcal{L}/\delta\psi = \partial\mathcal{L}/\partial\psi - (-1)^p D\partial\mathcal{L}/\partial D\psi, \tag{5}$$

$$V_A \equiv \delta\mathcal{L}/\delta\xi^A = \partial\mathcal{L}/\partial\xi^A - D\partial\mathcal{L}/\partial D\xi^A, \tag{6}$$

$$V_{AB} \equiv \delta\mathcal{L}/\delta\Omega^{AB} = T_{AB}\psi \wedge \partial\mathcal{L}/\partial D\psi + \partial\mathcal{L}/\partial D\xi^{[A} \times \xi_{B]} + D\partial\mathcal{L}/\partial\mathcal{F}^{AB}. \tag{7}$$

The symmetry transformations in DGT consist of the diffeomorphism transformations and the dS transformations. For the diffeomorphism transformations, they can be promoted to a gauge-invariant version [16,17], namely, the parallel transports in the fiber bundle with the gauge group as the structure group. The action of an infinitesimal parallel transport on a variable is a gauge-covariant Lie derivative (the gauge-covariant Lie derivative has been used in the metric-affine gauge theory of gravity [18]) $L_v \equiv v\rfloor D + Dv\rfloor$, where v is the vector field, which generates the infinitesimal parallel transport, and $\rfloor$ denotes a contraction, for example, $(v\rfloor\psi)_{\mu_2\cdots\mu_p} = v^{\mu_1}\psi_{\mu_1\mu_2\cdots\mu_p}$. Put $\delta = L_v$ in Equation (3), utilize the arbitrariness of v, then one obtains the chain rule:

$$\begin{aligned} v\rfloor\mathcal{L} \;=\;& (v\rfloor\psi) \wedge \partial\mathcal{L}/\partial\psi + (v\rfloor D\psi) \wedge \partial\mathcal{L}/\partial D\psi + (v\rfloor D\xi^A) \times \partial\mathcal{L}/\partial D\xi^A \\ & + (v\rfloor\mathcal{F}^{AB}) \wedge \partial\mathcal{L}/\partial\mathcal{F}^{AB}, \end{aligned} \tag{8}$$

and the first Noether identity:

$$(v\rfloor D\psi) \wedge V_\psi + (-1)^p(v\rfloor\psi) \wedge DV_\psi + (v\rfloor D\xi^A) \times V_A + (v\rfloor\mathcal{F}^{AB}) \wedge V_{AB} = 0. \tag{9}$$

On the other hand, the dS transformations are defined as vertical isomorphisms on the fiber bundle. The actions of an infinitesimal dS transformation on the fundamental variables are as follows:

$$\delta\psi = B^{AB}T_{AB}\psi, \quad \delta\xi^A = B^{AB}\xi_B, \quad \delta\Omega^{AB} = -DB^{AB}, \tag{10}$$

where $B^A{}_B$ is a dS algebra-valued function, which generates the infinitesimal dS transformation. Substitute Equation (10) and $\delta\mathcal{L} = 0$ into Equation (4), and make use of Equation (7) and the arbitrariness of B^{AB}, one arrives at the second Noether identity:

$$DV_{AB} = -T_{AB}\psi \wedge V_\psi - V_{[A} \times \xi_{B]}. \tag{11}$$

The above analyses are so general that they do not require the existence of a metric field. In the special case with a metric field being defined, $\xi^A\xi_A$ equating to a positive constant, and $p = 0$, the above analyses coincide with those in Ref. [16].

2.2. Reduction in the Lorentz Gauges

Consider the case with $\xi^A\xi_A = l^2$, where l is a positive function. Then, one may define the projector $\tilde{\delta}^A{}_B = \delta^A{}_B - \xi^A\xi_B/l^2$, the generalized tetrad $\tilde{D}\xi^A = \tilde{\delta}^A{}_B D\xi^B$, and a symmetric rank-2 tensor (this formula has been given in Refs. [11,19], and is different from that originally proposed by Stelle and West [7] by a factor $(l_0/l)^2$, where l_0 is the vacuum expectation value of l),

$$g_{\mu\nu} = \eta_{AB}(\widetilde{D}_\mu \xi^A)(\widetilde{D}_\nu \xi^B), \tag{12}$$

which is a localization of the dS metric, $\mathring{g}_{\mu\nu} = \eta_{AB}(d_\mu \mathring{\xi}^A)(d_\nu \mathring{\xi}^B)$, where η_{AB} is the 5d Minkowski metric, and $\mathring{\xi}^A$ are the 5d Minkowski coordinates on the 4d dS space.

Though Equation (12) seems less natural than the choice $g^*_{\mu\nu} = \eta_{AB}(D_\nu \xi^A)(D_\nu \xi^B)$, it coincides with another natural identification (15) (the relation between Equations (12) and (15) is discussed below in this Section). If $g_{\mu\nu}$ is non-degenerate, it is a metric field with Lorentz signature, and one may define $\widetilde{D}^\mu \xi_A \equiv g^{\mu\nu} \widetilde{D}_\nu \xi_A$. Put $v^\mu = \widetilde{D}^\mu \xi_A$ in Equation (9) and utilize $(\widetilde{D}_\mu \xi^A)(\widetilde{D}^\mu \xi_B) = \widetilde{\delta}^A{}_B$, one obtains:

$$\begin{aligned} \widetilde{V}_A \;=\; & -(\widetilde{D}\xi_A \rfloor D\psi) \wedge V_\psi - (-1)^p (\widetilde{D}\xi_A \rfloor \psi) \wedge DV_\psi - (\widetilde{D}\xi_A \rfloor d\ln l) \times V_C \xi^C \\ & -(\widetilde{D}\xi_A \rfloor \mathcal{F}^{CD}) \wedge V_{CD}, \end{aligned} \tag{13}$$

where $\widetilde{V}_A = \widetilde{\delta}^B{}_A V_B$. When l is a constant, Equation (13) implies that the ξ^A equation ($\widetilde{V}_A = 0$ for this case) can be deduced from the other field equations ($V_\psi = 0$ and $V_{CD} = 0$), as pointed out in Ref. [19]. Substitute Equation (13) into Equation (11), and make use of $\widetilde{V}_{[A} \times \xi_{B]} = V_{[A} \times \xi_{B]}$ and $\widetilde{D}\xi_{[A} \times \xi_{B]} = D\xi_{[A} \times \xi_{B]}$, one attains

$$\begin{aligned} DV_{AB} \;=\; & -T_{AB}\psi \wedge V_\psi + (D\xi_{[A} \times \xi_{B]} \rfloor D\psi) \wedge V_\psi + (-1)^p (D\xi_{[A} \times \xi_{B]} \rfloor \psi) \wedge DV_\psi \\ & +(D\xi_{[A} \times \xi_{B]} \rfloor d\ln l) \times V_C \xi^C + (D\xi_{[A} \times \xi_{B]} \rfloor \mathcal{F}^{CD}) \wedge V_{CD}. \end{aligned} \tag{14}$$

When l is a constant, Equation (14) coincides with the corresponding result in Ref. [16]. As shown later in this Section, Equation (14) unifies the two Noether identities in PGT.

To see this, let us define the Lorentz gauges by the condition $\xi^A = \delta^A{}_4 l$ [7]. If $h^A{}_B \in SO(1,4)$ preserves these gauges, then $h^A{}_B = \mathrm{diag}(h^\alpha{}_\beta, 1)$, where $h^\alpha{}_\beta$ belongs to the Lorentz group $SO(1,3)$. In the Lorentz gauges, $\Omega^\alpha{}_\beta$ transforms as a Lorentz connection, and $\Omega^\alpha{}_4$ transforms as a co-tetrad field. Therefore, one may identify $\Omega^\alpha{}_\beta$ as the spacetime connection $\Gamma^\alpha{}_\beta$, and $\Omega^\alpha{}_4$ as the co-tetrad field e^α divided by some quantity with the dimension of length, a natural choice for which is l. Resultantly, Ω^{AB} is identified with a combination of geometric quantities as follows:

$$\Omega^{AB} = \begin{pmatrix} \Gamma^{\alpha\beta} & l^{-1} e^\alpha \\ -l^{-1} e^\beta & 0 \end{pmatrix}. \tag{15}$$

In the case with constant l, this formula is given in Refs. [7,20], and, in the case with varying l, it is given in Refs. [10,19]. In the Lorentz gauges, $\widetilde{D}\xi^4 = 0$, $\widetilde{D}\xi^\alpha = \Omega^\alpha{}_4 l = e^\alpha$ (where Equation (15) is used), and so $g_{\mu\nu}$ defined by Equation (12) satisfies $g_{\mu\nu} = \eta_{\alpha\beta} e^\alpha{}_\mu e^\beta{}_\nu$, implying that Equation (12) coincides with Equation (15). Moreover, according to Equation (15), one finds the expression for $\mathcal{F}^{AB}$ in the Lorentz gauges as follows [19]:

$$\mathcal{F}^{AB} = \begin{pmatrix} R^{\alpha\beta} - l^{-2} e^\alpha \wedge e^\beta & l^{-1}[S^\alpha - d\ln l \wedge e^\alpha] \\ -l^{-1}[S^\beta - d\ln l \wedge e^\beta] & 0 \end{pmatrix}, \tag{16}$$

where $R^\alpha{}_\beta = d\Gamma^\alpha{}_\beta + \Gamma^\alpha{}_\gamma \wedge \Gamma^\gamma{}_\beta$ is the spacetime curvature, and $S^\alpha = de^\alpha + \Gamma^\alpha{}_\beta \wedge e^\beta$ is the spacetime torsion.

Now one can interpret the results in Section 2.1 in the Lorentz gauges. In those gauges, $D\psi = D^\Gamma \psi + 2l^{-1} e^\alpha T_{\alpha 4} \wedge \psi$, $D\xi^\alpha = e^\alpha$, $D\xi^4 = dl$, and so Equation (2) becomes:

$$\mathcal{L} = \mathcal{L}^L(\psi, D^\Gamma \psi, l, dl, e^\alpha, R^{\alpha\beta}, S^\alpha), \tag{17}$$

where $D^\Gamma \psi = d\psi + \Gamma^{\alpha\beta} T_{\alpha\beta} \wedge \psi$. It is the same as a Lagrangian 4-form in PGT [21], with the fundamental variables being ψ, l, $\Gamma^{\alpha\beta}$, and e^α. The relations between the variational derivatives with respect to the PGT variables and those with respect to the DGT variables can be deduced from the following equality:

$$\delta\xi^A \times V_A + 2\delta\Omega^{\alpha 4} \wedge V_{\alpha 4} = \delta l \times \Sigma_l + \delta e^\alpha \wedge \Sigma_\alpha, \tag{18}$$

where $\Sigma_l \equiv \delta\mathcal{L}^L/\delta l$ and $\Sigma_\alpha \equiv \delta\mathcal{L}^L/\delta e^\alpha$. Explicitly, the relations are:

$$\Sigma_\psi \equiv \delta\mathcal{L}^L/\delta\psi = V_\psi, \tag{19}$$

$$\Sigma_l = V_4 - 2l^{-2}e^\alpha \wedge V_{\alpha 4}, \tag{20}$$

$$\Sigma_{\alpha\beta} \equiv \delta\mathcal{L}^L/\delta\Gamma^{\alpha\beta} = V_{\alpha\beta}, \tag{21}$$

$$\Sigma_\alpha = 2l^{-1}V_{\alpha 4}. \tag{22}$$

It is remarkable that the DGT variational derivative V_{AB} unifies the two PGT variational derivatives $\Sigma_{\alpha\beta}$ and Σ_α. With the help of Equations (19)–(22), the $\alpha\beta$ components and $\alpha 4$ components of Equation (14) are found to be:

$$D^\Gamma\Sigma_{\alpha\beta} = -T_{\alpha\beta}\psi \wedge \Sigma_\psi + e_{[\alpha} \wedge \Sigma_{\beta]}, \tag{23}$$

$$\begin{aligned} D^\Gamma\Sigma_\alpha &= D^\Gamma_\alpha\psi \wedge \Sigma_\psi + (-1)^p(e_\alpha\rfloor\psi) \wedge D^\Gamma\Sigma_\psi + \partial_\alpha l \times \Sigma_l \\ &\quad + (e_\alpha\rfloor R^{\beta\gamma}) \wedge \Sigma_{\beta\gamma} + (e_\alpha\rfloor S^\beta) \wedge \Sigma_\beta, \end{aligned} \tag{24}$$

which are just the two Noether identities in PGT [21], with both ψ and l as the matter fields; $\partial_\alpha l = e_\alpha\rfloor dl$ This completes our proof for the earlier statement that the DGT identity (14) unifies the two Noether identities in PGT.

3. Polynomial Models for DGT

3.1. Stelle–West Gravity

It is natural to require that the Lagrangian for DGT is regular with respect to the fundamental variables. The simplest regular Lagrangians are polynomial in the variables, and, in order to recover the EC theory, the polynomial Lagrangian should be at least linear in the gauge curvature. Moreover, to ensure $\mathcal{F}^{AB} = 0$ is naturally a vacuum solution, the polynomial Lagrangian should be at least quadratic in $\mathcal{F}^{AB}$ (when the Lagrangian is linear in $\mathcal{F}^{AB}$, one may add some 'constant term' (independent of $\mathcal{F}^{AB}$) to ensure $\mathcal{F}^{AB} = 0$ is a vacuum solution, but this way is not so natural). The general Lagrangian quadratic in $\mathcal{F}^{AB}$ reads:

$$\begin{aligned} \mathcal{L}^G &= (\kappa_1\,\epsilon_{ABCDE}\,\xi^E + \kappa_2\,\eta_{AC}\xi_B\xi_D + \kappa_3\,\eta_{AC}\eta_{BD})\mathcal{F}^{AB} \wedge \mathcal{F}^{CD} \\ &= \kappa_1\mathcal{L}^{SW} + \kappa_2(S^\alpha \wedge S_\alpha - 2S^\alpha \wedge d\ln l \wedge e_\alpha) \\ &\quad + \kappa_3[R^{\alpha\beta} \wedge R_{\alpha\beta} + d(2l^{-2}S^\alpha \wedge e_\alpha)], \end{aligned} \tag{25}$$

where the κ_1 term is the SW Lagrangian, the κ_2 and κ_3 terms are parity odd, and the κ_3 term is a sum of the Pontryagin and modified Nieh–Yan topological terms. This quadratic Lagrangian is a special case of the, at most, quadratic Lagrangian proposed in Refs. [10,22], and one should note that the quadratic Lagrangian satisfies the requirement mentioned above about the vacuum solution, while the, at most, quadratic Lagrangian does not always satisfy that requirement.

Among the three terms in Equation (25), the SW term is the only one that can be reduced to the EC Lagrangian in the case with positive and constant $\xi^A\xi_A$. Thus, the SW Lagrangian is the simplest choice for the gravitational Lagrangian, which (i) is regular with respect to the fundamental variables; (ii) can be reduced to the EC Lagrangian; (iii) ensures $\mathcal{F}^{AB} = 0$ is naturally a vacuum solution.

For the above reason, the gravitational Lagrangian is taken to be $\mathcal{L}^{\rm SW}$, i.e., put $\kappa_1 = 1$ and $\kappa_2 = \kappa_3 = 0$ in Equation (25). The SW Lagrangian 4-form $\mathcal{L}^{\rm SW}$ takes the same form as $\mathcal{L}^{\rm MM}$ in the first line of Equation (1), while ξ^A is not constrained by any condition. Substitute Equation (1) into Equations (6)–(7), make use of $\partial\mathcal{L}^{\rm SW}/\partial\mathcal{F}^{AB} = \epsilon_{ABCDE}\,\xi^E\mathcal{F}^{CD}$ and the Bianchi identity $D\mathcal{F}^{AB} = 0$, one obtains the gravitational field equations:

$$-\kappa\,\epsilon_{ABCDE}\,\mathcal{F}^{AB}\wedge\mathcal{F}^{CD} = \delta\mathcal{L}^{\rm m}/\delta\xi^E, \tag{26}$$

$$-\kappa\,\epsilon_{ABCDE}\,D\xi^E\wedge\mathcal{F}^{CD} = \delta\mathcal{L}^{\rm m}/\delta\Omega^{AB}, \tag{27}$$

where $\mathcal{L}^{\rm m}$ is the Lagrangian of the matter field coupled to the SW gravity, with κ as the coupling constant. In the vacuum case, Equation (27) is given in Ref. [22] by direct computation, while here, Equation (27) is obtained from the general Equation (7).

In the Lorentz gauges, $\mathcal{L}^{\rm SW}$ takes the same form as $\mathcal{L}^{\rm MM}$ in the second line of Equation (1), while l becomes a dynamical field. The gravitational field equations read:

$$-(\kappa/4)\epsilon_{\alpha\beta\gamma\delta}\,\epsilon^{\mu\nu\sigma\rho}e^{-1}R^{\alpha\beta}{}_{\mu\nu}R^{\gamma\delta}{}_{\sigma\rho} - 4\kappa\,l^{-2}R + 72\kappa\,l^{-4} = \delta S_{\rm m}/\delta l, \tag{28}$$

$$-\kappa\,\epsilon_{\alpha\beta\gamma\delta}\,\epsilon^{\mu\nu\sigma\rho}e^{-1}\partial_\nu l\times R^{\gamma\delta}{}_{\sigma\rho} + 8\kappa\,e_{[\alpha}{}^{\mu}e_{\beta]}{}^{\nu}\partial_\nu l^{-1} + 4\kappa\,l^{-1}T^{\mu}{}_{\alpha\beta} = \delta S_{\rm m}/\delta\Gamma^{\alpha\beta}{}_{\mu}, \tag{29}$$

$$-8\kappa\,l^{-1}(G^{\mu}{}_{\alpha} + \Lambda e_{\alpha}{}^{\mu}) = \delta S_{\rm m}/\delta e^{\alpha}{}_{\mu}, \tag{30}$$

where $e = \det(e^{\alpha}{}_{\mu})$, R is the scalar curvature, $G^{\mu}{}_{\alpha}$ is the Einstein tensor, $T^{\mu}{}_{\alpha\beta} = S^{\mu}{}_{\alpha\beta} + 2e_{[\alpha}{}^{\mu}S^{\nu}{}_{\beta]\nu}$, and $S_{\rm m}$ is the action of the matter field. Although when l is a constant $\mathcal{L}^{\rm SW}$ reduces to the EC Lagrangian with a cosmological constant and a GB topological term, the field equations do not reduce to those of EC with a cosmological constant. The reason lies in the existence of Equation (28), which is nontrivial, even when l is a constant. As a result, the coupling constant κ cannot be fixed by simply comparing Equations (29) and (30) with the EC equations. As shown below in Section 4.4, κ could be determined by a comparison between the theory and cosmological observations.

3.2. Polynomial dS Fluid

For the same reason of choosing a polynomial Lagrangian for DGT, I intend to use those matter sources with polynomial Lagrangian. It has been shown that the Lagrangian of fundamental fields can be reformulated into polynomial forms [14,15]. However, when describing the universe, it is more adequate to use a fluid as the matter source. The Lagrangian of an ordinary perfect fluid (PF) [23] can be written in a Lorentz-invariant form:

$$\mathcal{L}^{\rm PF}_{\mu\nu\rho\sigma} = -\epsilon_{\alpha\beta\gamma\delta}e^{\alpha}{}_{\mu}e^{\beta}{}_{\nu}e^{\gamma}{}_{\rho}e^{\delta}{}_{\sigma}\rho + \epsilon_{\alpha\beta\gamma\delta}J^{\alpha}e^{\beta}{}_{\nu}e^{\gamma}{}_{\rho}e^{\delta}{}_{\sigma}\wedge\partial_\mu\phi, \tag{31}$$

where ϕ is a scalar field, J^α is the particle number current which is Lorentz covariant and satisfies $J^\alpha J_\alpha < 0$, $\rho = \rho(n)$ is the energy density, and $n \equiv \sqrt{-J^\alpha J_\alpha}$ is the particle number density. The Lagrangian (31) is polynomial in the PGT variable $e^{\alpha}{}_{\mu}$, but it is not polynomial in the DGT variables when it is reformulated into a dS-invariant form, in which case the Lagrangian reads:

$$\begin{aligned}\mathcal{L}^{\rm PF}_{\mu\nu\rho\sigma} &= -\epsilon_{ABCDE}(D_\mu\xi^A)(D_\nu\xi^B)(D_\rho\xi^C)(D_\sigma\xi^D)(\xi^E/l)\,\rho \\ &\quad +\epsilon_{ABCDE}J^A(D_\nu\xi^B)(D_\rho\xi^C)(D_\sigma\xi^D)\wedge(\xi^E/l)\,\partial_\mu\phi,\end{aligned} \tag{32}$$

where J^A is a dS-covariant particle number current, which satisfies $J^AJ_A < 0$ and $J^A\xi_A = 0$, $\rho = \rho(n)$ and $n \equiv \sqrt{-J^AJ_A}$. Because l^{-1} appears in Equation (32), the Lagrangian is not polynomial in ξ^A.

A straightforward way to modify Equation (32) into a polynomial Lagrangian is to multiply it by l. In the Lorentz gauges, $J^4 = 0$, and one may define the invariant $J^\mu \equiv J^\alpha e_\alpha{}^\mu$. Then, the modified Lagrangian $\mathcal{L}'^{\rm PF}_{\mu\nu\rho\sigma} = -e\epsilon_{\mu\nu\rho\sigma}\rho l + e\epsilon_{\mu'\nu\rho\sigma}J^{\mu'}\wedge l\times\partial_\mu\phi$. It can be verified that this Lagrangian violates the particle number conservation law

$\nabla_\mu J^\mu = 0$, where ∇_μ is the linearly covariant, metric-compatible and torsion-free derivative. To preserve the particle number conservation, we may replace $l \times \partial_\mu \phi$ by $\partial_\mu(l\phi)$, and the corresponding dS-invariant Lagrangian is:

$$\begin{aligned} \mathcal{L}^{\rm DF}_{\mu\nu\rho\sigma} &= -\epsilon_{ABCDE}(D_\mu\xi^A)(D_\nu\xi^B)(D_\rho\xi^C)(D_\sigma\xi^D)\,\xi^E\rho(n) \\ &\quad +\epsilon_{ABCDE}J^A(D_\nu\xi^B)(D_\rho\xi^C)(D_\sigma\xi^D)\wedge\left(\frac{1}{4}D_\mu\xi^E\times\phi+\xi^E\partial_\mu\phi\right). \end{aligned} \tag{33}$$

The perfect fluid depicted by the above Lagrangian is called the polynomial dS fluid, or dS fluid (DF) for short. In the Lorentz gauges,

$$\begin{aligned} \mathcal{L}^{\rm DF}_{\mu\nu\rho\sigma} &= -e\epsilon_{\mu\nu\rho\sigma}\rho l + \epsilon_{\alpha\beta\gamma\delta}J^\alpha e^\beta{}_\nu e^\gamma{}_\rho e^\delta{}_\sigma \wedge (\partial_\mu l \times \phi + l\,\partial_\mu\phi) \\ &= -e\epsilon_{\mu\nu\rho\sigma}\rho l + e\epsilon_{\mu'\nu\rho\sigma}J^{\mu'}\wedge\partial_\mu(l\phi), \end{aligned} \tag{34}$$

which is equivalent to Equation (31) when l is a constant.

Define the Lagrangian function $\mathscr{L}_{\rm DF}$ by $\mathcal{L}^{\rm DF}_{\mu\nu\rho\sigma} = \mathscr{L}_{\rm DF}\, e\epsilon_{\mu\nu\rho\sigma}$, then $\mathscr{L}_{\rm DF} = -\rho l + J^\mu\partial_\mu(l\phi)$. To compare the polynomial dS fluid with the ordinary perfect fluid, let us consider a general model with the Lagrangian function:

$$\mathscr{L}_{\rm m} = -\rho l^k + J^\mu \partial_\mu(l^k\phi), \tag{35}$$

where $k \in \mathbb{R}$. When $k = 0$, it describes the ordinary perfect fluid; when $k = 1$, it describes the polynomial dS fluid. The variation of $S_{\rm m} \equiv \int dx^4 e\mathscr{L}_{\rm m}$ with respect to ϕ gives the particle number conservation law $\nabla_\mu J^\mu = 0$. The variation with respect to J^α yields $\partial_\mu(l^k\phi) = -\mu U_\mu l^k$, where $\mu \equiv d\rho/dn = (\rho+p)/n$ is the chemical potential, $p = p(n)$ is the pressure, and $U^\mu \equiv J^\mu/n$ is the 4-velocity of the fluid particle. Making use of these results, one may check that the on-shell Lagrangian function is equal to pl^k, and the variational derivatives:

$$\delta S_{\rm m}/\delta l = -k\rho l^{k-1}, \tag{36}$$

$$\delta S_{\rm m}/\delta \Gamma^{\alpha\beta}{}_\mu = 0, \tag{37}$$

$$\delta S_{\rm m}/\delta e^\alpha{}_\mu = (\rho+p)l^k U^\mu U_\alpha + pl^k e_\alpha{}^\mu. \tag{38}$$

One can see that $\delta S_{\rm m}/\delta l = 0$ for the ordinary perfect fluid, while $\delta S_{\rm m}/\delta l = -\rho$ for the polynomial dS fluid.

Finally, it should be noted that the polynomial dS fluid does not support a signature change corresponding to $\xi^A\xi_A$ varying from negative to positive. The reason is that when $\xi^A\xi_A < 0$, there exists no J^A. which satisfies $J^A J_A < 0$ and $J^A\xi_A = 0$.

4. Cosmological Solutions

4.1. Field Equations for the Universe

In this Section, the coupling system of the SW gravity and the fluid model (35) is analyzed in the homogenous, isotropic, parity-invariant and spatially flat universe characterized by the following ansatz [13]:

$$e^0{}_\mu = d_\mu t, \quad e^i{}_\mu = a\, d_\mu x^i, \tag{39}$$

$$S^0{}_{\mu\nu} = 0, \quad S^i{}_{\mu\nu} = b\, e^0{}_\mu \wedge e^i{}_\nu, \tag{40}$$

where a and b are functions of the cosmic time t, and $i = 1,2,3$. On account of Equations (39) and (40), the Lorentz connection $\Gamma^{\alpha\beta}{}_\mu$ and curvature $R^{\alpha\beta}{}_{\mu\nu}$ can be calculated [13]. Further, assume that $U^\mu = e_0{}^\mu$, then $U_\mu = -e^0{}_\mu$, and so $U_\alpha = -\delta^0{}_\alpha$. Now, the reduced form of each term of Equations (28)–(30) can be attained. In particular,

$$\epsilon_{\alpha\beta\gamma\delta}\,\epsilon^{\mu\nu\sigma\rho}e^{-1}R^{\alpha\beta}{}_{\mu\nu}R^{\gamma\delta}{}_{\sigma\rho} = 96(ha)^{\cdot}a^{-1}h^2, \tag{41}$$

$$R = 6[(ha)^{\cdot}a^{-1} + h^2], \tag{42}$$

$$\epsilon_{0i\gamma\delta}\, \epsilon^{\mu\nu\sigma\rho} e^{-1} \partial_\nu l \times R^{\gamma\delta}{}_{\sigma\rho} = -4h^2 \dot{l}\, e_i{}^\mu, \tag{43}$$

$$\epsilon_{ij\gamma\delta}\, \epsilon^{\mu\nu\sigma\rho} e^{-1} \partial_\nu l \times R^{\gamma\delta}{}_{\sigma\rho} = 0, \tag{44}$$

$$T^\mu{}_{0i} = -2b\, e_i{}^\mu, \;\; T^\mu{}_{ij} = 0, \tag{45}$$

$$G^\mu{}_0 = -3h^2 e_0{}^\mu, \tag{46}$$

$$G^\mu{}_i = -[2(ha)^{\cdot}a^{-1} + h^2] e_i{}^\mu, \tag{47}$$

$$\delta S_{\mathrm{m}}/\delta e^0{}_\mu = -\rho l^k e_0{}^\mu, \tag{48}$$

$$\delta S_{\mathrm{m}}/\delta e^i{}_\mu = p l^k e_i{}^\mu, \tag{49}$$

where dot on top of a quantity or being a superscript denotes the differentiation with respect to t, and $h = \dot{a}/a - b$. Substitution of the above equations into Equations (28)–(30) leads to:

$$(ha)^{\cdot}a^{-1}(h^2 + l^{-2}) + l^{-2}(h^2 - \Lambda) = k\rho l^{k-1}/24\kappa, \tag{50}$$

$$(h^2 + l^{-2})\dot{l} - 2b\, l^{-1} = 0, \tag{51}$$

$$8\kappa l^{-1}(-3h^2 + \Lambda) = \rho l^k, \tag{52}$$

$$8\kappa l^{-1}[-2(ha)^{\cdot}a^{-1} - h^2 + \Lambda] = -p l^k, \tag{53}$$

which constitute the field equations for the universe.

Generally, if the requirement of parity invariance is removed, then the ansatz (40) should be replaced by [24]:

$$S^0{}_{\mu\nu} = 0, \; S^i{}_{\mu\nu} = b(t) e^0{}_\mu \wedge e^i{}_\nu + c(t) \epsilon^i{}_{jk} e^j{}_\mu \wedge e^k{}_\nu. \tag{54}$$

Correspondingly, Equations (50)–(53) are generalized to be:

$$(ha)^{\cdot}a^{-1}(h^2 - c^2 + l^{-2}) - 2hc(ca)^{\cdot}a^{-1} + (h^2 - c^2 - 3l^{-2})l^{-2} = \frac{k}{24\kappa}\rho l^{k-1}, \tag{55}$$

$$\left(h^2 - c^2 + l^{-2}\right)\dot{l} - 2bl^{-1} = 0, \tag{56}$$

$$c\left(h\dot{l} + l^{-1}\right) = 0, \tag{57}$$

$$-3(h^2 - c^2) + \Lambda = \frac{l^{k+1}}{8\kappa}\rho, \tag{58}$$

$$-\left[2(ha)^{\cdot}a^{-1} + h^2 - c^2\right] + \Lambda = -\frac{l^{k+1}}{8\kappa}p. \tag{59}$$

When $c = 0$, the above equations reduce to Equations (50)–(53). In virtue of Equation (57), there are two branches of solutions—one is parity even ($c = 0$) and the other is parity odd ($c \neq 0$), which satisfies $h\dot{l} + l^{-1} = 0$. In this paper, only the parity-even case is considered.

4.2. The Vacuum Solution

In the vacuum, $\rho = p = 0$, then Equations (50)–(53) read:

$$(ha)^{\cdot}a^{-1}(h^2 + l^{-2}) + l^{-2}(h^2 - \Lambda) = 0, \tag{60}$$

$$(h^2 + l^{-2})\dot{l} - 2bl^{-1} = 0, \tag{61}$$

$$-3h^2 + \Lambda = 0, \tag{62}$$

$$-2(ha)^{\cdot}a^{-1} - h^2 + \Lambda = 0. \tag{63}$$

It can be shown that Equations (60) and (63) can be deduced from Equations (61) and (62), and the solution for the latter reads:

$$a/a_0 = (l/l_0)e^{\int_{t_0}^{t}(\pm l^{-1})dt}, \tag{64}$$

$$b = l^{-1}\dot{l}, \tag{65}$$

where l is an arbitrary positive function, and a_0 and l_0 are the values of a and l at some moment t_0. In particular, if l is a constant, then:

$$a = a_0 e^{H(t-t_0)},\ b = 0, \tag{66}$$

where $H = \dot{a}/a = \pm l^{-1}$ is a constant. This solution is just the dS space, which describes an inflationary universe.

4.3. The Material Solution

In the general case with matter, let us first derive the continuity equation from the field Equations (50)–(53). Rewrite Equation (52) as:

$$h^2 = l^{-2} - \rho l^{k+1}/24\kappa. \tag{67}$$

Substituting Equation (67) into Equation (53) yields:

$$(ha)^{\cdot}a^{-1} = l^{-2} + (\rho + 3p)l^{k+1}/48\kappa. \tag{68}$$

Multiply Equation (68) by $2h$, making use of Equation (67) and $h = \dot{a}/a - b$, one gets:

$$2h\dot{h} = (\rho + p)l^{k+1}\dot{a}a^{-1}/8\kappa - 2b(ha)^{\cdot}a^{-1}, \tag{69}$$

in which, according to Equations (50), (51), and (67),

$$2b(ha)^{\cdot}a^{-1} = \dot{l}[(k+1)\rho l^{k}/24\kappa + 2l^{-3}]. \tag{70}$$

Differentiate Equation (67) with respect to t, and compare it with Equations (69) and (70), one arrives at the continuity equation:

$$\dot{\rho} + 3(\rho + p)\dot{a}a^{-1} = 0, \tag{71}$$

which is, unexpectedly, the same as the usual one. Suppose that $p = w\rho$, where w is a constant. Then, Equation (71) has the solution:

$$\rho = \rho_0(a/a_0)^{-3(1+w)}, \tag{72}$$

where a_0 and ρ_0 are the values of a and ρ at some moment t_0.

Now, one can solve Equations (50)–(52), while Equation (53) is replaced by Equation (71) with the solution (72). First, substitute Equations (67) and (68) into Equation (50), one finds:

$$\rho l^{k+3} = 48\kappa(3w - k - 1)/(3w + 1). \tag{73}$$

Assume that $\kappa < 0$, then according to the above relation, $\rho l^{k+3} > 0$ implies $(3w - k - 1)/(3w + 1) < 0$. The only concern are the cases with $k = 0, 1$, and assume that $k + 1 > -1$, then $\rho l^{k+3} > 0$ constrains w by:

$$-\frac{1}{3} < w < \frac{k+1}{3}. \tag{74}$$

For the ordinary fluid ($k = 0$), the pure radiation ($w = 1/3$) cannot exist. In fact, on account of Equation (73), $\rho l^3 = 0$ in this case, which is unreasonable. This problem is similar to that appeared in Ref. [13]. On the other hand, for the dS fluid ($k = 1$), Equation (74) becomes $-1/3 < w < 2/3$, which contains both the cases with pure matter ($w = 0$) and pure radiation ($w = 1/3$). Generally, the combination of Equations (72) and (73) yields:

$$l = l_0(a/a_0)^{\frac{3(w+1)}{k+3}}, \tag{75}$$

where l_0 is the value of l when $t = t_0$, and is related to ρ_0 by Equation (73).

Second, substituting Equation (67) into Equation (51), and utilizing Equations (73) and (75), one obtains:

$$b = \frac{3(w+1)(k+2)}{(3w+1)(k+3)}\dot{a}\, a^{-1}, \tag{76}$$

and hence,

$$h = \frac{3w-2k-3}{(3w+1)(k+3)}\dot{a}\, a^{-1}. \tag{77}$$

Third, substitution of Equations (73) and (77) into Equation (52) leads to:

$$\dot{a}a^{-1} = H_0(l_0/l), \tag{78}$$

where $H_0 \equiv (\dot{a}\, a^{-1})_{t_0}$ is the Hubble constant, being related to l_0 by:

$$H_0 = \sqrt{\frac{3w+1}{-3w+2k+3}} \times (k+3)l_0^{-1}. \tag{79}$$

Here, note that Equation (74) implies that $3w+1 > 0$, $-3w+k+1 > 0$, $k+1 > -1$, and so $-3w+2k+3 > 0$. In virtue of Equations (75), (76), and (78), one has:

$$b = b_0(a_0/a)^{\frac{3(w+1)}{k+3}}, \tag{80}$$

where b_0 is related to H_0 by Equation (76). Moreover, substitute Equation (75) into Equation (78) and solve the resulting equation, one attains:

$$(a/a_0)^{\frac{3(w+1)}{k+3}} - 1 = \frac{3(w+1)}{k+3} \times H_0(t-t_0). \tag{81}$$

In conclusion, the solutions for the field Equations (50)–(53) are given by Equations (72), (75), (80), and (81), with the independent constants a_0, H_0, and t_0.

4.4. Comparison with Observations

If k is specified, one can determine the value of the coupling constant κ from the observed values of $H_0 = 67.4\,\text{km} \times \text{s}^{-1} \times \text{Mpc}^{-1}$ and $\Omega_0 \equiv 8\pi\rho_0/3H_0^2 = 0.315$ [25]. For example, put $k = 1$, then according to Equation (79) (with $w = 0$), one has:

$$l_0 = 4/\sqrt{5}H_0 = 8.19 \times 10^{17}\,\text{s}. \tag{82}$$

Substitution of Equation (82) and $\rho_0 = 3H_0^2\Omega_0/8\pi = 1.79 \times 10^{-37}\,\text{s}^{-2}$ into Equation (73) yields:

$$\kappa = -\rho_0 l_0^4/96 = -8.41 \times 10^{32}\,\text{s}^2. \tag{83}$$

This value is an important reference for the future work, which will explore the viability of the model in the solar system scale.

Furthermore, the deceleration parameter $q \equiv -a\ddot{a}/\dot{a}^2$ derived from the above models can be compared with the observed one. With the help of Equations (78) and (75), one finds $\dot{a} \sim a^{(k-3w)/(k+3)}$, then $\ddot{a} = \frac{k-3w}{k+3} \times \dot{a}^2 a^{-1}$, and so:

$$q = \frac{3w-k}{k+3}. \tag{84}$$

Putting $w = 0$, one can find that the universe accelerates ($q < 0$) if $k > 0$, linearly expands ($q = 0$) if $k = 0$, and decelerates ($q > 0$) if $k < 0$. In particular, for the model with an ordinary fluid ($k = 0$), the universe expands linearly (this result is different from that of Ref. [13], where the cosmological solution describes a decelerating universe; as stated before, the gravitational Lagrangian in Ref. [13] is equal to $(-l/2)\mathcal{L}^{\mathrm{SW}}$, which is not equivalent to $\mathcal{L}^{\mathrm{SW}}$); while for the model with a dS fluid ($k = 1$), the universe accelerates with $q = -1/4$, which is consistent with the observational result $-1 \leq q_0 < 0$ [26–28], where q_0 is the present-day value of q. It should be noted that Equation (84) implies that q is a constant when w is a constant, and so the models cannot describe the transition from deceleration to acceleration when w is a constant.

5. Remarks

It is shown that the requirement of regular Lagrangian may be crucial for DGT, as it is shown that the SW gravity coupled with an ordinary perfect fluid (whose Lagrangian is not regular with respect to ξ^A when $\xi^A\xi_A = 0$) does not permit a radiation epoch and the acceleration of the universe, while the SW gravity coupled with a polynomial dS fluid (whose Lagrangian is regular with respect to ξ^A) is out of these problems. Yet, in the latter model, there exists the problem that it cannot describe the transition from deceleration to acceleration in the matter epoch. Actually, only the parity-even branch of the model is analyzed here. One may further analyze the parity-odd branch and check whether the transition problem exists in that case.

Moreover, there are two possible ways to refine the present model. The first is to modify the gravitational part to be the general quadratic model (25), which is a special case of the, at most, quadratic model proposed in Refs. [10,22], but the coupling of which with the polynomial dS fluid is unexplored. It is unknown whether the effect of the κ_2 term could solve the problem encountered in the SW gravity.

The second way is to modify the matter part. Although the Lagrangian of the polynomial dS fluid is regular with respect to ξ^A, it is not regular with respect to J^A when $\xi^A\xi_A = 0$, in which case there should be $J^A J_A \geq 0$, and so the number density $n \equiv \sqrt{-J^A J_A}$ is not regular. One could find a new fluid model whose Lagrangian is regular with respect to all variables, based on the polynomial models for fundamental fields proposed in Refs. [14,15].

Moreover, the present study may be extended to the inflationary epoch. As was shown in Section 4.2, in the vacuum, the theory contains the dS solution, which describes an inflationary universe. Then there should be a transition from the inflationary epoch to the radiation epoch. As usual, this might be achieved by introducing a particle production rate Γ given by some quantum theory. In GR, with the help of energy conservation, the contribution of Γ to the effective pressure p_{eff} can be derived [29,30]. As indicated in Section 4.3, energy conservation in the present theory takes the same form as that in GR, and so it could be believed that the derivation of p_{eff} also applies to the present theory. Replacing p by p_{eff} in the dS fluid, one could further explore the corresponding dynamics.

Funding: This research is supported by the National Natural Science Foundation for Young Scientists of China under Grant No. 12005307.

Data Availability Statement: Not applicable.

Acknowledgments: I would like to thank S.-D. Liang and Z.-B. Li for their abiding help. Furthermore, thanks are due to the anonymous referees who asked enlightening questions. Lastly, I owe my thanks to my parents and wife.

Conflicts of Interest: The author declares no conflict of interest.

References

1. Kibble, T.W.B. Lorentz invariance and the gravitational field. *J. Math. Phys.* **1961**, *2*, 212–221. [CrossRef]
2. Sciama, D.W. On the analogy between charge and spin in general relativity. In *Infeld Festschrift: Recent Developments in General Relativity. A Collection of Papers Dedicated to Leopold Infeld;* Pergamon Press: Oxford, UK, 1962; pp. 415–439

3. Blagojević, M.; Hehl, F.W. (Eds.) *Gauge Theories of Gravitation. A Reader with Commentaries*; Imperial College Press: London, UK, 2013.
4. Ponomarev, V.N.; Barvinsky, A.O.; Obukhov, Y.N. *Gauge Approach and Quantization Methods in Gravity Theory*; Nauka: Moscow, Russia, 2017. Available online: http://en.ibrae.ac.ru/pubtext/259/ (accessed on 10 September 2022).
5. Mielke, E.W. *Geometrodynamics of Gauge Fields*; Springer International Publishing Switzerland: Cham, Switzerland, 2017. [CrossRef]
6. Stelle, K.S.; West, P.C. De Sitter gauge invariance and the geometry of the Einstein–Cartan theory. *J. Phys. A Math. Theor.* **1979**, *12*, L205–L210. [CrossRef]
7. Stelle, K.S.; West, P.C. Spontaneously broken de Sitter symmetry and the gravitational holonomy group. *Phys. Rev. D* **1980**, *21*, 1466–1488. [CrossRef]
8. MacDowell, S.W.; Mansouri, F. Unified geometric theory of gravity and supergravity. *Phys. Rev. Lett.* **1977**, *38*, 739–742. [CrossRef]
9. West, P.C. A geometric gravity Lagrangian. *Phys. Lett. B* **1978**, *76*, 569–570. [CrossRef]
10. Westman, H.; Złośnik, T. Exploring Cartan gravity with dynamical symmetry breaking. *Class. Quant. Grav.* **2014**, *31*, 95004. [CrossRef]
11. Magueijo, J.; Rodríguez-Vázquez, M.; Westman, H.; Złośnik, T. Cosmological signature change in Cartan Gravity with dynamical symmetry breaking. *Phys. Rev. D* **2014**, *89*, 63542. [CrossRef]
12. Westman, H.; Złośnik, T. An introduction to the physics of Cartan gravity. *Ann. Phys.* **2015**, *361*, 330–376. [CrossRef]
13. Alexander, S.; Cortês, M.; Liddle, A.; Magueijo, J.; Sims, R.; Smolin, L. The cosmology of minimal varying Lambda theories. *Phys. Rev. D***2019**, *100*, 83507. [CrossRef]
14. Pagels, H.R. Gravitational gauge fields and the cosmological constant. *Phys. Rev. D* **1984**, *29*, 1690–1698. [CrossRef]
15. Westman, H.; Złośnik, T. Cartan gravity, matter fields, and the gauge principle. *Ann. Phys.* **2013**, *334*, 157–197. [CrossRef]
16. Lu, J.-A. Energy, momentum and angular momentum conservation in de Sitter gravity. *Class. Quantum Grav.* **2016**, *33*, 155009. [CrossRef]
17. Hehl, F.W.; von der Heyde, P.; Kerlick, G.D.; Nester, J.M. General relativity with spin and torsion: Foundations and prospects. *Rev. Mod. Phys.* **1976**, *48*, 393–416. [CrossRef]
18. Hehl, F.W.; McCrea, J.D.; Mielke, E.W.; Ne'eman, Y. Metric-affine gauge theory of gravity: Field equations, Noether identities, world spinors, and breaking of dilation invariance. *Phys. Rep.* **1995**, *258*, 1–171. [CrossRef]
19. Lu, J.-A.; Huang, C.-G. Kaluza–Klein-type models of de Sitter and Poincaré gauge theories of gravity. *Class. Quantum Grav.* **2013**, *30*, 145004. [CrossRef]
20. Guo, H.-Y. The local de Sitter invariance. *Kexue Tongbao (Chin. Sci. Bull.)* **1976**, *21*, 31–34.
21. Obukhov, Y.N. Poincaré gauge gravity: Selected topics. *Int. J. Geom. Meth. Mod. Phys.* **2006**, *3*, 95–138. [CrossRef]
22. Westman, H.; Złośnik, T. Gravity, Cartan geometry, and idealized waywisers. *arXiv* **2012**, arXiv:1203.5709. [CrossRef]
23. Brown, J.D. Action functionals for relativistic perfect fluids. *Class. Quant. Grav.* **1993**, *10*, 1579–1606. [CrossRef]
24. Magueijo, J.; Złośnik, T. Parity violating Friedmann Universes. *Phys. Rev. D.* **2019**, *100*, 84036. [CrossRef]
25. Planck Collaboration. Planck 2018 results. VI. Cosmological parameters. *Astron. Astrophys.* **2020**, *641*, A6. [CrossRef]
26. Riess, A.G.; Filippenko, A.V.; Challis, P.; Clocchiatti, A.; Diercks, A.; Garnavich, P.M.; Gilliard, R.L.; Hogan, C.J.; Jsa, S.; Kirshner, R.P. Observational evidence from supernovae for an accelerating universe and a cosmological constant. *Astron. J.* **1998**, *116*, 1009–1038. [CrossRef]
27. Schmidt, B.; Suntzeff, N.B.; Phillips, M.M.; Schommer, R.A.; Clocchiatti, A.; Kirshner, R.P.; Garnavich, P.; Challis, P.; Leibundgut, B.; Spyromilio, J.; et al. The high-Z supernova search: Measuring cosmic deceleration and global curvature of the universe using type IA supernovae. *Astrophys. J.* **1998**, *507*, 46–63. [CrossRef]
28. Perlmutter, S.; Aldering, G.; Goldhaber, G.; Knop, R.A.; Nugent, P.; Castro, P.G.; Deustua, S.; Fabbro, S.; Goobar, A.; Hook, I.M.; et al. Measurements of Omega and Lambda from 42 high redshift supernovae. *Astrophys. J.* **1999**, *517*, 565–586. [CrossRef]
29. Prigogine, I.; Geheniau, J.; Gunzig, E.; Nardone, P. Thermodynamics and cosmology. *Gen. Relativ. Gravit.* **1989**, *21*, 767–776. [CrossRef]
30. Calvão, M.O.; Lima, J.A.S.; Wagal, I. On the thermodynamics of matter creation in cosmology. *Phys. Lett. A* **1992**, *162*, 223–226. [CrossRef]

physics

MDPI

Article

Generalized Extended Uncertainty Principle Black Holes: Shadow and Lensing in the Macro- and Microscopic Realms

Nikko John Leo S. Lobos [1,2,*] and Reggie C. Pantig [2,3]

1 Mathematics and Physics Department, Technological Institute of the Philippines, 363 Pascual Casal St., Manila 1001, Philippines
2 Physics Department, De La Salle University, 2401 Taft Avenue, Manila 1004, Philippines
3 Physics Department, Mapúa Institute of Technology, 658 Muralla St., Intramuros, Manila 1002, Philippines
* Correspondence: njllobos.mp@tip.edu.ph

Abstract: Motivated by the recent study about the extended uncertainty principle (EUP) black holes, we present in this study its extension called the generalized extended uncertainty principle (GEUP) black holes. In particular, we investigated the GEUP effects on astrophysical and quantum black holes. First, we derive the expression for the shadow radius to investigate its behavior as perceived by a static observer located near and far from the black hole. Constraints to the large fundamental length scale, L_*, up to two standard deviations level were also found using the Event Horizont Telescope (EHT) data: for black hole Sgr. A*, $L_* = 5.716 \times 10^{10}$ m, while for M87* black hole, $L_* = 3.264 \times 10^{13}$ m. Under the GEUP effect, the value of the shadow radius behaves the same way as in the Schwarzschild case due to a static observer, and the effect only emerges if the mass, M, of the black hole is around the order of magnitude of L_* (or the Planck length, $l_{\rm Pl}$). In addition, the GEUP effect increases the shadow radius for astrophysical black holes, but the reverse happens for quantum black holes. We also explored GEUP effects to the weak and strong deflection angles as an alternative analysis. For both realms, a time-like particle gives a higher value for the weak deflection angle. Similar to the shadow, the deviation is seen when the values of L_* and M are close. The strong deflection angle gives more sensitivity to GEUP deviation at smaller masses in the astrophysical scenario. However, the weak deflection angle is a better probe in the micro world.

Keywords: black hole; strong gravitational lensing; weak gravitational lensing; shadow cast; Gauss–Bonnet theorem; generalized extended uncertainty principle

Citation: Lobos, N.J.L.S.; Pantig, R.C. Generalized Extended Uncertainty Principle Black Holes: Shadow and Lensing in the Macro- and Microscopic Realms. *Physics* **2022**, *4*, 1318–1330. https://doi.org/10.3390/physics4040084

Received: 1 August 2022
Accepted: 8 October 2022
Published: 28 October 2022

1. Introduction

Black hole theory has never been more exciting than before when the Event Horizon Telescope (EHT) Collaboration revealed the first image of a black hole in M87 galaxy [1], and more recently, the black hole Sgr. A* in our galaxy [2]. These pictures, with very special algorithms, provided further evidence that black holes exist in nature. Black holes are compact objects with gravity so strong that not even light can escape its gravitational grip.

Black hole solutions are found by solving the Einstein field equation, and the simplest black hole model that is static and spherically symmetric was found by Karl Schwarzschild [3] (see [4] for English translation). Later on, the metric of a spinning black hole, which is static and axisymmetric, was found by Roy Kerr [5]. Conceptually, black holes are massive objects where all the mass is concentrated into a point, thus giving the object an infinite density. In essence, there is no doubt that there must be some interplay between gravity and quantum mechanics in these extreme regions. Indeed, black holes are laboratories where one can probe the quantum nature of gravity [6].

Central to the microscopic realm is the Heisenberg uncertainty principle (HUP), which states that

$$\Delta x \, \Delta p \geq \hbar/2, \tag{1}$$

which is derived from the commutation relation of the position, $\hat{x}$, and momentum, $\hat{p}$, operators, with $\hbar$ the reduced Planck's constant. That is, $[\hat{x}, \hat{p}] = i\hbar$. Equation (1) can provide limitations in testing predictions, but nonetheless a hypothetical energy probe can still detect very short distance scales. The main problem is that, beyond the Planck length, $l_{\rm Pl}$, there is no guarantee that the spacetime observed is still smooth. Such a chaotic spacetime in the microscopic realm is called the quantum foam [7]. It is only then that the HUP must be modified to accommodate the Planck length, and the most accepted modification is called the generalized uncertainty principle (GUP) [8–10], which adds uncertainty quadratic in momentum:

$$\Delta x\,\Delta p \geq 1 + \beta\, l_{\rm Pl}^2\, \Delta p^2, \tag{2}$$

where β is a dimensionless quantity usually taken as unity and can be either positive or negative [11].

As nature is fond of symmetry and duality, similar to the yin-yang symbol, it is only natural to suspect that if there is a minimum fundamental length, there must be a large fundamental length scale in our Universe. Hence, the GUP is naturally extended [12], to include the large fundamental length, L_*, through a quadratic correction in the position uncertainty. That is,

$$\Delta x\,\Delta p \geq 1 + \alpha\, \Delta x^2/L_*^2, \tag{3}$$

which is commonly called the extended uncertainty principle (EUP), with α being another dimensionless constant. Equation (3) was also derived from first principles in Ref. [13]. While GUP is commonly analyzed in the literature due to its vast application in the microscopic world [14], the application of EUP seems to be dearth in the literature. For instance, the analysis of EUP effects on the thermodynamics of Friedmann–Robertson–Walker (FRW) Universe [15] was analyzed long ago and a year later applied to the geometry of de Sitter (dS) and anti-de Sitter (AdS) spacetime [16]. The effects of the EUP correction has also been studied in Rindler and cosmological horizons [17], relativistic Coulomb potential [18], bound-state solutions of the two-dimensional Dirac equation with Aharonov–Bohm–Coulomb interaction [19], Jüttner gas [20]. With the help of the GUP and EUP parameters, bounds for the Hubble parameter's value were also studied to resolve the Hubble tension [21]. It is only recently that EUP correction has been applied in the context of black holes [22], with $r_h \sim \Delta x$ given the gravitons are considered the quantum particles inside such confinement. Since then, various studies have explored the black hole with EUP correction; see Refs. [23–31].

We are motivated to continue the analysis of Ref. [22] and further investigate the most general form of the uncertainty principle [32],

$$\Delta x\,\Delta p \geq 1 + \beta\, l_{\rm Pl}^2\, \Delta p^2 + \alpha\, \Delta x^2/L_*^2, \tag{4}$$

as been applied to the shadow cast and gravitational lensing of astrophysical black holes and quantum black holes [33,34]. To this end. the black hole metric that contains the GEUP correction must be expressed as (in time and cylindrical space coordiantes) [22]

$$ds^2 = -A(r)dt^2 + B(r)dr^2 + C(r)d\theta^2 + D(r)d\phi^2, \tag{5}$$

where

$$\begin{aligned} A(r) &= 1 - \frac{2\mathcal{M}}{r}, \quad B(r) = A(r)^{-1}, \\ C(r) &= r^2, \quad D(r) = r^2 \sin^2\theta. \end{aligned} \tag{6}$$

With the GEUP correction in Equation (4), the mass, M, of the black hole corrected to [22]:

$$\mathcal{M} = M\left(1 + \frac{4\alpha M^2}{L_*^2} + \frac{\beta\hbar}{2M^2}\right). \tag{7}$$

Here, we first show $\hbar$ to emphasize the quantum correction for quantum particles. Note that, since M is geometrized, one can relate $\hbar$ to the Planck length representing the known minimal length $l_{\rm Pl} = 1.616 \times 10^{-35}$ m. Furthermore, $\alpha = \beta = 1$, and L_*'s value is estimated based on the observational constraints from the EHT in Section 2. First, we explore the behavior of the shadow radius of the object being considered (i.e., supermassive black hole (SMBH) for macroscopic and some elementary particles for the microscopic realm). Shadows are important since they can reveal imprints that allow one to test gravity theories in the strong field regime; shadows were first studied in Ref. [35]. In 1979, Luminet gave the formula for the angular radius of the shadow [36]. Then several studies have explored the shadows of quantum black holes [37–44]. In this paper, we are also interested in probing the GEUP effects using the strong and weak deflection angles. Gravitational lensing is one of the most successful tools as it verified Einstein's general theory of relativity in 1919 [45] through the Sun's solar eclipse. Since then, it has been crucial in probing various tests of gravitation theories. Several tools have been developed [46–48], and in 2008, the Gauss–Bonnet theorem on the optical geometries in asymptotically flat spacetimes was developed [49]. It was extended by Werner [50] to include stationary spacetimes in the Finsler–Randers type optical geometry on Nazim's osculating Riemannian manifolds. Ishihara and others then found a way to extend the Gauss-Bonnet theorem (GBT) to incorporate finite distance effects [51,52], which also applies to non-asymptotic spacetimes. Finally, instead of using points at infinity as integration domain for the GBT, the study in [53] used the photonsphere to naturally find an alternative to the Ishihara method, which also accommodates the deflection angle of massive particles. For recent works about quantum black holes' deflection angles, see Refs. [24,29,54–58].

The paper is organized as follows. Section 2 is devoted to exploring the shadow behavior of the GEUP black hole and microscopic entities been viewed as quantum black holes. In Section 3, the Gauss–Bonnet theorem is used to study the weak deflection angle of the mentioned objects. Section 4 considers the strong deflection angle as a generalization of the weak deflection angle studied in Section 3. Then, in Section 5, we formulate the conclusion based on the results of the prior Sections. In this paper, geometrized units are used wherein $G = c = 1$, with G being the gravitaion constant and c the speed of light, and the metric signature $(-,+,+,+)$; hence, $\hbar$ in Equation (7) can be replaced by the Planck length.

2. Shadow and Constraints to the Large Fundamental Length Scale

In this Section, we study the shadow of the GEUP black hole. Thanks to r and t independence of the metric, such symmetry allows us to analyze light-like geodesics along the equatorial plane ($\theta = \pi/2$) without compromising generality. Thus, $D(r) = C(r)$ in the metric (5). These geodesics can be derived through the Lagrangian,

$$\mathcal{L} = \frac{1}{2}[-A(r)\dot{t} + B(r)\dot{r} + C(r)\dot{\phi}]. \tag{8}$$

Here on, the dot denotes the time derivation.

Through the variational principle, the Euler–Lagrange equation gives two constants of motion

$$E = A(r)\frac{dt}{d\lambda}, \quad L = C(r)\frac{d\phi}{d\lambda}, \tag{9}$$

from where one can define the impact parameter as

$$b \equiv \frac{L}{E} = \frac{C(r)}{A(r)}\frac{d\phi}{dt}. \tag{10}$$

Here, λ denotes the affine parameter defined by $\tau = \mu\lambda$, where τ is the proper time and μ is the particle's rest mass.

For light-like geodesics, the metric can be set as $ds^2 = 0$, and using Equation (9), one obtains the orbit equation:

$$\left(\frac{dr}{d\phi}\right)^2 = \frac{C(r)}{B(r)}\left(\frac{h(r)^2}{b^2} - 1\right), \tag{11}$$

where by definition [59],

$$h(r)^2 = \frac{C(r)}{A(r)}. \tag{12}$$

Through the above equation, we can obtain the location of the photonsphere by taking $h'(r) = 0$, where the prime denotes r-derivation. To this end, since the mass M is just imbued with quantum correction, the location of the photonsphere is

$$r_{\mathrm{ph}} = 3\mathcal{M}. \tag{13}$$

Our concern in this Section is how the observer will perceive the GEUP black hole at near and far away locations. Let the observer be at the coordinates $(t_{\mathrm{obs}}, r_{\mathrm{obs}}, \theta_{\mathrm{obs}} = \pi/2, \phi_{\mathrm{obs}} = 0)$. Then, the observer can construct [60] the relation,

$$\tan(\alpha_{\mathrm{sh}}) = \lim_{\Delta x \to 0} \frac{\Delta y}{\Delta x} = \left(\frac{C(r)}{B(r)}\right)^{1/2} \frac{d\phi}{dr}\bigg|_{r=r_{\mathrm{obs}}}, \tag{14}$$

which can be rewritten as

$$\sin^2(\alpha_{\mathrm{sh}}) = \frac{b_{\mathrm{crit}}^2}{h(r_{\mathrm{obs}})^2}, \tag{15}$$

where b_{crit} is a function of the photonsphere given in Equation (13). A spacetime may have a different expression for $h(r)$, thus for a general spacetime, the critical impact parameter reads [61]:

$$b_{\mathrm{crit}}^2 = \frac{h(r_{\mathrm{ph}})}{\left[B'(r_{\mathrm{ph}})C(r_{\mathrm{ph}}) - B(r_{\mathrm{ph}})C'(r_{\mathrm{ph}})\right]}\Big[h(r_{\mathrm{ph}})B'(r_{\mathrm{ph}})C(r_{\mathrm{ph}}) - h(r_{\mathrm{ph}})B(r_{\mathrm{ph}})C'(r_{\mathrm{ph}}) - 2h'(r_{\mathrm{ph}})B(r_{\mathrm{ph}})C(r_{\mathrm{ph}})\Big], \tag{16}$$

and, for the GEUP black hole, one finds:

$$b_{\mathrm{crit}}^2 = 27\mathcal{M}^2. \tag{17}$$

Finally, one obtains the behavior of the shadow radius, applicable for both macroscopic and quantum black holes:

$$R_{\mathrm{sh}} = 3\mathcal{M}\sqrt{3\left(1 - \frac{2\mathcal{M}}{r_{\mathrm{obs}}}\right)}. \tag{18}$$

Note that this expression is valid even when the static observer is near the black hole. In addition, if $r_{\mathrm{obs}} \to \infty$, Equation (18) can be approximated with $R_{\mathrm{sh}} = 3\sqrt{3}\mathcal{M}$.

Let us start first with astrophysical black holes, such as Sgr. A* and M87*, and discuss some observational constraints. According to Refs. [1,2], the mass ($M_\odot$ denotes the mass of the Sun), distance from Earth, and angular shadow diameter of M87* are [62]: $M_{\mathrm{M87^*}} = 6.5 \pm 0.90 \times 10^9\ M_\odot$, $D = 16.8$ Mpc, and $\alpha_{\mathrm{M87^*}} = 42 \pm 3$ μas, respectively. For Sgr. A*, these values are: $M_{\mathrm{Sgr.\ A^*}} = 4.3 \pm 0.013 \times 10^6\ M_\odot$ (Very Large Telescope Interferometer, VLTI), $D = 8277 \pm 33$ pc, and $\alpha_{\mathrm{Sgr.\ A^*}} = 48.7 \pm 7$ μas (EHT), respectively [1,2]. The diameter of the shadow size using these empirical data and in units of the black hole mass can be calculated using

$$d_{\mathrm{sh}} = D\theta/M. \tag{19}$$

Then, the diameter of the shadow image of M87* and Sgr. A* are $d_{\mathrm{sh}}^{\mathrm{M87^*}} = (11 \pm 1.5)M$, and $d_{\mathrm{sh}}^{\mathrm{Sgr.\ A^*}} = (9.5 \pm 1.4)M$, respectively. Meanwhile, the theoretical shadow diameter can be obtained as $d_{\rm sh}^{\rm theo} = 2R_{\rm sh}$. The observational constraints' results are plotted in Figure 1.

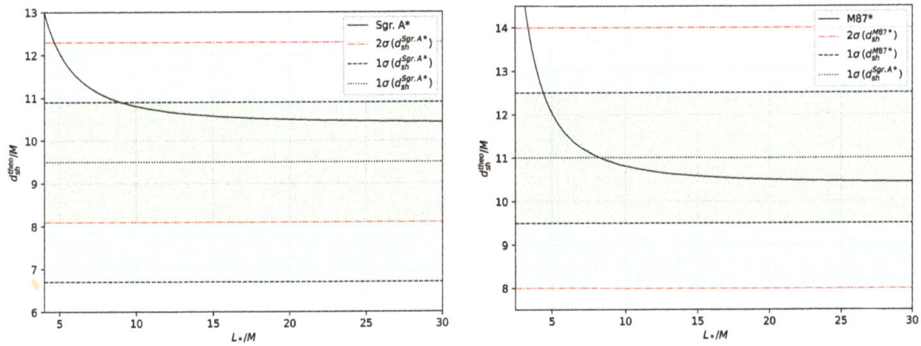

Figure 1. Observational contraints for the M-normalized theoretical shadow diameter for various M-normalized fundamental length scales, L_*, for black holes Sgr. A* and M87*, where M is the black hole mass. (**Left**): one standard deviation (1σ) for $L_* \sim 5.716 \times 10^{10}$ m, 2σ for $L_* \sim 2.985 \times 10^{10}$ m. (**Right**): 1σ for $L_* \sim 4.224 \times 10^{13}$ m, 2σ for $L_* \sim 3.264 \times 10^{13}$ m. At the mean, $L_* \sim 7.950 \times 10^{13}$ m.

Theoretically, let us now consider how the static observer perceives the shadow radius at different locations in the radial coordinate for different values of L_*. In the literature, only the case of $r_{\rm obs} \to \infty$ were considered [22,29].

In Figure 2, left plot, the dashed line is the Schwarzschild case for both SMBHs, which overlaps the shadow radius coming from empirical data [1,2] shown for comparison. Note that the GEUP effect merely increases the shadow radius while the trend of the behaviour of the curve is the same as in the Schwarzschild case. In the right plot, one can see how the shadow radius behaves due to the GEUP effect. For instance, deviations begin to manifest if the value of L_* is close to the mass of the black hole, which is also visible from the green line as soon as L_* comparable to the Hubble length is used. In this scenario, the effect of the parameters in the microscopic realm does not even manifest.

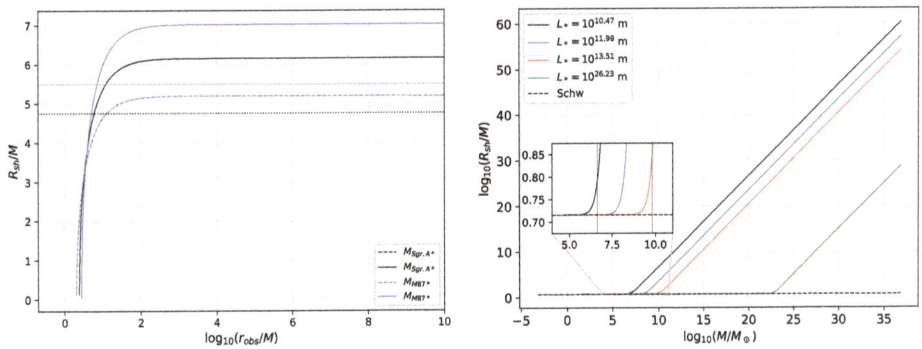

Figure 2. (**Left**): the shadow radius of Sgr. A* and M87* with observer location dependency. The dashed line represents the Schwarzschild case and the solid line for the general extended uncertainty principle (GEUP) case. The horizontal black and blue dotted lines represent the shadow radius of Sgr. A* and M87* based on the Event Horizon Telescope (EHT) data [1,2]. (**Right**): the shadow radius as a function of the black hole mass. The black and blue vertical lines in the inset plot represent the mass of the Sgr. A* and M87*, respectively. "Schw" denotes the Schwarzild case and $M_\odot$ denotes the mass of the Sun.

Equation (18) also admits analysis for quantum black holes. The results are plotted in Figure 3.

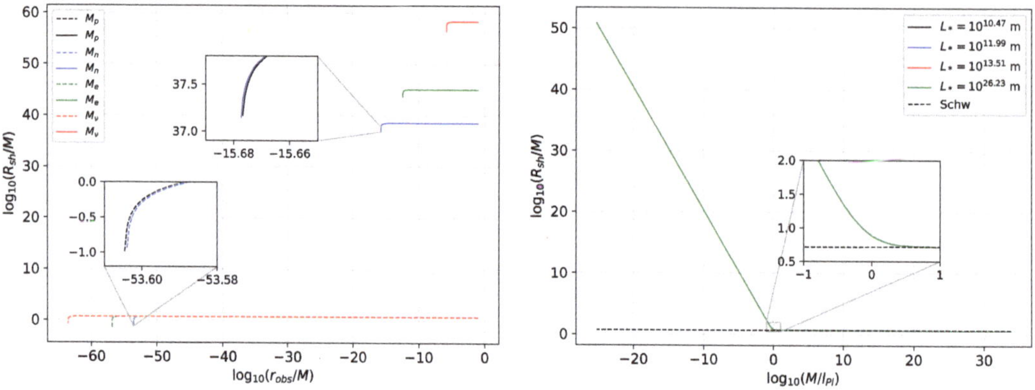

Figure 3. (**Left**): Observer-dependent shadow radius of some elementary particles: proton (p), neutron (n), electron (e), and neutrino (ν). (**Right**): The shadow radius plotted under the assumption that the observer/detector is at $r_{\text{obs}} \gg M$, where M is the quantum black hole's mass, for different values of L_*. The overlapping of these lines means that L_* has no effect in the microscopic realm. l_{Pl} denotes the Planck mass.

The left plot shows the case where the static observer may be represented by a detector that can probe masses as small as the proton, neutron, electron, and neutrino [63], where their geometrized masses are used. The dashed and solid lines represent the Schwarzschild and GEUP cases, respectively. The right plot reveals that L_* is indeed irrelevant in the microscopic realm. Nonetheless, with the GUP correction, the plot reveals the detector's position where the shadow of the particle manifests. Take, for example, the neutrino. Without GUP correction, the shadow radius is around 10^{-63} order of magnitude for a wide range of detector locations. The GUP correction lessens this range and makes the shadow radius larger. For instance, if the detector is at $r = 1.59 \times 10^{-67}$ m, then the shadow radius is around $R_{\text{sh}} = 5.03 \times 10^{-6}$ m. Note how the shadow radius of these particles levels at greater distances. Finally, one observes that, without GUP correction, the shadow radii are nearly identical to each other. With the GUP correction, we have seen that, as the mass of the particle decreases, the shadow radius tends to increase while the range where a detector can observe it decreases.

3. Weak Deflection Angle

In this Section, we explore a different phenomenon and examine the effect of the GEUP correction on the weak deflection angle by black holes in the macroscopic and microscopic realms. To do so, we use the GBT. Consider the domain $(D_a, \bar{g})$ (where $a = 1, 2, ..., N$ and $\bar{g}$ is the optical geometry metric) that is connected over an osculating Riemannian manifold $(\mathcal{M}, \bar{g})$ along some boundaries, and let κ_g be the geodesic curvature of the boundary ∂D_a. Then the GBT states that [49,64]

$$\iint_{D_a} K dS + \sum_{a=1}^{N} \int_{\partial D_a} \kappa_g d\ell + \sum_{a=1}^{N} \theta_a = 2\pi\chi(D_a), \tag{20}$$

where $\chi(D_a)$ is the Euler characteristic, K is the Gaussian optical curvature, $dS = \sqrt{\bar{g}} dr d\phi$, ℓ denotes the line element, and θ_a is the exterior angle at the Nth vertex.

Although the spacetime herein is asymptotically flat under the GEUP correction, we used the generalized GBT that considers non-asymptotically flat spacetime and massive particle deflection. In Ref. [53], the photonsphere radius $r_{\rm ph}$ is the one considered as part of the quadrilateral for integration domain. It is shown that the weak deflection angle,

$$\hat{\alpha} = \iint_{{}^{R}_{r_{\rm ph}}\square^{S}_{r_{\rm ph}}} K dS + \phi_{\rm RS}, \tag{21}$$

where integral is taken over through $r_{\rm ph} \to {\rm S} \to {\rm R} \to r_{\rm ph}$ Here, S and R are the radial positions of the source and receiver, respectively, and $\phi_{\rm RS}$ is the coordinate position angle between the source and the receiver defined as $\phi_{\rm RS} = \phi_{\rm R} - \phi_{\rm S}$. g is the determinant of the Jacobi metric in static and spherically symmetric spacetime:

$$dl^2 = g_{ij}dx^i dx^j = (E^2 - \mu^2 A(r))\left(\frac{B(r)}{A(r)}dr^2 + \frac{C(r)}{A(r)}d\Omega^2\right). \tag{22}$$

Here, E is the energy of the massive particle defined by

$$E = \mu/\sqrt{1 - v^2}, \tag{23}$$

where v is the particle's velocity. As only the equatorial plane is considered here due to spherical symmetry, the determinant of the Jacobi metric reads:

$$g = \frac{B(r)C(r)}{A(r)^2}(E^2 - \mu^2 A(r))^2. \tag{24}$$

Following Ref. [53], one obtains the final expression for the weak deflection angle:

$$\hat{\alpha} \sim \frac{\mathcal{M}(v^2+1)}{bv^2}\left(\sqrt{1 - b^2 u_{\rm R}^2} + \sqrt{1 - b^2 u_{\rm S}^2}\right) \tag{25}$$

which also involves the finite distance $u_{\rm S}$ and $u_{\rm R}$. The obtained expression for $\hat{\alpha}$ can still be further approximated as soon as $b^2u^2 \sim 0$:

$$\hat{\alpha} \sim \frac{2\mathcal{M}(v^2+1)}{bv^2} \tag{26}$$

For the case of photons, when $v = 1$, one finds:

$$\hat{\alpha} \sim 4\mathcal{M}/b. \tag{27}$$

The weak deflection angle result is usually applied to SMBH. As soon as $\hat{\alpha}$ is usually plotted against the impact parameter b/M, in Figure 4 we are interested how $\hat{\alpha}$ changes as the black hole mass under the effect of GEUP varies. Without the GEUP correction, the plot would only represent straight lines. From Figure 4, one observes that similar to the shadow radius, the deviation occurs when M is close to the value of L_*. The time-like deflection also produces a higher value of $\hat{\alpha}$, and the lower the impact parameter, the greater the deflection. Note that in this plot, $b = 10M$ is still in the regime for weak deflection angle since this is higher than the critical impact parameter, $b_{\rm crit} = 3\sqrt{3}\mathcal{M}$. We use this information for the weak deflection for quantum black holes and strong deflection angle.

As a final remark to the plots is that showing how $\hat{\alpha}$ changes as the mass M varies has its shortcomings since b/M is constant. For instance, if one considers the mass of the Earth, $b = 1000M$ equals 4.4 m, which is too small compared to the radius of the Earth (6371 km). Thus, the line plot in Figure 4 may have its range of validity relative to the chosen value of the impact parameter. Such a result has a critical implication as far as the GEUP model in this study is concerned. One can verify that if the dimensional reduction is used in the metric in Equation (7) to calculate $\hat{\alpha}$, that is when $l_{\rm Pl} = 1$ [22], one can observe a very high value for $\hat{\alpha}$ for low mass compact objects (such as Earth, for example).

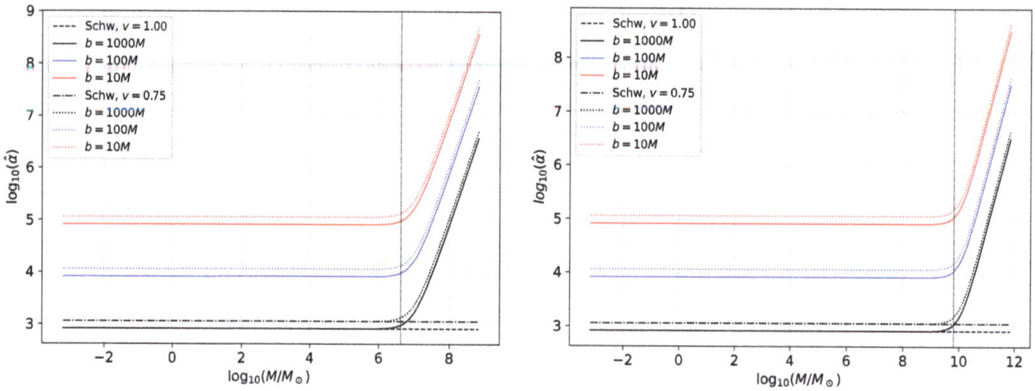

Figure 4. Weak deflection angle by Sgr. A* (**Left**), and M87* (**Right**) for different values of impact parameter. The vertical dotted line is the mass of the black hole considered. See text for details.

We also apply the weak deflection angle for quantum black holes [33,34]. We do this by plotting $\hat{\alpha}$ versus $\log_{10}(M/l_{\rm Pl})$ in Figure 5. Let us note that when one geometrizes the Planck mass, the Planck length is obtained, so, for simplicity, $\hat{\alpha}$ is plotted in terms of $M/l_{\rm Pl}$ in Figure 5. Qualitatively, from Figure 5, one observes the same features as those are known for the weak deflection for astrophysical black holes. Here, one can see that the deviation begins to manifest when the $\log_{10}(M/l_{\rm Pl}) \sim 0$, and these are the masses that are comparable with $l_{\rm Pl}$ ($\sim$2.176 $\times$ 10^{-8} kg in metric units). In this case, $\hat{\alpha} \sim$ 114,815 µas and can be detectable if one directs a photon at an impact parameter of $b \sim 1.62 \times 10^{-32}$ m. Such particle is still massive, and its physical dimension may cause a collision instead of a deflection. Weak deflection may occur unless the particle is compressed to allow such a small value for the impact parameter. In the plot shown, the vertical dotted line represents the neutrino's mass. One can see that $\hat{\alpha} \sim 3.89 \times 10^{60}$µas for $b = 1000M$. Such a large weak deflection angle can be made smaller by increasing b. However, the main obstacle in this case is that one cannot observe neutrinos at rest.

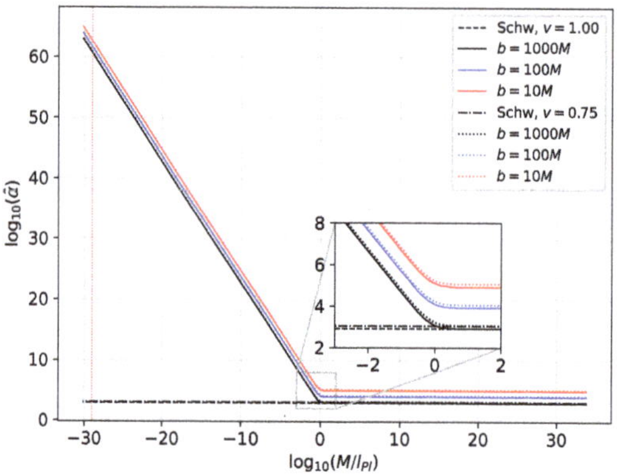

Figure 5. Weak deflection angle by quantum black hole. The red vertical dotted line corresponds to the mass of neutrino.

4. Strong Deflection Angle

Near the black hole region, specifically in the critical impact parameter, the deflection angle is described by the strong deflection expression as shown in Refs. [48,65,66]. The photonsphere region is crucial in strong deflection calculation; hence, we use Equation (13). Following Refs. [48,65,66], one obtains the strong deflection angle to read:

$$\hat{\alpha}_{\rm str} = -\bar{a}\ln(b_0/b_{\rm crit} - 1) + \bar{b} + \mathcal{O}(b - b_{\rm crit}), \tag{28}$$

where $\bar{a}$ and $\bar{b}$ are the coefficients of strong deflection and b_0 and $b_{\rm crit}$ correspond to the impact parameters evaluated at the closest approach and critical impact parameter, respectively. The coefficients of the strong deflection are calculated based on Ref. [65], namely:

$$\bar{a} = \sqrt{\frac{2B(r_{\rm ph})C(r_{\rm ph})}{C''(r_{\rm ph})A(r_{\rm ph}) - A''(r_{\rm ph})C(r_{\rm ph})}}, \tag{29}$$

and

$$\bar{b} = \bar{a}\ln\left[r_{\rm ph}\left(\frac{C''(r_{\rm ph})}{C(r_{\rm ph})} - \frac{A''(r_{\rm ph})}{A(r_{\rm ph})}\right)\right] + I_{\rm R}(r_{\rm ph}) - \pi, \tag{30}$$

where $A(r_{\rm ph})$, $B(r_{\rm ph})$, and $C(r_{\rm ph})$ are metric functions evaluated at the photon sphere region, and $I_{\rm R}$ denotes the regular integral evaluated from 0 to 1. The double prime signifies second derivative with respect to r evaluated at the photonsphere, $r \to r_{\rm ph}$.

The second term in Equation (30) can be calculated using the procedure illustrated in [65,66], where

$$I_{\rm R}(r_{\rm ph}) = \int_0^1 \left[\frac{2(1 - A_{\rm ph})\sqrt{A(z, r_{\rm ph})B(z, r_{\rm ph})}}{A'(z, r_{\rm ph})C(z, r_{\rm ph})\sqrt{A_{\rm ph}/C_{\rm ph} - A(z, r_{\rm ph})/C(z, r_{\rm ph})}}\right] dz, \tag{31}$$

and $A(z, r_{\rm ph})$, $B(z, r_{\rm ph})$, and $C(z, r_{\rm ph})$ are metric functions $A(r)$, $B(r)$, and $C(r)$ evaluated using the new variable [65],

$$z \equiv 1 - r_{\rm ph}/r. \tag{32}$$

Let us express Equation (32) in terms of r and substitute it to the metric functions. Applying the expression in Equations (29)–(31) to the black hole metric (5), one finds:

$$\bar{a} = 1, \tag{33}$$

and

$$\bar{b} = \ln\left[216(7 - 4\sqrt{3})\right] - \pi. \tag{34}$$

When α and β are set to zero, the Schwarzschild expression is retrieved for strong deflection [67]:

$$\hat{\alpha}_{\rm str} = -\ln[b/b_{\rm crit} - 1] - 0.40023, \tag{35}$$

with the critical impact parameter from Equation (16) [61], resulting to Equation (17). In choosing the value of b it is essential to note that the ratio, $b/b_{\rm crit}$, must not be significantly far from 1. Equation (35) diverges for $b_{\rm crit} = b$. This shows that the photonsphere captures particles in this region. In the plots shown below in Figures 6 and 7, b (in units of M) is chosen to be slightly larger than $b_{\rm crit} = 3\sqrt{3}\mathcal{M}$. We plot the strong deflection angle demonstarting how GEUP affects astrophysical and quantum black holes.

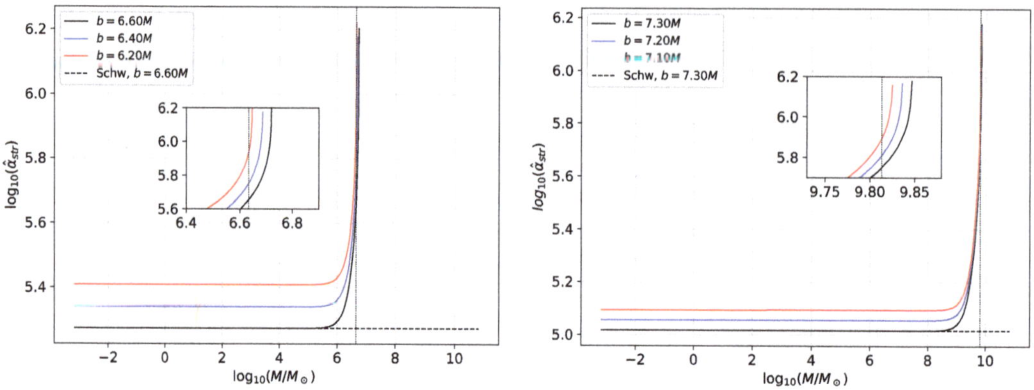

Figure 6. Behavior of strong deflection angle by Sgr. A* (**Left**) and M87* (**Right**). The black vertical dotted line is the corresponding mass of the supermassive black hole (SMBH). See text for details.

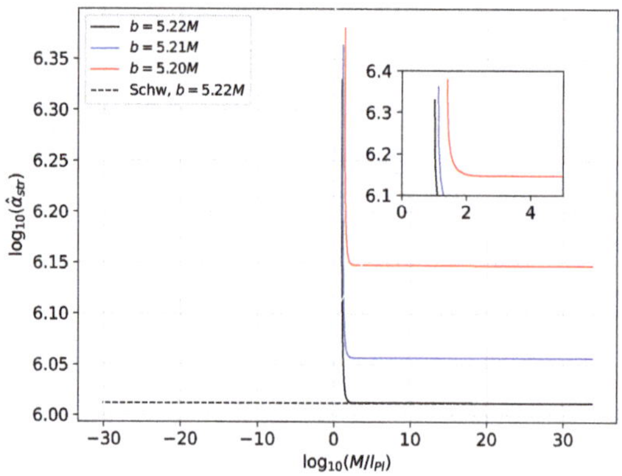

Figure 7. Strong deflection angle by quantum black holes.

Figure 6 shows that the strong deflection angle curves are steeper than the weak deflection angle. While one observes the same feature of the low impact parameter producing higher deflection angle, one can see that the deviations due to the GEUP in the strong deflection regime occur early (at lower mass) than that obsreved for the weak deflection angle (cf. Figure 4), thus providing with an enhanced detectability.

Due to Equation (35), there is some value for mass M where the strong deflection ceases, and this value is near the value of the GEUP parameters L_* and l_{Pl} (see also Figure 7). Without the influence of GEUP, the strong deflection angle seems to have no limit for any values of mass M (as shown by the dashed black line). The same feature can be observed for quantum black holes. Again, while strong deflection is theoretically possible for small particles, a problem in its detectability is looming in the impact parameters, b, since it might be small compared to the particle's physical dimension.

5. Conclusions

While the effects of the generalized (GUP) and extended (EUP) uncertainty principles are commonly analyzed separately in the literature, our study in this paper is about unifying these two quantum corrections as applied to black hole physics. Motivated by the study of Ref. [22], we investigated the effect of GEUP on the shadow and lensing for astrophysical black holes and very small particles viewed as quantum black holes.

We first find constraints to the values of the fundamental length scale, L_*, using astrophysical data from the Event Horizon Telescope (EHT) Collaboration. For the two standard deviations level of uncertainty, we found an upper bound, $L_* \sim 2.985 \times 10^{10}$ m, for Sgr. A* and $L_* \sim 3.264 \times 10^{13}$ m for M87* black holes. Interestingly, for M87*, there is a value for L_* which crosses the mean of the shadow diameter, which is $L_* \sim 7.950 \times 10^{13}$ m. We note that this order of magnitudes agrees with the constraints of gravitational lensing observables, position, magnification, and differential time delays in [23]. We also examined how the shadow radius behaves based on the position of the observer from the GEUP black hole. The results obtained indicate that a black hole with GEUP generally follows the same pattern for the shadow radius curve as that retained. in the Schwarzschild case. In particular, the GEUP parameter L_* generally increases the shadow radius for black hole masses with the same order of magnitude as L_*. We also did not find any influence of GUP on the shadow of astrophysical black holes. Shadows for quantum black holes are also investigated. Here, as the quantum black hole's mass, M, under GEUP correction decreases, we found that quantum black hole's corresponding shadow increases. The position of detectors also affects the radius of such shadows. Lastly, it is shown that L_* does not affect the quantum black hole's shadow.

Alternatively, we probe more into the effects of GEUP by considering the strong and weak deflection angles. For the weak deflection angle, the main result indicates that deviation caused by GEUP occurs when the masses are comparable to the fundamental length scales. Such a deviation occurs early at a strong deflection angle. Furthermore, due to the fundamental length scales, there is a limitation for the occurrence of strong deflection angle. For example, if hypothetically the deflection angle by a neutrino is observed, then strong deflection cannot be applied due to the limitations imposed by the Planck length, l_{Pl}.

Nonetheless, the weak deflection angle is still a better probe since it can be applied for relatively high impact parameters. The drawback is that measurement may not be possible due to the quantum nature of a particle. Finally, as far as the GEUP model in this study is concerned, the strong and weak deflection angles cannot probe whether L_* affects the quantum realm, and vice versa. Lowering the value of L_* may give an interesting result, but it may have some implications in the astrophysical phenomena that might be ruled out by observation. In theory, this direction is worth investigating.

Author Contributions: Conceptualization, R.C.P.; methodology, R.C.P. and N.J.L.S.L.; software, N.J.L.S.L.; validation, R.C.P.; formal analysis, R.C.P. and N.J.L.S.L.; investigation, R.C.P. and N.J.L.S.L.; resources, R.C.P. and N.J.L.S.L.; data curation, R.C.P. and N.J.L.S.L.; writing—original draft preparation, N.J.L.S.L.; writing—review and editing, R.C.P.; visualization, R.C.P.; supervision, R.C.P. All authors have read and agreed to the published version of the manuscript.

Funding: This research received no external funding.

Data Availability Statement: Not applicable.

Acknowledgments: We would like to thank the anonymous reviewers for their useful comments and suggestions that helped to improve the paper. R.C.P. would like to acknowledge support by the COST Action CA18108—Quantum gravity phenomenology in the multi-messenger approach (QG-MM).

Conflicts of Interest: The authors declare no conflict of interest.

References

1. Akiyama, K. et al. [Event Horizon Telescope Collaboration]. First M87 Event Horizon Telescope results. I. The shadow of the supermassive black hole. *Astrophys. J. Lett.* **2019**, *875*, L1. [CrossRef]
2. Akiyama, K. et al. [Event Horizon Telescope Collaboration]. First Sagittarius A* event Horizon Telescope results. I. The shadow of the supermassive black hole in the center of the Milky Way. *Astrophys. J. Lett.* **2022**, *930*, L12. [CrossRef]
3. Schwarzschild, K. Über das Gravitationsfeld eines Massenpunktes nach der Einsteinschen Theorie. *Sitzungsber. Preuss. Akad. Wiss. Berl. Math. Phys.* **1916**, *1916*, 189–196. Available online: https://www.jp-petit.org/Schwarzschild-1916-exterior-de.pdf (accessed on 1 October 2022).

4. Schwarzschild, K. "Golden Oldie": On the gravitational field of a mass point according to Einstein's theory. *Gen. Relat. Gravit.* **2003**, *35*, 951–959. [CrossRef]
5. Kerr, R.P. Gravitational field of a spinning mass as an example of algebraically special metrics. *Phys. Rev. Lett.* **1963**, *11*, 237–238. [CrossRef]
6. Rovelli, C. *Quantum Gravity*; Cambridge University Press: Cambridge, MA, USA, 2004. [CrossRef]
7. Wheeler, J.A. Geons. *Phys. Rev.* **1955**, *97*, 511–536. [CrossRef]
8. Maggiore, M. Quantum groups, gravity, and the generalized uncertainty principle. *Phys. Rev. D* **1994**, *49*, 5182–5187. [CrossRef] [PubMed]
9. Maggiore, M. A generalized uncertainty principle in quantum gravity. *Phys. Lett. B* **1993**, *304*, 65–69. [CrossRef]
10. Maggiore, M. The algebraic structure of the generalized uncertainty principle. *Phys. Lett. B* **1993**, *319*, 83–86. [CrossRef]
11. Amati, D.; Ciafaloni, M.; Veneziano, G. Can spacetime be probed below the string size? *Phys. Lett. B* **1989**, *216*, 41–47. [CrossRef]
12. Bambi, C.; Urban, F.R. Natural extension of the generalized uncertainty principle. *Class. Quant. Grav.* **2008**, *25*, 095006. [CrossRef]
13. Costa Filho, R.N.; Braga, J.P.M.; Lira, J.H.S.; Andrade, J.S., Jr. Extended uncertainty from first principles. *Phys. Lett. B* **2016**, *755*, 367–370. [CrossRef]
14. Tawfik, A.; Diab, A. Generalized uncertainty principle: Approaches and applications. *Int. J. Mod. Phys. D* **2014**, *23*, 1430025. [CrossRef]
15. Zhu, T.; Ren, J.-R.; Li, M.-F. Influence of generalized and extended uncertainty principle on thermodynamics of FRW universe. *Phys. Lett. B* **2009**, *674*, 204–209. [CrossRef]
16. Mignemi, S. Extended uncertainty principle and the geometry of (anti)-de-sitter space. *Mod. Phys. Lett. A* **2010**, *25*, 1697–1703. [CrossRef]
17. Dabrowski, M.P.; Wagner, F. Extended uncertainty principle for Rindler and cosmological horizons. *Eur. Phys. J. C* **2019**, *79*, 716. [CrossRef]
18. Hamil, B.; Merad, M.; Birkandan, T. Effects of extended uncertainty principle on the relativistic Coulomb potential. *Int. J. Mod. Phys. A* **2021**, *36*, 2150018. [CrossRef]
19. Hamil, B.; Merad, M.; Birkandan, T. Bound-state solutions of the two-dimensional Dirac equation with Aharonov–Bohm-Coulomb interaction in the presence of extended uncertainty principle. *Phys. Scr.* **2020**, *95*, 105307. [CrossRef]
20. Moradpour, H.; Aghababaei, S.; Ziaie, A.H. A Note on effects of generalized and extended uncertainty principles on Jüttner gas. *Symmetry* **2021**, *13*, 213. [CrossRef]
21. Aghababaei, S.; Moradpour, H.; Vagenas, E.C. Hubble tension bounds the GUP and EUP parameters. *Eur. Phys. J. Plus* **2021**, *136*, 997. [CrossRef]
22. Mureika, J.R. Extended Uncertainty Principle black holes. *Phys. Lett. B* **2019**, *789*, 88–92. [CrossRef]
23. Lu, X.; Xie, Y. Probing an Extended Uncertainty Principle black hole with gravitational lensings. *Mod. Phys. Lett. A* **2019**, *34*, 1950152. [CrossRef]
24. Kumaran, Y.; Övgün, A. Weak deflection angle of extended uncertainty principle black holes. *Chin. Phys. C* **2020**, *44*, 025101. [CrossRef]
25. Cheng, H.; Zhong, Y. Instability of a black hole with $f(R)$ global monopole under extended uncertainty principle. *Chin. Phys. C* **2021**, *45*, 105102. [CrossRef]
26. Hassanabadi, H.; Chung, W.S.; Lütfüoğlu, B.C.; Maghsoodi, E. Effects of a new extended uncertainty principle on Schwarzschild and Reissner–Nordström black holes thermodynamics. *Int. J. Mod. Phys. A* **2021**, *36*, 2150036. [CrossRef]
27. Hamil, B.; Lütfüoğlu, B.C. The effect of higher-order extended uncertainty principle on the black hole thermodynamics. *EPL (Europhys. Lett.)* **2021**, *134*, 50007. [CrossRef]
28. Ökcü, O.; Aydiner, E. Investigating bounds on the extended uncertainty principle metric through astrophysical tests. *EPL (Europhys. Lett.)* **2022**, *138*, 39002. [CrossRef]
29. Pantig, R.C.; Yu, P.K.; Rodulfo, E.T.; Övgün, A. Shadow and weak deflection angle of extended uncertainty principle black hole surrounded with dark matter. *Ann. Phys.* **2022**, *436*, 168722. [CrossRef]
30. Hamil, B.; Lütfüoğlu, B.C.; Dahbi, L. EUP-corrected thermodynamics of BTZ black hole. *Int. J. Mod. Phys. A* **2022**, *37*, 2250130. [CrossRef]
31. Chen, H.; Hassanabadi, H.; Lütfüoğlu, B.C.; Long, Z.W. Quantum corrections to the quasinormal modes of the Schwarzschild black hole. *arXiv* **2022**, arXiv:2203.03464. [CrossRef]
32. Kempf, A.; Mangano, G.; Mann, R.B. Hilbert space representation of the minimal length uncertainty relation. *Phys. Rev. D* **1995**, *52*, 1108–1118. [CrossRef] [PubMed]
33. Calmet, X.; Carr, B.; Winstanley, E. *Quantum Black Holes*; Springer Briefs in Physics; Springer: Berlin/Heidelberg, Germany, 2014. [CrossRef]
34. Hawking, S. Gravitationally collapsed objects of very low mass. *Mon. Not. R. Astron. Soc.* **1971**, *152*, 75–78. [CrossRef]
35. Synge, J.L. The escape of photons from gravitationally intense stars. *Mon. Not. R. Astron. Soc.* **1966**, *131*, 463–466. [CrossRef]
36. Luminet, J.P. Image of a spherical black hole with thin accretion disk. *Astron. Astrophys.* **1979**, *75*, 228–235. Available online: https://adsabs.harvard.edu/full/1979A%26A....75..228L (accessed on 1 October 2022).
37. Konoplya, R.A. Quantum corrected black holes: Quasinormal modes, scattering, shadows. *Phys. Lett. B* **2020**, *804*, 135363. [CrossRef]

38. Hu, Z.; Zhong, Z.; Li, P.C.; Guo, M.; Chen, B. QED effect on a black hole shadow. *Phys. Rev. D* **2021**, *103*, 044057. [CrossRef]
39. Tamburini, F.; Feleppa, F.; Thidé, B. Constraining the Generalized Uncertainty Principle with the light twisted by rotating black holes and M87*. *Phys. Lett. B* **2022**, *826*, 136894. [CrossRef]
40. Devi, S.; S, A.N.; Chakrabarti, S.; Majhi, B.R. Shadow of quantum extended Kruskal black hole and its super-radiance property. *arXiv* **2021**, arXiv:2105.11847. [CrossRef]
41. Anacleto, M.A.; Campos, J.A.V.; Brito, F.A.; Passos, E. Quasinormal modes and shadow of a Schwarzschild black hole with GUP. *Ann. Phys.* **2021**, *434*, 168662. [CrossRef]
42. Xu, Z.; Tang, M. Testing the quantum effects near the event horizon with respect to the black hole shadow. *Chin. Phys. C* **2022**, *46*, 085101. [CrossRef]
43. Karmakar, R.; Gogoi, D.J.; Goswami, U.D. Quasinormal modes and thermodynamic properties of GUP-corrected Schwarzschild black hole surrounded by quintessence. *arXiv* **2022**, arXiv:2206.09081. [CrossRef]
44. Rayimbaev, J.; Pantig, R.C.; Övgün, A.; Abdujabbarov, A.; Demir, D. Quasiperiodic oscillations, weak field lensing and shadow cast around black holes in symmergent gravity. *arXiv* **2022**, arXiv:2206.06599. [CrossRef]
45. Dyson, F.W.; Eddington, A.S.; Davidson, C. A Determination of the deflection of light by the Sun's gravitational field, from observations made at the total eclipse of May 29, 1919. *Philos. Trans. R. Soc. Lond. A* **1920**, *220*, 291–333. [CrossRef]
46. Virbhadra, K.S.; Ellis, G.F.R. Schwarzschild black hole lensing. *Phys. Rev. D* **2000**, *62*, 084003. [CrossRef]
47. Bozza, V.; Capozziello, S.; Iovane, G.; Scarpetta, G. Strong field limit of black hole gravitational lensing. *Gen. Rel. Grav.* **2001**, *33*, 1535–1548. [CrossRef]
48. Bozza, V. Gravitational lensing in the strong field limit. *Phys. Rev. D* **2002**, *66*, 103001. [CrossRef]
49. Gibbons, G.W.; Werner, M.C. Applications of the Gauss-Bonnet theorem to gravitational lensing. *Class. Quantum Gravity* **2008**, *25*, 235009. [CrossRef]
50. Werner, M.C. Gravitational lensing in the Kerr-Randers optical geometry. *Gen. Rel. Grav.* **2012**, *44*, 3047–3057. [CrossRef]
51. Ishihara, A.; Suzuki, Y.; Ono, T.; Kitamura, T.; Asada, H. Gravitational bending angle of light for finite distance and the Gauss-Bonnet theorem. *Phys. Rev. D* **2016**, *94*, 084015. [CrossRef]
52. Ishihara, A.; Suzuki, Y.; Ono, T.; Asada, H. Finite-distance corrections to the gravitational bending angle of light in the strong deflection limit. *Phys. Rev. D* **2017**, *95*, 044017. [CrossRef]
53. Li, Z.; Zhang, G.; Övgün, A. Circular orbit of a particle and weak gravitational lensing. *Phys. Rev. D* **2020**, *101*, 124058. [CrossRef]
54. Xu, C.; Yang, Y. Determination of bending angle of light deflection subject to possible weak and strong quantum gravity effects. *Int. J. Mod. Phys. A* **2020**, *35*, 2050188. [CrossRef]
55. Zhang, R.; Lin, J.; Huang, Q. Strong gravitational lensing for the quantum-modified Schwarzschild black hole. *Int. J. Theor. Phys.* **2021**, *60*, 387–396. [CrossRef]
56. Fu, Q.M.; Zhang, X. Gravitational lensing by a black hole in effective loop quantum gravity. *Phys. Rev. D* **2022**, *105*, 064020. [CrossRef]
57. Lu, X.; Xie, Y. Gravitational lensing by a quantum deformed Schwarzschild black hole. *Eur. Phys. J. C* **2021**, *81*, 627. [CrossRef]
58. Jusufi, K.; Övgün, A.; Banerjee, A.; Sakallı, t.I. Gravitational lensing by wormholes supported by electromagnetic, scalar, and quantum effects. *Eur. Phys. J. Plus* **2019**, *134*, 428. [CrossRef]
59. Perlick, V.; Tsupko, O.Y.; Bisnovatyi-Kogan, G.S. Influence of a plasma on the shadow of a spherically symmetric black hole. *Phys. Rev. D* **2015**, *92*, 104031. [CrossRef]
60. Perlick, V.; Tsupko, O.Y.; Bisnovatyi-Kogan, G.S. Black hole shadow in an expanding universe with a cosmological constant. *Phys. Rev. D* **2018**, *97*, 104062. [CrossRef]
61. Pantig, R.C.; Övgün, A. Testing dynamical torsion effects on the charged black hole's shadow, deflection angle and greybody with M87* and Sgr A* from EHT. *arXiv* **2022**, arXiv:2206.02161. [CrossRef]
62. Nandi, K.K.; Izmailov, R.N.; Yanbekov, A.A.; Shayakhmetov, A.A. Ring-down gravitational waves and lensing observables: How far can a wormhole mimic those of a black hole? *Phys. Rev. D* **2017**, *95*, 104011. [CrossRef]
63. Eitel, K. Direct neutrino mass experiments. *Nucl. Phys. B Proc. Suppl.* **2005**, *143*, 197–204. [CrossRef]
64. do Carmo, M. *Riemannian Geometry*; Birkhäuser: Boston, MA, USA, 1992.
65. Tsukamoto, N. Deflection angle in the strong deflection limit in a general asymptotically flat, static, spherically symmetric spacetime. *Phys. Rev. D* **2017**, *95*, 064035. [CrossRef]
66. Zhao, S.S.; Xie, Y. Strong deflection gravitational lensing by a modified Hayward black hole. *Eur. Phys. J. C* **2017**, *77*, 1–10. [CrossRef]
67. Bisnovatyi-Kogan, G.S.; Tsupko, O.Y. Strong gravitational lensing by Schwarzschild black holes. *Astrophysics* **2008**, *51*, 99–111. [CrossRef]

physics

MDPI

Article

The Cosmology of a Non-Minimally Coupled $f(R,T)$ Gravitation

Değer Sofuoğlu [1], Rishi Kumar Tiwari [2], Amare Abebe [3,4,*], Alnadhief H. A. Alfedeel [5,6] and Eltegani I. Hassan [5]

1 Department of Physics, Faculty of Science, Istanbul University, Vezneciler/Fatih, Istanbul 34134, Turkey
2 Department of Mathematics, Govt. Model Science College, Rewa 486003, Madhya Pradesh, India
3 Centre for Space Research, North-West University, Mahikeng 2745, South Africa
4 National Institute for Theoretical and Computational Sciences (NITheCS), South Africa
5 Department of Mathematics and Statistics, Imam Mohammad Ibn Saud Islamic University (IMSIU), Riyadh 13318, Saudi Arabia
6 Department of Physics, Faculty of Science, University of Khartoum, P.O. Box 321, Khartoum 11115, Sudan
* Correspondence: amareabebe.gidelew@nwu.ac.za

Abstract: A non-minimally coupled cosmological scenario is considered in the context of $f(R,T) = f_1(R) + f_2(R)f_3(T)$ gravity (with R being the Ricci scalar and T the trace of the energy-momentum tensor) in the background of the flat Friedmann–Robertson–Walker (FRW) model. The field equations of this modified theory are solved using a time-dependent deceleration parameter for a dust. The behavior of the model is analyzed taking into account constraints from recent observed values the deceleration parameter. It is shown that the analyzed models can explain the transition from the decelerating phase to the accelerating one in the expansion of the universe, by staying true to the results of the observable universe. It is shown that the models are dominated by a quintessence-like cosmological dark fluid at the late universe.

Keywords: $f(R,T)$ theory; deceleration parameter; non-minimal coupling; dark energy

Citation: Sofuoğlu, D.; Tiwari, R.K.; Abebe, A.; Alfedeel, A.H.A.; Hassan, E.I. The Cosmology of a Non-Minimally Coupled $f(R,T)$ Gravitation. *Physics* **2022**, *4*, 1348–1358. https://doi.org/10.3390/physics4040086

Received: 12 August 2022
Accepted: 18 October 2022
Published: 7 November 2022

1. Introduction

The cosmological portrait of the universe has been changed by some indications of recent type Ia supernovae data [1–3] and the results of Planck Collaboration [4]. The revolutionary sign of these observations is that the expansion of the universe is currently accelerating. Various theories have been developed in the literature to explain this cosmic acceleration. It is believed that the cause of this acceleration is an energy called dark energy, which cannot be explained by the baryonic matter distribution. This dark energy is now known to have a large proportion of about 70% of the total energy distribution in the universe. In Λ-cold-dark-matter (ΛCDM) cosmology, this dark energy is usually explained by adding the cosmological constant Λ to the field equations of the general theory of relativity (GR).

However, such a cosmological scenario is pregnant with some cosmological problems [5], so some alternative models have been proposed [6–8]. In these alternative cosmologies, the model that behaves similar to the ΛCDM model is obtained without using the cosmological constant. The main reason here is the need for a cosmological model that can give the results of the observable universe, but on the other hand, will keep the problems brought by ΛCDM at bay. For example, it is one of the consequences of such a need to take the matter–energy content of the universe as the scalar field as exotic matter in Einstein's field equations that can produce enough negative pressure to accelerate the expansion of the universe; see Refs. [9–14].

In this context, modified gravity theories that serve this purpose are of great interest in current cosmological studies. The $f(R,T)$ theory of gravity is one of the most popular of these modified gravitational theories [15]. Here, the gravitational Lagrangian is given by an arbitrary function of the Ricci scalar, R, and the trace, T, of the energy–momentum tensor, the dependence of which can be induced by exotic imperfect fluids or quantum effects [15].

Since it was proposed in 2011, many researchers have performed a wide variety of research on this theory. Some to be mentioned are as follows. Xu et al. [16] studied quantum cosmology effects of the gravitational interaction described by the model. Friedmann–Robertson–Walker (FRW) model was examined by Myrzakulov [17]. Sharif and Zubair [18] obtained exact solutions of the field equations of $f(R,T)$ theory using the anisotropic behavior of spacetime for exponential and power expansion laws. Moraes et al. examined the transition from deceleration to acceleration in this theory [19]. Shamir [20] studied the locally rotationally symmetric (LRS) Bianchi type-I model. A string cosmological model was considered by Sharma and Singh [21] for Bianchi type-II universe. In order to understand the dynamic behavior of the anisotropic universe in $f(R,T)$ gravity, a large-scale search was made for the Bianchi-type VI_h model [22,23] by Mishra et al. Tiwari and Sofuoğlu [24] investigated the cosmological implications of a quadratically varying decelariton parameter in locally-rotationally-symmetric (LRS) Bianchi type-I model. Tiwari et al. studied Bianchi type-I universe taking into account time dependent gravitational and cosmological parameters [25]. Evolution of axially symmetric anisotropic sources was investigated in $f(R,T)$ theory by Zubair and Noureen [26]. Alfedeel and Tiwari showed that the generalized Friedman equation's exact solution for the average scale factor involves the hypergeometric function considering a novel approach [27]. An accelerating model studied in the presence of varying cosmological term by Tiwari et al. [28]. A cosmological model with variable deceleration parameter in $f(R,T)$ theory was constructed by Tiwari et al. [29]. Sahoo et al. [30] studied on Bianchi type-I universe taking bulk viscous fluid. Moraes and Sahoo [31] considered nonminimal coupling between geometry and matter in this theory. Sharma et al. [32] examined the existing of non-minimal matter–geometry interaction in Bianchi type-I model. Tiwari et al. [33] studied a non-minimal cosmological model in the presence of a varying deceleration parameter.

In the current study, inspired by the above discussion, we consider the $f(R,T)$ modified theory of gravity in the background of flat FRW universe by considering a variable deceleration parameter to investigate the phase change (from decelerating to accelerating expansion phase) in the expansion of the universe. For the choice of a particular case of the non-minimally function $f(R,T) = f_1(R) + f_2(R)f_3(T)$, exact solution of the field equations has been obtained. In Section 2, a basic formalism of $f(R,T)$ theory is presented, the solutions of the field equations are obtained in Section 3, and the conclusions are given in Section 4.

2. $f(R,T)$ Gravity

Throughout this section, we review the basic derivation of the $f(R,T)$ theory of gravity. Let us start by introducing the the action of $f(R,T)$ gravity is defined by Harko et al. [15]:

$$S = \int \sqrt{-g}\, d^4x \left(\frac{1}{16\pi G} f(R,T) + L_{\mathrm{m}} \right), \tag{1}$$

where $f(R,T)$ is an arbitrary function of R and $T = g^{ij}T_{ij}$ the trace of the energy–momentum tensor, T_{ij}; Latin letters $i,j,k,l,\ldots$ denote 4-dimensional tensor indices and take on the values 0 (time), 1, 2, and 3 (space); $g = \det|g_{ij}|$ is the determinant of the metric tensor, g_{ij}; G is Newtonian constant of gravity, and L_{m} is the matter Lagrangian. Accordingly, the energy–momentum tensor, T_{ij}, is defined as

$$T_{ij} = -\frac{2}{\sqrt{-g}} \frac{\delta(\sqrt{-g}L_{\mathrm{m}})}{\delta g^{ij}} = L_{\mathrm{m}} g_{ij} - 2\frac{\delta L_{\mathrm{m}}}{\delta g^{ij}}\,. \tag{2}$$

Here, we assume that L_{m} is a function the metric tensor, g_{ij}, rather than of its derivatives. By varying the action S in Equation (1) with respect to the metric tensor, g_{ij}, the modified gravitational field equations for $f(R,T)$ gravity reads:

$$f_R(R,T)R_{ij} - \frac{1}{2}f(R,T)g_{ij} - (\nabla_i\nabla_j - g_{ij}\Box)f_R(R,T) = 8\pi T_{ij} - f_T(R,T)(T_{ij} + \Theta_{ij})\,, \tag{3}$$

where ∇_i is the covariant derivative, $\Box \equiv \nabla^i \nabla_i$ is the d'Alembertian operator and the fractions moved to a linear form better visible in the text. Please confirm. $f_R(R,T) = \partial f(R,T)/\partial R$, $f_T(R,T) = \partial f(R,T)/\partial T$, $\Theta_{ij} = g^{ab}\delta T_{ab}/\delta g^{ij}$. Contracting Equation (3) with metric tensor g^{ij} produces

$$f_R(R,T)R + 3\Box f_R(R,T) - 2f(R,T) = 8\pi T - f_T(R,T)(T+\Theta)\,, \tag{4}$$

where $\Theta = g^{ij}\Theta_{ij}$. Upon using the matter Lagrangian L_{m}, the energy–momentum tensor of matter is given by

$$T_{ij} = (\rho + p)u_i u_j - p g_{ij}\,, \tag{5}$$

where ρ, p, u_i are the fluid energy density, the pressure of the fluid and the fluid 4-velocity, respectively. Further, u_i is time-like quantity that satisfies $u^i u_i = 1$ and $u^i \nabla_j u_i = 0$. The variation of stress energy of perfect fluid is obtained by following Shamir [34] argument where the matter Lagrangian $L_{\mathrm{m}} = -p$ is assumed, thus

$$\Theta_{ij} = -2T_{ij} - p g_{ij}\,. \tag{6}$$

Substituting Equation (6) into Equation (3), the field equations take the form

$$f_R(R,T)R_{ij} - \frac{1}{2}f(R,T)g_{ij} - (\nabla_i\nabla_j - g_{ij}\Box)f_R(R,T) = 8\pi T_{ij} + f_T(R,T)(T_{ij} + p g_{ij})\,. \tag{7}$$

Equation (7) leads to the modified $f(R)$ and GR theories of gravity when $f(R,T) = f(R)$ and $f(R,T) = R$, respectively. Ref. [15] investigated three different functional forms of $f(R,T)$. More justification about the choice of $f(R,T)$ is given in [35]. These forms are given by

$$f(R,T) = \begin{cases} R + 2f(T), \\ f_1(R) + f_2(T), \\ f_1(R) + f_2(R)f_3(T). \end{cases}$$

In this paper, we adopt the last functional form,

$$f(R,T) = f_1(R) + f_2(R)f_3(T) = R + \lambda R T\,, \tag{8}$$

thus transforming Equation (7) into the following form:

$$R_{ij} - \frac{1}{2}g_{ij}R \;=\; +8\pi T_{ij} - \lambda[g_{ij}\Box - \nabla_i\nabla_j]T - \lambda T[R_{ij} - \frac{1}{2}g_{ij}R] + \lambda R[T_{ij} + p g_{ij}]\;, \tag{9}$$

or, alternatively,

$$R_{ij} - \frac{1}{2}g_{ij}R \;=\; \frac{8\pi + \lambda R}{(1+\lambda T)}T_{ij} - \frac{\lambda}{(1+\lambda T)}[g_{ij}\Box - \nabla_i\nabla_j]\,T + \frac{\lambda R}{(1+\lambda T)}p g_{ij}\,. \tag{10}$$

Here, λ is the coupling parameter of the model and vanishes automatically for GR.

The right-hand side of Equation (10) can be viewed as a total-effective energy momentum tensor, T^t_{ij},

$$T^t_{ij} = T_{ij} + T^f_{ij}\,, \tag{11}$$

with T^f_{ij} defined as

$$T^f_{ij} = \frac{\lambda}{1+\lambda T}(R T_{ij} + R p g_{ij} - [g_{ij}\Box - \nabla_i\nabla_j]T)\,, \tag{12}$$

showing the contribution term from $f(R,T)$. The limiting case $\lambda = 0$ in Equation (9) gives standard GR results.

The FRW Metric and Field Equations

The homogeneous and isotropic flat FRW universe is given as

$$ds^2 = dt^2 - a^2(t)(dx^2 + dy^2 + dz^2), \tag{13}$$

where $a(t)$ is the scale factor. In the flat FRW background, the non-minimally coupled $f(R,T)$ gravity for $T = \rho - 3p$, the time-time and space-space components of the modified field equations of Equation (9) gives the following Friedmann equations:

$$3H^2 = 8\pi\rho - 3\lambda H(\dot{\rho} - 3\dot{p}) - 3\lambda H^2(\rho - 3p) - 6\lambda(\dot{H} + 2H^2)(p + \rho)\,, \tag{14}$$

$$2\dot{H} + 3H^2 = -8\pi p - \lambda(2\dot{H} + 3H^2)(\rho - 3p) - \lambda(\ddot{T} - 2H\dot{T})\,. \tag{15}$$

Here, H is the Hubble constant and the dot denotes time derivation.

As soon as $\dot{T} = \dot{\rho} - 3\dot{p}$ and $\ddot{T} = \ddot{\rho} - 3\ddot{p}$, Equation (15) becomes:

$$2\dot{H} + 3H^2 = -8\pi p - \lambda(2\dot{H} + 3H^2)(\rho - 3p) - \lambda[\ddot{\rho} - 3\ddot{p} - 3H(\dot{\rho} - 3\dot{p})]\,. \tag{16}$$

Equations (14) and (16) are the generalized Friedmann equation in $f(R,T)$ theory of gravity. These equations cannot be solved since they contain $\lambda, \dot{\rho}, \dot{p}, \ddot{\rho}$ and $\ddot{p}$. On the other hand, Equation (9) gives the Bianchi identity as

$$-8\pi\nabla^j T_{ij} = \frac{\lambda R}{2}(\nabla_i T) + \lambda(\nabla^j R)[T_{ij} + pg_{ij}] + \lambda R[\nabla^j T_{ij} + \nabla_j p]\,, \tag{17}$$

which gives

$$(8\pi + \lambda R)\dot{\rho} + 3H(p + \rho) = -\frac{\lambda R}{2}(\dot{\rho} - \dot{p}) - \lambda\dot{R}(\rho + p)\,. \tag{18}$$

Assuming that $p = w\rho$, Equation (17) can be re-arranged for $\dot{\rho}$ as

$$\dot{\rho} = -\frac{3H(1+w) + \lambda\dot{R}}{8\pi + \frac{1}{2}\lambda R(3-w)}\rho\,, \tag{19}$$

upon differentiating with respect to time yields:

$$\ddot{\rho} = \left\{-\frac{3\dot{H}(1+w) + \lambda\ddot{R}}{8\pi + \frac{1}{2}\lambda R(3-w)} + \frac{[3H(1+w) + \lambda\dot{R}(5-w)][3H(1+w) + \lambda\dot{R}]}{[8\pi + \frac{1}{2}\lambda R(3-w)]^2}\right\}\rho\,, \tag{20}$$

where for a flat FRW metric,

$$R = -6(\dot{H} + 2H^2)\,, \tag{21}$$
$$\dot{R} = -6(\ddot{H} + 4H\dot{H})\,, \tag{22}$$
$$\ddot{R} = -6(\dddot{H} + 4\dot{H}^2 + 4H\ddot{H}) \tag{23}$$

are the Ricci scalar and its time deviatives. The generalized Friedmann Equations (14) and (16) now read:

$$3H^2 = 8\pi\rho - 3\lambda(1-3w)H[\dot{\rho} + H\rho] - 6\lambda(1+w)(\dot{H} + 2H^2)\rho\,, \tag{24}$$

$$2\dot{H} + 3H^2 = -8\pi w\rho - \lambda(1-3w)[(2\dot{H} + 3H^2)\rho - 2H\dot{\rho} + \ddot{\rho}]\,. \tag{25}$$

Subtracting Equations (24) and (25) one from another produces the generalized Raychaudhri equation,

$$\begin{aligned} 2\dot{H} = &-8\pi(1+w)\rho - \lambda(1-3w)[(2\dot{H} + 3H^2)\rho - 2H\dot{\rho} + \ddot{\rho}] \\ &+3\lambda(1-3w)H[\dot{\rho} + H\rho] - 6\lambda(1+w)(\dot{H} + 2H^2)\rho\,. \end{aligned} \tag{26}$$

Having known the value of $\dot{\rho}$ and $\ddot{\rho}$ (from Equations (19) and (20), respectively) and H, Equation (26) is solved in Section 3 below to obtain an expression for the energy density ρ of the matter content of universe directly. In this model, the Hubble parameter, H, and deceleration parameter (DP) are defined, respectively, as

$$H \equiv \frac{\dot{a}}{a} \quad \text{and} \quad q \equiv -1 - \frac{\dot{H}}{H^2} \,. \tag{27}$$

Using Equation (9), Equations (24) and (25) give the total-effective density, ρ^t, and the total-effective pressure, p^t:

$$\rho^t = \rho - 3\lambda(1-3w)H(\dot{\rho} + H\rho) + 6\lambda(1+w)(\dot{H} + 2H^2)\rho \,, \tag{28}$$

$$p^t = w\rho - \lambda(1-3w)[(2\dot{H} + 3H^2)\rho - 2H\dot{\rho} + \ddot{\rho}] \,. \tag{29}$$

From Equations (28) and (29), with the help of Equation (11), one obtains the density ρ in terms of H, $\dot{H}$, $\dot{\rho}$, and $\ddot{\rho}$:

$$\rho = \frac{1}{4}\left[\frac{2\dot{H} - 5\lambda H(1-3w)\dot{\rho} + \lambda(1-3w)\ddot{\rho}}{-2\pi(1+w) + \lambda(1+3w)\dot{H} + 3\lambda(1+w)H^2}\right] . \tag{30}$$

3. Solutions of the $f(R,T)$ Field Equations

To solve the system of Equations (20)–(24) containing two equations and three unknowns (a, ρ, and p), one more equation is needed. Since the Type Ia supernova observations and various astronomical observations [1,2,36,37] indicated that the universe is accelerating, a time-dependent DP is needed that can explain the transition from deceleration expansion in the past at $z \geq 1$ to acceleration expansion at present. In concordance with this argument, many parametrization have proposed that DP is time-dependent to study various problems in cosmology [38–40]. For instance, for Berman [41] and Gomide [42], the law of variation for Hubble parameter that yields a constant DP. Ref. [43] introduced a linear function of the Hubble parameter, and well motivated by [44,45]. Motivated by the above discussion, in this paper, we adopt a generalization form of deceleration parameter that is introduced in Equation (27) as a function of Hubble parameter:

$$q = \alpha - \frac{\beta}{H^2}, \tag{31}$$

where α is a dimensionless constant, while the other constant, β, has the dimensions of H^2. Using this relation along with Equation (27) for solving the scale factor and the Hubble parameter, one obtains:

$$a = \left\{\sinh\left[\sqrt{(1+\alpha)\beta}\; t + c\right]\right\}^{\frac{1}{1+\alpha}}, \tag{32}$$

$$H = \sqrt{\frac{\beta}{1+\alpha}}\coth\left[\sqrt{(1+\alpha)\beta}\; t + c\right], \tag{33}$$

where c is the constant of integration.

Substituting the values of H, R, and $\dot{R}$ into Equation (19) gives:

$$\dot{\rho} = \frac{3A\coth(\tau)[1 + w + 4B\lambda(2A - B)\text{cosech}^2(\tau)]}{8\pi + A\lambda(w-3)[A\coth^2(\tau) - 3B\text{cosech}^2(\tau)]}\rho \,, \tag{34}$$

where $A = \sqrt{\frac{\beta}{1+\alpha}}$ and $B = \sqrt{(1+\alpha)\beta}$.

Integrating Equation (34) results into the following expression for energy density:

$$\rho = \rho_0 \exp\left\{ \int \frac{3A\coth(\tau)[1 + w + 4B\lambda(2A - B)\mathrm{cosech}^2(\tau)]}{8\pi + \lambda(w-3)[A^2\coth^2(\tau) - 3AB\mathrm{cosech}^2(\tau)]} dt \right\}, \tag{35}$$

where $\tau \equiv \sqrt{(1+\alpha)\beta}\, t + c$, and ρ_0 is a constant of integration. It is worth mentioning that the processes of obtaining a simplified expression for the energy density ρ from Equations (20)–(25) is not straightforward as soon as it depends on $R, \dot{R}, \dot{\rho}$, and $\ddot{\rho}$. If $\lambda = 0$, Equation (35) give the GR limit as

$$\rho = \frac{3(1+w)\rho_0}{8\pi} \sqrt{\frac{\beta}{1+\alpha}} \sinh(\tau) . \tag{36}$$

It is always viable to write the explicit expression of ρ using its temporal derivatives, but the expression is large and complex. Instead, it is possible to calculate the integral of Equation (34) or Equation (35) for $w = 0$, which gives the following expression for the energy density ρ:

$$\rho = \rho_0 \left[(\sinh(\tau))^{-1} \right]^{\frac{3A}{4B(-9\lambda A^2 + 4\pi)}} \left[\frac{9\lambda A(-2A\cosh^2\tau + B)}{\sinh^2\tau} + 8\pi \right]^{\frac{-72\lambda A^2 B + 32\pi B + 9A}{-12B(-9\lambda A^2 + 4\pi)}} . \tag{37}$$

To show the graphical representation of the model, let us first write the expressions of the parameters in terms of redshift, z. Using the relation $a = (1+z)^{-1}$, one obtains:

$$q = \alpha - \frac{b}{h^2} = \alpha - \frac{b(1+\alpha)}{1 + (1+z)^{2+2\alpha}}, \tag{38}$$

$$h = \sqrt{(1+\alpha)^{-1}} \sqrt{(1+z)^{2(1+\alpha)} + 1}, \tag{39}$$

$$\rho = \rho_0 (1+z)^{\frac{3(1+\alpha)A}{4B(-9\lambda A^2 + 4\pi)}} \times \left[(1+z)^{2\alpha} \left(9\lambda A\Big((-2A + B)[z^2 + z + 1] + (-2A + 7B)z - 2 \Big) + 8\pi \right) \right]^{\frac{-72\lambda A^2 B + 32\pi B + 9A}{-12B(-9\lambda A^2 + 4\pi)}}, \tag{40}$$

where $h = H/H_0$ is the normalized expansion rate and $b = \beta/H_0^2$ is a normalized constant with H_0 being the Hubble constant.

To plot the graphs, different values of the constants α and b were selected considering the observable universe as initial conditions ($z = 0$). We take into account the results of three different observations for the current values of the deceleration parameter, namely $q_0 = -0.54$ [46], $q_0 = -0.73$ [47] and $q_0 = -0.81$ [48]; the models obtained for each of the observational values called Model 1 (M1), Model 2 (M2), and Model 3 (M3), respectively. In what follows, the graphs are plotted for the corresponding three different α and b pairs: $\alpha = 0.4761$, $b = 1.3903$ for M1, $\alpha = 0.5685$, $b = 1.6557$ for M2, and $\alpha = 0.6054$, $b = 1.7633$ for M3.

From Figure 1, one can see that the normalized Hubble parameter, h, has a larger value at high redshift zone and it is smaller at the low redshift zone for all the models. Figure 2 shows that while the sign of the deceleration parameter was initially positive, it became negative in the late universe for each model. This sign change indicates that the universe has moved from its decelerating expansion in the past to its current accelerating expansion. It is seen that the transition from slowing expansion to accelerating expansion takes place at almost $z = 1/2$ for three of the models.

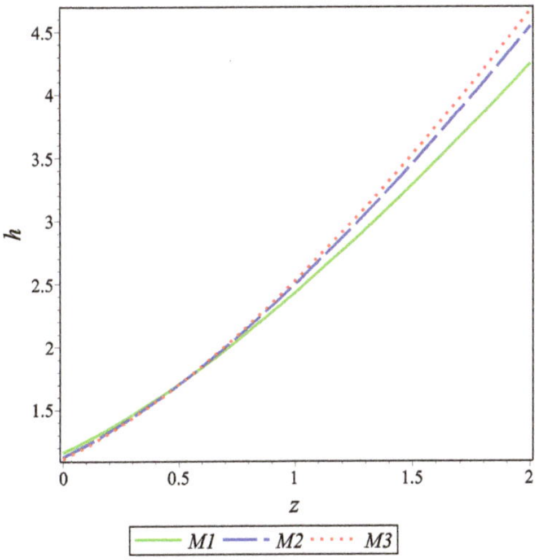

Figure 1. The normalized expansion rate, h (39), versus redshift, z, for three observational models M1, M2, and M3. See text for details.

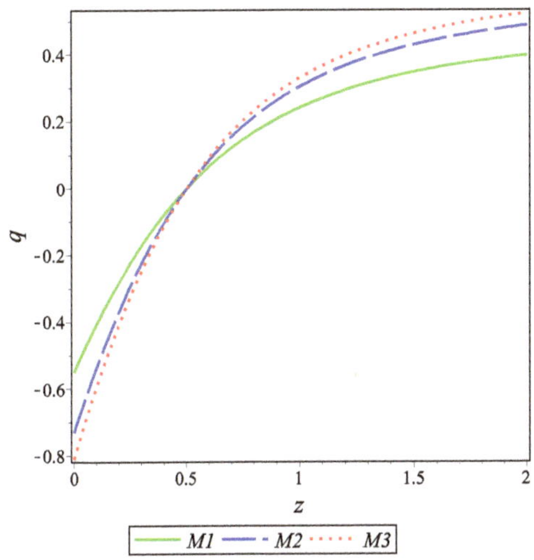

Figure 2. The deceleration parameter, q (38), versus redshift, z, for three observational models M1, M2, and M3. See text for details.

Figure 3 shows that the energy density ρ decreases from the high redshift region to the low redshift region and always remains positive.

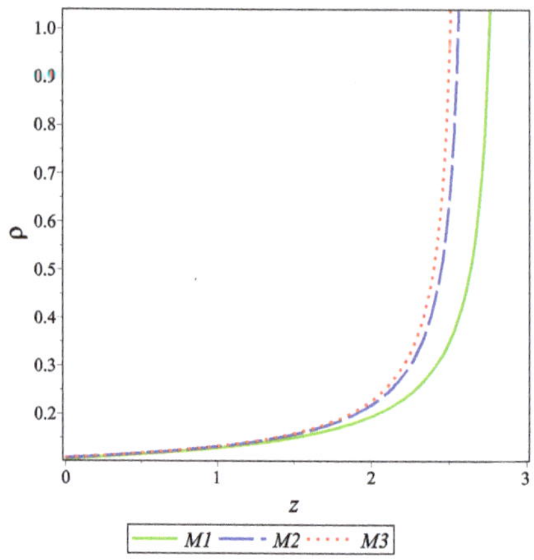

Figure 3. The energy density, ρ (40), versus redshift, z, with $\lambda = 0.1$ and $\rho_0 = 1$ for three observational models M1, M2, and M3. See text for details.

Now, let us use the relation

$$w^t = \frac{1}{3}(2q - 1) \tag{41}$$

between the deceleration parameter and total-effective equation of state (EoS) parameter, w^t, to obtain

$$w^t = \frac{1}{3}\left[2\alpha - \frac{2b(1+\alpha)}{(1+z)^{2+2\alpha}} - 1\right]. \tag{42}$$

Figure 4 shows the evolution of the total-effective EoS parameter in redshift for each model. It is seen that while the EoS parameter has positive values at high redshift regions, it decreases and takes negative values at low redshift regions. Current values of w^t indicate that the models are dominated by a quintessence-like dark fluid currently.

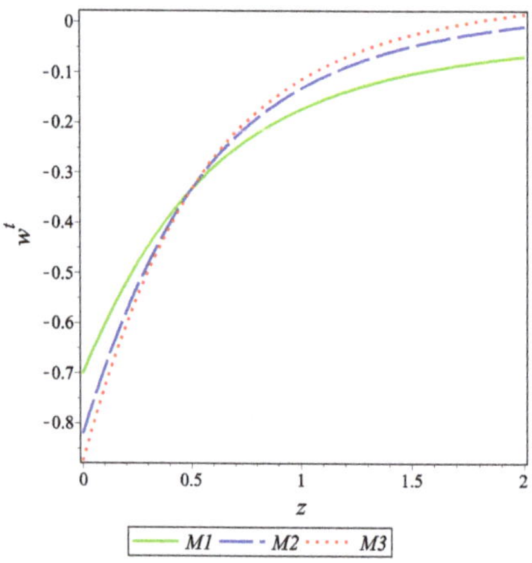

Figure 4. The The equation of state (EoS) parameter, w^t (42), versus redshift, z, for three observational models M1, M2, and M3. See text for details.

4. Conclusions

In this study, we investigated a non-minimally coupled cosmological model in the context of $f(R,T)$ theory for the flat Friedmann–Robertson–Walker (FRW) metric. For the choice of the function $f(R,T)$ in the form of $f(R,T) = f_1(R) + f_2(R)f_3(T)$ with $f_1(R) = f_2(R) = R$ and $f_3(T) = \lambda T$, the solutions of the field equations were solved under the assumption of a time-dependent deceleration parameter, which can explain the evolution of the expansion of the universe from beginning to current epoch. Results obtained were discussed by means of their graphs in redshift space, by taking into account three different observed values of the deceleration parameter as three different observational models (M1, M2, and M3).

The evolution of the deceleration parameters shows that the phase change in the expansion of the universe occurs at almost $z = 1/2$ redshift, and the deceleration parameter current values (q_0) are -0.55, -0.729, and -0.81 for M1, M2, and M3 models, respectively. These values of q_0 are consistent with the observational values of the deceleration parameter.

The total or effective equation of state parameter, w^t, is found to have positive values initially, but continue their evolution by taking negative values, for each model. The current values of w^t are less than $-1/3$ and greater than -1. These values tell us that each model is dominated by a quintessence-like dark fluid.

As a next step, it is worth constraining the parameter space of these models with current and upcoming astronomical data, as well as studying the cosmological perturbations in the context of these models to analyze large-scale structure formation scenarios.

Author Contributions: Conceptualization, D.S and R.K.T.; methodology, D.S., R.K.T. and A.A.; validation, D.S., A.A. and A.H.A.A.; formal analysis, D.S., A.A., A.H.A.A. and E.I.H.; writing—original draft preparation, D.S.; writing—review and editing, A.A. and A.H.A.A. All authors have read and agreed to the published version of the manuscript.

Funding: The authors extend their appreciation to the Deanship of Scientific Research at Imam Mohammad Ibn Saud Islamic University (Riyadh, Saudi Arabia) for funding this work through Research Group no. RG-21-09-18.

Data Availability Statement: The data used in the study are available from the corresponding references.

Conflicts of Interest: The authors declare no conflict of interest.

References

1. Perlmutter, S.; Aldering, G.; Goldhaber, G.; Knop, R.A.; Nugent, P.; Castro, P.G.; Deustua, S.; Fabbro, S.; Goobar, A.; Groom, D.E.; et al. Measurements of Ω and Λ from 42 high-redshift supernovae. *Astrophys. J.* **1999**, *517*, 565–586. [CrossRef]
2. Riess, A.G.; Filippenko, A.V.; Challis, P.; Clocchiatti, A.; Diercks, A.; Garnavich, P.M.; Gilliland, R.L.; Hogan, C.J.; Jha, S.; Kirshner, R.P.; et al. Observational evidence from supernovae for an accelerating universe and a cosmological constant. *Astron. J.* **1998**, *116*, 1009–1038. [CrossRef]
3. Riess, A.G.; Kirshner, R.P.; Schmidt, B.P.; Jha, S.; Challis, P.; Garnavich, P.M.; Esin, A.A.; Carpenter, C.; Grashius, R.; Schild, R.E.; et al. *BVRI* light curves for 22 Type Ia supernovae. *Astron. J.* **1999**, *117*, 707–724. [CrossRef]
4. Ade, P.A.R.; Aghanim, N.; Arnaud, M.; Ashdown, M.; Aumont, J.; Baccigalupi, C.; Banday, R.B.; Barriero, R.B.; Bartlett, G.; Bartolo, N.; et al. [Planck Collaboration] Planck 2015 results. XIII. Cosmological parameters. *Astron. Astrophys.* **2016**, *594*, A13.
5. Burgess, C.P. The cosmological constant problem: Why it's hard to get dark energy from micro-physics. *arXiv* **2013**, arXiv:1309.4133. [CrossRef]
6. Luongo, O.; Tommasisni, D. Modeling dark energy through an Ising fluid with network interactions. *Int. J. Mod. Phys. D* **2014**, *23*, 1450023. [CrossRef]
7. Luongo, O.; Quevedo, H. Cosmographic study of the universe's specific heat: A landscape for cosmology. *Gen. Relativ. Gravit.* **2014**, *46*, 1649. [CrossRef]
8. Luongo, O.; Muccino, M. Speeding up the universe using dust with pressure. *Phys. Rev. D* **2018**, *98*, 103520. [CrossRef]
9. Caldwell, R.R.; Dave, R.; Steinhardt, P.J. Cosmological imprint of an energy component with general equation of state. *Phys. Rev. Lett.* **1998**, *80*, 1582–1585. [CrossRef]
10. Tsujikawa, S. Quintessence: A review. *Class. Quant. Grav.* **2013**, *30*, 214003. [CrossRef]
11. Moraes, P.H.R.S.; Santos, J.R.L. Two scalar field cosmology from coupled one-field models. *Phys. Rev. D* **2014**, *89*, 083516. [CrossRef]

12. Khurshudyan, M.; Chubaryan, E.; Pourhassan, B. Interacting Quintessence Models of Dark Energy. *Int. J. Theor. Phys.* **2014**, *53*, 2370–2378. [CrossRef]
13. Khurshudyan, M.; Pourhassan, B.; Myrzakulov, R.; Chattopadhyay, S. An effective quintessence field with a power-law potential. *Astrophys. Space Sci.* **2015**, *356*, 383–391. [CrossRef]
14. Umar Farooq, M.; Jamil, M.; Debnath, U. Dynamics of interacting phantom and quintessence dark energies. *Astrophys. Space Sci.* **2011**, *334*, 243–248. [CrossRef]
15. Harko, T.; Lobo, F.S.N.; Nojiri, S.; Odintsov, S.D. $f(R,T)$ gravity. *Phys. Rev. D* **2011**, *84*, 024020. [CrossRef]
16. Xu, M.-X.; Harko, T.; Liang, S.-D. Quantum cosmology of $f(R,T)$ gravity. *Eur. Phys. J. C* **2016**, *76*, 449. [CrossRef]
17. Myrzakulov, R. FRW cosmology in $f(R,T)$ gravity. *Eur. Phys. J. C* **2012**, *72*, 2203. [CrossRef]
18. Sharif, M.; Zubair, M. Study of Bianchi I anisotropic model in $f(R,T)$ gravity. *Astrophys. Space Sci.* **2014**, *349*, 457–465. [CrossRef]
19. Moraes, P.H.R.S.; Ribeiro, G.; Correa, R.A.C. A transition from a decelerated to an accelerated phase of the universe expansion from the simplest non-trivial polynomial function of T in the $f(R,T)$ formalism. *Astrophys. Space Sci.* **2016**, *361*, 227. [CrossRef]
20. Shamir, M.F. Locally rotationally symmetric Bianchi type I cosmology in $f(R,T)$ gravity. *Eur. Phys. J.* **2015**, *75*, 354. [CrossRef]
21. Sharma, N.K.; Singh, J.K. Bianchi type-II String cosmological model with magnetic field in $f(R,T)$ gravity. *Int J. Theor. Phys.* **2014**, *53*, 2912–2922. [CrossRef]
22. Mishra, B.; Tarai, S.; Pacif, S.K.J. Dynamics of Bianchi VI_h universe with bulk viscous fluid in modified gravity. *Int. J. Geom. Mod. Phys.* **2018**, *15*, 1850036. [CrossRef]
23. Mishra, B.; Tarai, S.; Tripathy, S.K. Dynamical features of an anisotropic cosmological model. *Int. J. Phys.* **2018**, *92*, 1199–1206. [CrossRef]
24. Tiwari, R.K.; Sofuoğlu, D. Quadratically varying deceleration parameter in $f(R,T)$ gravity. *Int. J. Geom. Methods Mod. Phys.* **2020**, *17*, 2030003. [CrossRef]
25. Tiwari, R.K.; Beesham, A.; Singh, R.; Tiwari, L.K. Time varying G and Λ cosmology in $f(R,T)$ gravity theory. *Astrophys. Space Sci.* **2017**, *362*, 143. [CrossRef]
26. Zubair, M.; Noureen, I. Evolution of axially symmetric anisotropic sources in $f(R,T)$ gravity. *Eur. Phys. J. C* **2015**, *75*, 265. [CrossRef]
27. Alfedeel, A.H.A.; Tiwari, R.K. A novel approach to Bianchi type–I cosmological model in $f(R,T)$ gravity. *Indian J. Phys.* **2021**, *96*, 1877–1885. [CrossRef]
28. Tiwari, R.K.; Sofuoğlu, D.; Mishra, S.K. Accelerating universe with varying Λ in $f(R,T)$ theory of gravity. *New Astron.* **2021**, *83*, 101476. [CrossRef]
29. Tiwari, R.K.; Beesham, A.; Shukla, B. Cosmological model with variable deceleration parameter in $f(R,T)$ modified gravity. *Int. J. Geom. Meth. Mod. Phys.* **2018**, *15*, 1850115. [CrossRef]
30. Sahoo, P.K.; Sahoo, P.; Bishi, B.K. Anisotropic cosmological models in $f(R,T)$ gravity with variable deceleration parameter. *Int. J. Geom. Meth. Mod. Phys.* **2017**, *14*, 1750097. [CrossRef]
31. Moraes, P.H.R.S.; Sahoo, P.K. The simplest non-minimal matter-geometry coupling in the $f(R,T)$ cosmology. *Eur. Phys. J. C* **2017**, *77*, 480. [CrossRef]
32. Sharma, L.K.; Yadav, A.K.; Sahoo, P.K.; Singh, B.K. Non-minimal matter-geometry coupling in Bianchi I space-time. *Results Phys.* **2018**, *10*, 738–742. [CrossRef]
33. Tiwari, R.K.; Sofuoglu, D.; Isik, R.; Shukla, B.K.; Baysazan, E. Non-minimally coupled transit cosmology in $f(R,T)$ gravity. *Int. J. Geom. Meth. Mod. Phys.* **2022**, *19*, 2250118. [CrossRef]
34. Shamir, M.F. Bianchi type-I cosmology in $f(R,T)$ gravity. *J. Exp. Theor. Phys.* **2014**, *119*, 242–250. [CrossRef]
35. Fisher, S.B.; Carlson, E.D. Reexamining $f(R,T)$ gravity. *Phys. Rev. D* **2019**, *100*, 064059. [CrossRef]
36. Perlmutter, S.; Gabi, S.; Goldhaber, G.; Goobar, A.; Groom, D.E.; Hook, I.M.; Kim, A.G.; Kim, M.Y.; Lee, J.C.; Pain, R.; et al. Measurements of the cosmological parameters Ω and Λ from the first seven supernovae at $z \geq 0.35$. *Astrophys. J.* **1997**, *483*, 565–581. [CrossRef]
37. Perlmutter, S.; Aldering, G.; Della Valle, M.; Deustua, S.; Ellis, R.S.; Fabbro, S.; Fruchter, A.; Goldhaber, G.; Goobar, A.; Groom, D.E.; et al. Discovery of a supernova explosion at half the age of the Universe. *Nature* **1998**, *391*, 51–54. [CrossRef]
38. Tiwari, R.K.; Beesham, A.; Shukla, B.K. Cosmological models with viscous fluid and variable deceleration parameter. *Eur. Phys. J. Plus* **2017**, *132*, 20. [CrossRef]
39. Tiwari, R.K.; Beesham, A.; Shukla, B.K. Scenario of a two-fluid FRW cosmological model with dark energy. *Eur. Phys. J. Plus* **2017**, *132*, 126. [CrossRef]
40. Tiwari, R.K.; Beesham, A.; Shukla, B.K. Behaviour of the cosmological model with variable deceleration parameter. *Eur. Phys. J. Plus* **2016**, *131*, 447. [CrossRef]
41. Berman, M.S. A special law of variation for Hubble's parameter. *Nuovo Cim. B* **1983**, *74*, 182–186. [CrossRef]
42. Berman, M.S.; de Mello Gomide, F. Cosmological models with constant deceleration parameter. *Gen. Relativ. Gravit.* **1988**, *20*, 191–198. [CrossRef]
43. Tiwari, R.K.; Singh, R.; Shukla, B.K. A cosmological model with variable deceleration parameter. *Afric. Rev. Phys.* **2015**, *10*, 395–402. Available online: http://lamp.ictp.it/index.php/aphysrev/article/view/1137/460 (accessed on 15 October 2022). [CrossRef]

44. Pradhan, A.; Goswami, G.K.; Beesham, A.; Dixit, A. An FLRW interacting dark energy model of the Universe. *New Astron.* **2020**, *78*, 101368. [CrossRef]
45. Goswami, G.K.; Pradhan, A.; Beesham, A. A dark energy quintessence model of the universe. *Mod. Phys. Lett. A* **2020**, *35*, 2050002. [CrossRef]
46. Giostri, R.; Vargas dos Santos, M.; Waga, I.; Reis, R.R.R.; Calvão, M.O.; Lago, B.L. From cosmic deceleration to acceleration: New constraints from SN Ia and BAO/CMB. *J. Cosmol. Astropart. Phys.* **2012**, 3, 27. [CrossRef]
47. Cunha, J.V. Kinematic constraints to the transition redshift from supernovae type Ia union data. *Phys. Rev. D* **2009**, *79*, 047301. [CrossRef]
48. Rapetti, D.; Allen, S.W.; Amin, M.A.; Blandford, R.D. A kinematical approach to dark energy studies. *Mon. Not. R. Astron. Soc.* **2007**, *375*, 1510. [CrossRef]

physics

MDPI

Article

Exploring Quantum Geometry Created by Quantum Matter

Abhay Ashtekar

Physics Department and Institute for Gravitation and the Cosmos, The Pennsylvania State University, University Park, PA 16802, USA; ava1@psu.edu

Abstract: Exactly soluble models can serve as excellent tools to explore conceptual issues in non-perturbative quantum gravity. In perturbative approaches, it is only the two radiative modes of the linearized gravitational field that are quantized. The goal of this investigation is to probe the 'Coulombic' aspects of quantum geometry that are governed entirely by matter sources. Since there are no gravitational waves in three dimensions, 3-dimensional (3-d) gravity coupled to matter provides an ideal arena for this task. The analysis presented here reveals novel aspects of quantum gravity that bring out limitations of classical and semi-classical theories in unforeseen regimes: non-linearities of general relativity can magnify small quantum fluctuations in the matter sector to large effects in the gravitational sector. Finally, this analysis leads to thought experiments that bring out rather starkly why understanding of the nature of physical reality depends sensitively on the theoretical lens with which it is probed. As theories become richer, new scales emerge, triggering novel effects that could not be imagined before. The model provides a concise realization of this well-known chain.

Keywords: non-perturbative quantum gravity; exactly soluble models; quantum nature of Coulombic interaction; trans-Planckian frequencies

Citation: Ashtekar, A. Exploring Quantum Geometry Created by Quantum Matter. *Physics* **2022**, *4*, 1384–1402. https://doi.org/10.3390/physics4040089

Received: 4 October 2022
Accepted: 4 November 2022
Published: 8 December 2022

1. Introduction

This paper has several different motivations stemming from various aspects of quantum gravity. A number of bold ideas have been put forward in 4-dimensional (4-d) quantum gravity that appear to be plausible from one perspective but puzzling, and even unsettling, from another. Exactly soluble models with local degrees of freedom (sometimes called midi-superspace in the literature) can be used to examine these issues *non-perturbatively*. Of course, one has to make simplifying assumptions to arrive at these models. Therefore, one can miss important aspects of the full theory and has to exercise due caution in assessing the conclusions reached. Nonetheless, these models can provide pointers and bring to the forefront new conceptual issues in a crisp manner.

One such issue concerns the potential implications of 'trans-Planckian modes' associated with extremely high frequency fields that arise, e.g., in the analysis of Hawking radiation. Over the years, there have been suggestions that an adequate handling of these modes may lead to novel physics by, e.g., having to abandon local Lorentz invariance (see, e.g., [1]). Exactly soluble models provide a natural arena to analyze this issue. Models discussed here, for example, have unforeseen effects induced by high frequency modes. It is then natural to ask if they lead to a loss of local Lorentz invariance in the final non-perturbative theory or if this symmetry is preserved by the new physics. Another example comes from the widely held view that the usual counting of states in quantum field theory becomes inadequate at high frequencies and the correct counting would lead to a holographic picture in which physics in the bulk is fully captured by states residing on the boundary (see, e.g., [2–4] for early works and [5] for a recent review). Is this view supported in exactly soluble models?

A third example comes from another commonly held view first introduced by Wheeler [6] that there is a 'space-time foam' at the microscopic level because of the perpetual quantum fluctuations of the gravitational field. In Wheeler's paradigm, these fluctuations are to emerge as novel non-perturbative effects associated with the quantum nature of geometry, rather than from the ultraviolet behavior of graviton-mediated interactions in perturbative approaches. This view then leads one to visualize the 'non-perturbative ground state' of full quantum gravity as having a very different micro-structure to that suggested by the smooth, tame geometry of Minkowski space and the associated perturbative vacuum. Is this idea borne out in exactly soluble models? If so, the standard positive energy theorem of classical general relativity may not extend to non-perturbative quantum gravity; the energy density in the ground state could well be Planck density with either sign. Is this what happens?

Exactly soluble models can also be well-suited to probe an issue at the opposite corner of the theory: semi-classical gravity. This is a mathematically well-defined theory in which gravity/geometry is treated classically and matter fields quantum mechanically, and the two are related by semi-classical Einstein's equations in which the right-hand side is provided by the expectation value of stress-energy of quantum matter. It is generally assumed that the theory would provide a good approximation to the predictions of quantum gravity so long as the space-time curvature is low compared to the Planck scale. Exactly soluble models can shed light on the validity of this conjecture. Can one arrive at this theory from full quantum gravity in the appropriate regime? Are there potential surprises about its domain of validity?

Another motivation for this work comes from the ongoing debate on whether quantum matter can genuinely inject quantum information into the gravitational field or if it is only the radiative, or 'true', degrees of the gravitational field that are quantum mechanical (see, e.g., [7–10]). The soluble models discussed here can be thought of as three-dimensional general relativity coupled to the Maxwell (or Klein–Gordon) fields. The three-dimensional viewpoint provides a conceptually clean arena to probe this issue since the gravitational field is now completely determined by matter; there are no gravitons (or gravitational waves) in three dimensions. Nonetheless, quantum geometry does exhibit specific and quintessentially quantum features. In fact, the nonlinear coupling between matter and gravity can magnify small quantum fluctuations in the matter sector to large quantum effects in space-time geometry, leading to consequences that could not have been foreseen classically.

Finally, these soluble models also suggest a thought experiment that yields more general conceptual insights. The understanding of physical reality is deeply influenced by the theoretical paradigm one chooses to work in. Indeed, possibilities envisaged and experiments designed to test subsequent hypotheses are dictated in this paradigm. For the specific models we discuss, two constants of nature play a fundamental role: Planck's constant, $\hbar$, which dictates the size of quantum effects, and Newton's constant, G, which governs gravitational phenomena. We will find that 'switching them on and off' results in drastic changes in our understanding of the underlying physical reality.

The material is organized as follows. Section 2 discusses the classical 'midi-superspace models' and, Section 3, the unforeseen quantum gravity effects in these models. Section 4 summarizes the main results and puts them in a broader context. The key mathematical results presented here were summarized in a letter [11] quite some time ago. The current paper provides detailed derivations that have been been circulated only in private communications so far and examines the results from different angles, putting them in a broader conceptual perspective. Since significant time has lapsed, the paper also discusses related results in other that have appeared since (in particular, in [7–10,12–18]) as well as the above mentioned thought experiment, which has not appeared in the literature.

In the main discussion, 2 + 1 dimensional general relativity coupled with a scalar or Maxwell field is used. Space-time metric g_{ab} has the signature $(-,+,+)$ and its derivative operator is denoted by ∇, its Riemann tensor is defined by $R_{abc}{}^{d}k_d = 2\nabla_{[a}\nabla_{b]}k_c$, its Ricci tensor by $R_{ac} = R_{abc}{}^{b}$, its scalar curvature by $R = g^{ac}R_{ac}$, and its Einstein tensor by G_{ab}. We set $c = 1$ but display the dependence of various quantities on G and $\hbar$. For the convenience of the reader, let us note that, in 3-d, Newton's constant has dimensions $[G] \sim M^{-1}$. Therefore, unlike in 4-d general relativity, there is a natural mass scale in the classical theory. However, one cannot associate a given mass $M_\circ$ with a length; there is no notion of 'Schwarzschild radius' of $M_\circ$. It is also helpful to keep in mind the differences that arise in quantum gravity from the more familiar 4-d situation. While there is a natural notion of Planck length $([\hbar G] \sim L)$ and, therefore, of Planck frequency, there is no notion of Planck mass that would be relevant to quantum gravity. Section 2 briefly discusses how the 3-d models considered here arise from a symmetry reduction of 4-d general relativity, à la Kaluza–Klein. There, the 4-dimensional fields explicitly carry a prefix of 4.

2. Exactly Soluble Models

In this Section, we first recall the model of primary interest, and then briefly discuss other similar models that have been considered in the literature. These are often referred to as 'midi-superspace models' because they result from a symmetry reduction of 4-d general relativity but have an infinite number of true degrees of freedom (in contradistinction with the 'mini-superspace' of homogeneous cosmological models, which have only a finite number of degrees of freedom). While these models and their structure has been discussed in the literature, their salient features are reviewed here because the paper is intended for a broad audience. This discussion also enables us to fix the notation used.

As mentioned above, in most of this paper, we work primarily in three space-time dimensions and with the signature (−,+,+). Let us denote this space-time by (M, g_{ab}). M is assumed to be topologically R^3 and the metric g_{ab} to be asymptotically flat both at null and spatial infinity in the sense of Refs. [17–19], respectively. Matter fields are taken to be either Maxwell fields, F_{ab} or Klein–Gordon fields, Φ. In 3-d, there is a well-known duality between the two. Maxwell's equations imply that F_{ab} satisfies Maxwell's equations if and only if its the dual ${}^\star F_a := \frac{1}{2}\,\epsilon_a{}^{bc} F_{bc}$ is exact, i.e., ${}^\star F_a =: \nabla_a \Phi$, where Φ satisfies the Klein–Gordon equation, $\nabla^a \nabla_b \Phi = 0$, and $\epsilon_a{}^{bc}$ is the totally anti-symmetric, non-degerate tensor field, defined by g_{ab}. (Here, the fact that the topology of M is trivial has been used.) Interestingly, the dictionary ${}^\star F_a := \nabla_a \Phi$ translates the Maxwell stress-energy tensor, $T_{ab}^{\rm Max}$, to the Klein–Gordon stress-energy, $T_{ab}^{\rm KG}$:

$$\begin{aligned} T_{ab}^{\rm Max} &\equiv F_a{}^m\, F_{bm} - \frac{1}{4} g_{ab}\, F_{mn} F^{mn} = {}^\star F_a\, {}^\star F_b - \frac{1}{2} {}^\star F^m\, {}^\star F_m\, g_{ab} \\ &= \nabla_a \Phi\, \nabla_b \Phi - \frac{1}{2} \nabla^m \Phi \nabla_m \Phi\, g_{ab} \equiv T_{ab}^{\rm KG}\,. \end{aligned} \tag{1}$$

Therefore, the pair $(g_{ab},\, \Phi)$ satisfies the coupled Klein–Gordon *and* Einstein's equations if and only if the pair $(g_{ab},\, F_{ab})$ satisfies the Einstein–Maxwell equations. For notational simplicity, we work with a Klein–Gordon field Φ in calculations. However, conceptually, it is often more convenient to regard the source as a Maxwell field F_{ab} and its quantum excitations as photons.

Thus, the Klein–Gordon field Φ and the space-time metric satisfy the following coupled set of equations:

$$\Box \Phi = 0 \qquad \text{and} \qquad G_{ab} = 8\pi G\, T_{ab} \quad \Leftrightarrow \quad R_{ab} = 8\pi G\, \nabla_a \nabla_b \Phi\,. \tag{2}$$

Hamiltonian analysis of this system has also been carried out in some detail elsewhere [20]. However, because Equations (2) are rather complicated partial differential equations, one still does not have good control on the detailed properties of their solutions. However, it is known that equations simplify greatly if one restricts oneself to the axisymmetric sector of this system consisting of solutions, in which g_{ab} admits a rotational Killing field, $\partial_\theta \partial / \partial\theta$, with a regular axis, and Φ is Lie-dragged by the Killing field. Then, one can cast the space-time metric in the form

$$g_{ab}\, {\rm d}x^a\, {\rm d}x^b = e^{G\Gamma(R,T)}\, ({\rm d}T^2 + {\rm d}R^2) + R^2 {\rm d}\theta^2\,, \tag{3}$$

where $T \in (-\infty, \infty)$ and $0 \le R < \infty$. The chart is unique up to the transformation $T \to T + \text{const}$, and, by inspection, R is the norm of the rotational Killing field. Note that if the metric coefficient, $\Gamma(R,T)=0$, the metric becomes the Minkowski metric, $\mathring{g}_{ab}$. The second consequence of the assumption of axisymmetry is technically more powerful. It is straightforward to check that a scalar field Φ satisfies the Klein–Gordon equation, $\Box\Phi=0$, with respect to the metric g_{ab} if and only if it satisfies $\mathring{\Box}\Phi=0$ with respect to the Minkowski metric $\mathring{g}_{ab}$ where $\mathring{\Box}$ is the d'Alembertian defined by the Minkowski metric $\mathring{g}_{ab}$. Therefore, the two equations in Equation (2) now decouple: one can just solve the wave equation $\mathring{\Box}\Phi=0$ in Minkowski space, obtain a solution Φ, and then use it in $R_{ab}=8\pi G\,\nabla_a\Phi\nabla_b\Phi$ to determine the only unknown $\Gamma(R,T)$ in the metric physical g_{ab}. The final simplification is that this last equation can be readily solved to obtain

$$\Gamma(R,T)=\frac{1}{2}\int_0^R \mathrm{d}\bar{R}\,\bar{R}\;[(\partial_T\Phi)^2+(\partial_{\bar{R}}\Phi)^2](\bar{R},T)\,. \tag{4}$$

Note that the right side has a straightforward interpretation: it is just the energy of the scalar field Φ with respect to the *Minkowski metric* on a disc of radius R at time T. Therefore, in the axisymmetric case, there is a large class of asymptotically flat solutions on which one has excellent control [17,18]. This is our midi-superspace.

To summarize, in the midi-superspace of interest, to solve field Equation (2), one only needs an axisymmetric solution $\Phi(R,T)$ to the Klein–Gordon equation in *Minkowski space-time*, $(M,\mathring{g}_{ab})$. Given Φ, one can just write down the metric of Equation (3) such that the pair (g_{ab},Φ) solves the coupled Einstein–Klein–Gordon equations. This is the *general* solution in the axisymmetric sector of 3-d gravity coupled to a Klein–Gordon (or Maxwell) field. Finally, recall that a general solution, Φ of $\mathring{\Box}\Phi=0$ can be readily constructed by going to the frequency domain: it is given as an expansion on the basis $f^+_\omega(R,T):=J_0(\omega R)e^{-i\omega T}$ of positive frequency solutions, where J_0 is the zeroth order Bessel function of first kind:

$$\Phi(R,T)=\int_0^\infty \mathrm{d}\omega\,\left[\phi(\omega)f^+_\omega(R,T)\,+\,\bar{\phi}(\omega)\bar{f^+}_\omega(R,T)\right]. \tag{5}$$

This is the precise sense, in which the midi-superspace under consideration is an exactly soluble sector of 3-d gravity (and, as discussed below, also of 4-d vacuum gravity). For each choice of a regular function $\phi(\omega)$, one obtains a solution $\Phi(R,T)$ to $\mathring{\Box}\Phi=0$, which then determines the only unknown coefficient $\Gamma(R,T)$ in the metric (3) such that (Φ,g_{ab}) satisfy the coupled Einstein–Klein–Gordon system (2). This is a midi-superspace because the system has one local degree of freedom that is neatly coded in the function $\phi(\omega)$ in the above formulation.

To make the conceptual considerations sharper, it is often convenient to focus on solutions, in which $\Phi(R,T)$ has initial data of compact support on a Cauchy slice (although this restriction can be relaxed by allowing the much milder fall-off specified in [19]). It then follows that the $\Phi(R,T)$ vanishes outside some radius $R=R_\circ(T)$ for each T. Denote by M' the complement of the support of the given solution Φ. In the region M', one has

$$\begin{aligned}\Gamma(R,T) &= \frac{1}{2}\int_0^\infty \mathrm{d}R\,R\;[(\partial_T\Phi)^2+(\partial_R\Phi)^2](R,T)\\ &= \int_0^\infty \mathrm{d}\omega\,\omega\,|\phi(\omega)|^2 =: H_\circ(\phi)\,,\end{aligned} \tag{6}$$

where, in the first step, the initial integral over the interval $R\in(0,R_\circ(T))$ could be extended to that on $R\in(0,\infty)$ because $\Phi(R,T)$ vanishes for $R>R_\circ(T)$. Note that $H_\circ(\phi)$ is a constant because it equals the total, conserved energy of the solution Φ with respect to the Minkowski metric, $\mathring{g}_{ab}$; $H_\circ(\phi)$ is independent of T. Therefore, on M', the metric assumes the form,

$$g_{ab}\,\mathrm{d}x^a\,\mathrm{d}x^b\,|_{M'}\;=\;e^{GH_\circ}\,(\mathrm{d}T^2+\mathrm{d}R^2)\;+\;R^2\mathrm{d}\theta^2\,, \tag{7}$$

showing that the metric g_{ab} is flat in this region. This also follows more directly from the fact that the Ricci tensor R_{ab} vanishes in the absence of sources and, in 3-d, R_{abcd} is determined entirely by R_{ab}. However, unless $H_\circ = 0$—i.e., unless Φ vanishes identically—the physical metric g_{ab} is not globally flat. Also, although it is smooth everywhere on all of M, it has a 'conical structure with a deficit angle' in M'. Note also that g_{ab} admits a time translation Killing vector ∂/∂_T on M', in addition to the rotational Killing vector. The solution is asymptotically flat both at spatial and null infinity [17–19].

Hamiltonian analysis of this midi-superspace [19,21] reveals a striking result: The Hamiltonian $H(g,\Phi)$, i.e., the generator of the asymptotic time translation for the full system, is *not* the energy $H_\circ(\phi)$ of the Klein–Gordon field in Minkowski space but is *non-polynomially* related to it:

$$H(g,\Phi) = \frac{1}{4G}\left(1 - e^{-4GH_\circ(\phi)}\right). \tag{8}$$

This is the physical energy of the system in the Einstein–Klein–Gordon theory; as with the Arnowitt–Deser–Misner (ADM) energy in 4-d, it includes all contributions, including those from gravity. (While the Hamiltonian analysis refers to spatial infinity, H is also the past limit of the Bondi energy at null infinity [17].) H is manifestly positive and, in striking contrast to $H_\circ$, it is *bounded from above*. This property is *genuinely non-perturbative*: if one were to expand H in powers of G, to the leading order G^0, one would obtain $H_\circ$; if one truncates the series to any finite order, the expression would again be unbounded from above. It is only when one sums all the terms that one finds that H is bounded. For a flat conic space-time, this result was obtained by Henneaux already in 1983 [22] and for geometries sourced by $\mathcal{N}$ point particles, by Deser, Jackiw, and 't Hooft [23] in 1984. But note that positivity and boundedness of H is a general result in 3-d general relativity, so long as the matter satisfies the positive energy condition, $T_{ab}\, t_1^a\, t_2^b \geq 0$, for all time-like vectors t_1^a, t_2^a [19]. Finally, note that the non-perturbative form (8) is possible because the 3-d Newton's constant has physical dimensions of inverse mass.

In 4-d gravity, the ADM or the Bondi energy is expressible as a 2-surface 'charge-integral' at infinity that involves only geometrical/gravitational fields. In the 3-d case, the expression (8) of energy, on the other hand, involves $H_\circ$, which is a bulk integral involving just the matter fields. However, since it measures the deficit angle of the metric g_{ab} in M', it has a geometric interpretation (see [23] for geometries sourced by point particles). Moreover, as explained below, it can also be expressed as a *line integral involving only geometry*. Since the isometry group in the tangent space of any point of (M, g_{ab}) is the 3-d Lorentz group, there is a natural notion of $SU(1,1)$ spinors, λ_A, where the capital Latin letters indexes, e.g., $A, B, \ldots$, refer to the 2-dimensional spin space. Parallel transporting any λ_A at a point $p \in M$ along a closed curve (C) starting and ending at p yields another spinor λ'_A at p, related to the original λ_A via $\lambda'_A = U^{(C)}{}_A{}^B\,(p)\,\lambda_B$, where $U^{(C)}{}_A{}^B\,(p)$ is an $SU(1,1)$ rotation. Now, the trace, $\mathrm{Tr}\, U^{(C)} = U^{(C)}{}_A{}^A(p)$, is independent of the choice of point p on the curve. Furthermore, since the curvature of g_{ab} vanishes in M', $\mathrm{Tr}\, U^{(C)}$ is also independent of the choice of the curve (C) that lies entirely in M'. Let us restrict to such curves. Then, the trace is a function only of the total energy $H(g,\Phi)$ of the following system:

$$\mathrm{Tr}\, U^{(C)}(g) = 2\cos\left[\frac{\pi H(g,\phi)}{4G}\right]. \tag{9}$$

Consequently, there is an elementary thought experiment that one can make in the asymptotic region to determine H and, therefore, also the metric coefficient, g_{RR} (or G_{TT}), in the source-free region M': transport particles with spin $\frac{1}{2}$ around any closed loop C in M', measure the $SU(1,1)$ rotation $U_C(p)$ they undergo, and take the trace. Since Φ vanishes on M', this experiment can be carried out *without reference to matter fields*. This experiment will be used again in Section 4.

Finally, there is another striking feature of 3-d gravity: the absence of gravitational collapse leading to the formation of black holes. This is related to a fact noted above: in

3-d, there is no analog of the Schwarzschild radius associated with a given mass $M_\circ$. In the midi-superspace considered here, this feature is seen in detail: for regular initial data for Φ (e.g., of compact support), no matter how densely one packs the scalar field is packed, one obtains a non-singular solution g_{ab} on all of $M = R^3$ without any horizons.

Let us conclude with a number of remarks addressed primarily to the general relativity community.

Remarks

1. An important difference in the asymptotic structure of 4-d and 3-d gravity is the following. In 4-d, one can and does ask that all physical metrics g_{ab} of interest approach a *fixed* Minkowski metric g°_{ab} asymptotically and the information about the mass is encoded in the leading deviation. In 3-d, noted above, g_{ab} does not approach the fixed Minkowski metric even to the zeroth order because of the factor involving $H_\circ$ that varies from one space-time to another, depending on its energy content. This fact complicates the discussion of the asymptotic symmetry group in both spatial [19] and null [17] regimes. In particular, the notion of 'time translation' is now more subtle. This issue of time is discussed in [14,21].

2. The model presented here is a symmetry reduction of the celebrated Einstein–Rosen waves of 4-d, vacuum general relativity [24,25]. These solutions have cylindrical symmetry, i.e., two hypersurface orthogonal, commuting Killing vector fields; a spatial translation, for example, along the z direction and a rotation around the z axis. The Kaluza–Klein reduction is carried out for the translational Killing field with norm, e.g., e^ψ, and the 3-d scalar field Φ is given by $\Phi = \psi/4\pi G$ [17]. The rotational Killing field of ${}^4g_{ab}$ descends from the 4-manifold ${}^4M$ to the 3-manifold M and becomes a Killing field of the 3-metric g_{ab}. Einstein's vacuum equations ${}^4R_{ab} = 0$ in 4-d are equivalent to the 3-d field in Equation (2). For details, especially of the asymptotic analysis from a 4-d perspective, see Ref. [18]. In the context of Einstein–Rosen waves, $H_\circ$ is called the 'c-energy' and is sometimes interpreted as the energy 'per unit length along the z-axis' in these waves. Careful Hamiltonian analysis [19,21] shows that it is more appropriate to assign this interpretation to H.

3. There is an interesting variation [12] of the midi-superspace of 4-d Einstein–Rosen waves in which the space-time topology is $R^2 \times T^2$ rather than R^4 and, consequently, the corresponding $\mathcal{I}^+$ has topology $T^2 \times R$. In the Kaluza–Klein reduction, the 3-d space-time has topology $R^2 \times S^1$, rather than R^3. Locally, the structure of equations is the same as the one discussed in this Section, but there are global differences. In particular, there is a right and a left spatial infinity. This model has been analyzed from a Hamiltonian perspective in detail and then used as a point of departure for non-perturbative quantization in [15]. The canonical analysis follows the same lines as [21] but with interesting conceptual differences that arise from the difference in boundary conditions. Another interesting exactly soluble model is provided by Gowdy space-times, where the spatial topology is that of a 3-torus T^3. Here, the midi-superspace is significantly richer in that the 4-d Killing vector used in the Kaluza–Klein reduction is not hypersurface orthogonal. Consequently, there is an additional local degree of freedom in the 3-dimensional description. The phase space and non-perturbative canonical quantization of this model is discussed in [14] in the setting of 4-dimensional connection dynamics.

3. Non-Perturbative Quantum Theory

This Section is divided into three parts. In the first, the quantum framework is introduced; the second part discusses the unforeseen results on the nature of quantum geometry that follow from this framework; and the third part introduces a thought experiment that brings out the deep interplay between our notion of physical reality and the theoretical paradigm used to frame it.

3.1. The Framework

Classically, the model is exactly soluble because the matter field, Φ, and the dynamical space-time metric, g_{ab} could be decoupled. This procedure led us to encode the unconstrained degree of freedom of the total system in Φ satisfying $\mathring{\Box}\Phi = 0$, or, equivalently, in the freely specifiable function $\phi(\omega)$. Therefore, a natural strategy is to first focus on Φ and quantize it, and then investigate the nature of the quantum geometry $\hat{g}_{ab}$ it determines. This procedure makes the underlying mathematical structure simpler since the first step involves just quantum fields in 3-d Minkowski space-time (M, g°_{ab}).

Let us first collect the equations from the textbook quantization of the Klein–Gordon field $\Phi(R,T)$ in the 3-d Minkowski space-time (M, g°_{ab}) that are needed. The positive and negative frequency expansion (5) of the solution immediately leads to the 'field operator'

$$\hat{\Phi}(R,T) = \int_0^\infty \mathrm{d}\omega \, \left[\phi^+_\omega(R,T)\,\hat{A}(\omega) + \overline{\phi^+}_\omega(R,T)\,\hat{A}^\dagger(\omega)\right], \tag{10}$$

where $\hat{A}(\omega)$ and $\hat{A}^\dagger(\omega)$ are the annihilation and creation operators satisfying $[\hat{A}(\omega),\, \hat{A}^\dagger(\omega')]$ = $\hbar\,\delta(\omega,\omega')$, where $\delta(x,y)$ is there Dirac distribution. These operators act the symmetric Fock space, $\mathcal{F}$, where the underlying 1-particle Hilbert space, $\mathcal{H}$, is spanned by $\phi(\omega)$ or, equivalently, classical solutions, $\Phi(R,T)$, *which have a finite norm*:

$$||\phi||^2 := \frac{1}{\hbar}\int_0^\infty \mathrm{d}\omega \, |\phi(\omega)|^2 . \tag{11}$$

Note that while solutions $\Phi(R,T)$ and their 'frequency components' $\phi(\omega)$ are classical concepts that make no reference to $\hbar$, the presence of $\hbar$ is essential in expression (11) since the norm has to be dimensionless; (11) is a quintessentially quantum expression. The Hamiltonian operator $\hat{H}_\circ$ has the standard expression,

$$\hat{H}_\circ = \int_0^\infty \mathrm{d}\omega \, \omega \hat{A}^\dagger(\omega)\,\hat{A}(\omega) . \tag{12}$$

(There is no explicit factor of $\hbar$ because $[\hat{A}(\omega),\, \hat{A}^\dagger(\omega')] = \hbar\,\delta(\omega,\omega')$). Now, in the free field theory, coherent states play a key role in the discussion of the relation of quantum and classical theories: Given a classical solution $\mathring{\Phi}(R,T)$, or equivalently, $\mathring{\phi}(\omega)$, one has a normalized coherent state $|\Psi_{\mathring{\phi}}\rangle$:

$$\begin{aligned} |\Psi_{\mathring{\phi}}\rangle &= N\, e^{\frac{1}{\hbar}\int \mathrm{d}\omega\, \mathring{\phi}(\omega) \hat{A}^\dagger(\omega)}\, |0\rangle , \\ \text{where} \quad N &= e^{\frac{1}{2\hbar}\int \mathrm{d}\omega\, |\mathring{\phi}(\omega)|^2} \quad \text{is the normalization constant.} \end{aligned} \tag{13}$$

Recall that the expectation values of $\hat{\Phi}(R,T)$ and $\hat{H}$ in this state just yield the classical solution $\mathring{\Phi}(R,T)$ and its energy:

$$\langle\Psi_{\mathring{\phi}}|\hat{\Phi}(R,T)|\Psi_{\mathring{\phi}}\rangle = \Phi_\circ(R,T) \qquad \text{and} \qquad \langle\Psi_{\mathring{\phi}}|\hat{H}_\circ(R,T)|\Psi_{\mathring{\phi}}\rangle = H_\circ(\mathring{\phi}) . \tag{14}$$

Furthermore, the state $|\Psi_{\mathring{\phi}}\rangle$ is sharply peaked at the classical solution $\Phi_\circ(R,T)$ in the sense that the uncertainty in the field and its momentum is saturated and equally distributed in the appropriate sense at all times T. The uncertainty in the Hamiltonian $H_\circ$ is given by

$$(\Delta\hat{H}_\circ)^2 = \hbar \int_0^\infty \mathrm{d}\omega\, \omega^2\, |\mathring{\phi}(\omega)|^2 \quad \text{and} \quad \frac{(\Delta\hat{H}_\circ)}{(\langle\hat{H}\rangle)} \approx \frac{1}{\left(\frac{1}{\hbar}\int_0^\infty \mathrm{d}\omega\, |\mathring{\phi}|^2(\omega)\right)^{\frac{1}{2}}} = \frac{1}{\langle\hat{\mathcal{N}}\rangle^{\frac{1}{2}}} . \tag{15}$$

In the second equation, it is assumed that the profile $\mathring{\phi}(\omega)$ is sharply peaked at some fixed frequency to bring out the physical meaning of the right side and denoted the number operator on the Fock space, $\mathcal{F}$, by $\hat{\mathcal{N}}$. Thus, the relative uncertainty is proportional to the inverse of the square root of the expected number of 'photons' in the coherent state under consideration. The more intense the classical beam $\mathring{\phi}(\omega)$—i.e., the greater the value of $\int \mathrm{d}\omega\, |\mathring{\phi}|^2(\omega)$ relative to $\hbar$—the less the relative uncertainty in the quantum state and more trustworthy are the classical results vis a vis the correct quantum answer. Note that the frequency, at which $\mathring{\phi}(\omega)$ is peaked is irrelevant in this consideration; the classical approximation will continue to be excellent one even if the peak is shifted to arbitrarily high frequencies as long as $\int_0^\infty \mathrm{d}\omega\, |\mathring{\phi}|^2(\omega) \gg \hbar$. This is a standard result in quantum optics.

All these considerations referred only to the matter sector—either a scalar field Φ, or a Maxwell field, $F_{ab} = \epsilon_{abc}\nabla^c\Phi$—which, thanks to the integrability of the model, could be analyzed without knowing the physical metric g_{ab}. Now, in the classical theory, the physical metric g_{ab} is determined completely by Einstein's equations for any given matter field $\Phi(R,T)$. In the quantum theory, one can let the true degree of freedom reside in $\hat{\Phi}(R,T)$, represented as an operator (10) on $\mathcal{F}$. Therefore, the metric coefficient $g_{RR} = -g_{TT}$ in the asymptotic region, namely, $e^{GH_\circ}$ of Equation (7), also becomes an operator on $\mathcal{F}$:

$$\hat{g}_{RR}\,\big|_{\mathrm{asym}} := e^{G\hat{H}_\circ} = e^{G\int_0^\infty \mathrm{d}\omega\,\omega\,\hat{A}^\dagger(\omega)\,\hat{A}(\omega)}\,. \tag{16}$$

Similarly, the non-perturbative Hamiltonian H of the total system—matter *and* gravity—becomes the operator,

$$\hat{H} := \frac{1}{4G}\left(1 - e^{-4G\hat{H}_\circ}\right) \tag{17}$$

on $\mathcal{F}$. These operators encode the non-perturbative information contained in gravity–matter interactions.

Generally, non-trivial models are exactly soluble because they can be mapped to a trivial model. In the case under consideration, the Einstein–Klein–Gordon (or Einstein–Maxwell) system (g_{ab}, Φ) could be mapped to a free field, $\Phi(R,T)$, in Minkowski space. Non-triviality is then transferred to the map that *relates the two models*.The relation is now given by Equations (7) and (8) in the classical theory and Equations (16) and (17) in the quantum theory. As shown in Section 3.2, these equations have certain unforeseen consequences that bring out the non-triviality of the original Einstein-matter system.

Remarks

1. While defining various operators, one encounters the issue of factor ordering. For the free Hamiltonian $\hat{H}_\circ$ of the scalar field $\hat{\Phi}$ in Minkowski space, there is an unambiguous answer: normal ordering used in Equation (12). Relation (7) between the classical $H_\circ$ and metric coefficients $g_{RR} = -g_{TT}$ and relation (8) between $H_\circ$ and the Hamiltonian H of the full system led us to define quantum operators $\hat{g}_{RR}|_{\mathrm{asym}}$ and $\hat{H}$. However, it is natural to ask whether one could also use other choices. For example, could one not have normal ordered the operator *after* exponentiation and used $\hat{H}' = :\hat{H}:$ instead? The operator $\hat{H}'$ shares several attractive properties with $\hat{H}$. In particular, it annihilates the vacuum state $|0\rangle \in \mathcal{F}$ and its expectation value in coherent states $|\Psi_{\mathring{\phi}}\rangle$ also yields the value of the classical Hamiltonian $H(\mathring{\phi})$. However, in sharp contrast to $\hat{H}$, the spectrum of $\hat{H}'$ is the *entire real line*. Given that the classical observable H takes values only in the finite interval $[0, \frac{1}{4G}]$, $\hat{H}'$ is an inadmissible choice for the corresponding quantum observable. The choice made in Equations (16) and (17) is free of such drawbacks. In particular, the spectrum of $\hat{H}$ is precisely $[0, \frac{1}{4G}]$.

2. The 'field operator' $\hat{\Phi}(R,T)$ of Equation (10) is actually an operator-valued distribution. Consequently, there are technical subtleties in giving precise mathematical meaning to $\hat{g}_{RR} = -\hat{g}_{TT}$ in full space-time (discussed, e.g., in Ref. [21]). However, for conceptual issues that are at the forefront of the current analysis, one could bypass this issue by focusing only on the asymptotic region, where the classical matter fields $\Phi_\circ(R,T)$, at which the coherent states $|\Psi_{\mathring{\phi}}\rangle$ are peaked, vanish. More precisely, careful treatment yields the same results as those obtained in this Section by restricting the discussion to the asymptotic region from the beginning.

3. In the present framework, the true degree of freedom lies in the scalar field $\Phi(R,T)$. Therefore, it was natural to construct the Hilbert space of states for this field and represent geometric observables by appropriate operators thereon. However, one might be concerned that this procedure is somewhat unnatural from the perspective of general relativity and *only* the physical metric g_{ab} should be used in the quantization procedure. This is indeed

possible using a canonical quantization procedure, which refers only to the physical metric as was first discussed in the 4-d context of cylindrically symmetric waves in a remarkably early work by Kuchař [26], and later in the 3-d context in [27] (albeit, without realizing that the second is a Kaluza–Klein reduction of the first). However, these investigations overlooked subtle issues related to boundary conditions and the distinction between diffeomorphisms representing gauge and the true dynamics they imply. Careful handling of these issues is needed to put the Hamiltonian theory and the canonical quantization procedure on a firm footing. When this is done, the framework also brings out subtleties associated with the issue of time. Finally, the Hilbert space of quantum states can also be selected without having to introduce the Minkowski metric $\mathring{g}_{ab}$. For details on both of these points, see [21].

3.2. Unforeseen Quantum Gravity Effects

The Hilbert space of quantum states is the Fock space $\mathcal{F}$, constructed entirely from the matter sector. As shown above, if one uses observables—such as $\hat{\Phi}$ and $\hat{H}_\circ$—that refer only to the matter sector, then the classical theory is an excellent approximation if the system is in a coherent state $|\Psi_{\mathring{\phi}}\rangle$. In these states, the expectation values of quantum observables equal values of their classical counterparts and their quantum fluctuations are negligible provided that the expected number of 'photons', $\langle\Psi_{\mathring{\phi}}|\hat{\mathcal{N}}|\Psi_{\mathring{\phi}}\rangle$, is large, irrespective of the choice of $\mathring{\phi}$. Now, observables—such as $\hat{g}_{RR}$ that encodes the quantum metric and $\hat{H}$ that represents the Hamiltonian $\hat{H}$ of the full system, including gravity—are also represented by operators on $\mathcal{F}$. In the classical theory, values of these observables can be computed using the matter sector alone. Therefore, one's first expectation would be that the classical theory would again provide an excellent approximation for these observables if the system is in a coherent state $|\Psi_{\mathring{\phi}}\rangle$. However, as shown in this Subsection, this expectation is not borne out. Even though geometry is completely determined by matter and coherent states $|\Psi_{\mathring{\phi}}\rangle$ are sharply peaked on classical configurations $\mathring{\phi}$, quantum properties of gravitational observables in these states can be very different from their classical analogs. For brevity, from now on, the suffix 'asym' in $g_{RR}\,|_{\rm asym}$ is dropped since the subsequent discussion refers only to the asymptotic metric.

Let us begin with the action of the operator $\hat{g}_{RR} = e^{G\hat{H}_\circ}$ on coherent states:

$$\begin{aligned}
e^{G\hat{H}_\circ}|\Psi_{\mathring{\phi}}\rangle &= N\, e^{G\hat{H}_\circ}\, e^{\frac{1}{\hbar}\int_0^\infty \mathrm{d}\omega\, \mathring{\phi}(\omega)\, \hat{A}^\dagger(\omega)}\ |0\rangle \\
&= N\, e^{G\hat{H}_\circ} \sum_{n=0}^{\infty} \frac{1}{\hbar^n\, n!} \int_0^\infty \mathrm{d}\omega_1 \ldots \mathrm{d}\omega_n\ \mathring{\phi}(\omega_1)\ldots\mathring{\phi}(\omega_n)\ |\omega_1,\ldots\omega_n\rangle \\
&= N \sum_{n=0}^{\infty} \frac{1}{\hbar^n\, n!} \int_0^\infty \mathrm{d}\omega_1 \ldots \mathrm{d}\omega_n\ \mathring{\phi}(\omega_1)\ldots\mathring{\phi}(\omega_n)\ e^{G\hbar(\omega_1+\ldots+\omega_n)}\ |\omega_1,\ldots\omega_n\rangle \\
&= N \sum_{n=0}^{\infty} \frac{1}{\hbar^n\, n!} \int_0^\infty \mathrm{d}\omega_1 \ldots \mathrm{d}\omega_n\ \mathring{\varphi}(\omega_1)\ldots\mathring{\varphi}(\omega_n)\ \ |\omega_1,\ldots\omega_n\rangle\,,
\end{aligned} \tag{18}$$

where, as before, $N = e^{\frac{1}{2\hbar}\int \mathrm{d}\omega\, |\mathring{\phi}(\omega)|^2}$ is the normalization constant for the coherent state, and $\mathring{\varphi}(\omega)$ in the last step is given by $\mathring{\varphi}(\omega) = e^{G\hbar\omega}\, \mathring{\phi}(\omega)$. Using the definition of the normalized coherent state $|\Psi_{\mathring{\varphi}}\rangle$, one can express the result as

$$e^{G\hat{H}_\circ}|\Psi_{\mathring{\phi}}\rangle = e^{\frac{1}{2\hbar}\int_0^\infty \mathrm{d}\omega\, |\mathring{\phi}(\omega)|^2 (e^{2G\hbar\omega}-1)}\ |\Psi_{\mathring{\varphi}}\rangle\,. \tag{19}$$

Thus, the action *shifts* the peak of the coherent state from $\mathring{\phi}$ to $\mathring{\varphi}$ and multiplies it by a constant. Note that the shift rescales $\mathring{\phi}(\omega)$ by a factor that is *exponential* in $G\hbar\omega$. Finally, since the inner-product between these two coherent states is given by

$$\langle\Psi_{\mathring{\phi}}|\Psi_{\mathring{\varphi}}\rangle = e^{-\frac{1}{2\hbar}\int_0^\infty \mathrm{d}\omega |\mathring{\phi}(\omega)|^2 (1+e^{2G\hbar\omega}-2e^{G\hbar\omega})}\,, \tag{20}$$

the expectation value of $\hat{g}_{RR}|_{\rm asym}$ is given by

$$\langle\Psi_{\mathring{\phi}}|\,\hat{g}_{RR}\,|\Psi_{\mathring{\phi}}\rangle \;=\; e^{\frac{1}{\hbar}\int_0^\infty d\omega\,|\mathring{\phi}(\omega)|^2\,(e^{G\hbar\omega}-1)}\,, \tag{21}$$

in contrast to the classical value,

$$g_{RR} = e^{G\int_0^\infty d\omega\,\omega\,|\mathring{\phi}(\omega)|^2}\,. \tag{22}$$

There are several notable differences between the expectation value (21) and the classical value (22). First, even though the expectation value is evaluated in a coherent state, it depends explicitly on $\hbar$. Thus, unlike $\langle\Psi_{\mathring{\phi}}|\hat{\Phi}(R,T)|\Psi_{\mathring{\phi}}\rangle$ or $\langle\Psi_{\mathring{\phi}}|\hat{H}_\circ|\Psi_{\mathring{\phi}}\rangle$ in the matter sector, in the geometric sector, the expectation value (21) itself carries a signature of quantum effects. Second, since the matter sector knows only about $\hbar$ and c, one does not have a preferred frequency scale; with the availability of both G and $\hbar$ one does, and the Planck frequency is given by $\omega_{\rm Pl} = (G\hbar)^{-1}$. Therefore, one can examine the expectation value in various limits by assuming that $\mathring{\phi}(\omega)$ is sharply peaked at various frequencies $\mathring{\omega}$.

Let us first consider the low frequency limit, $G\hbar\mathring{\omega} \ll 1$. Then, the expectation value (21) can be approximated as

$$\langle\Psi_{\mathring{\phi}}|\,\hat{g}_{RR}\,|\Psi_{\mathring{\phi}}\rangle \;\approx\; e^{G\int_0^\infty d\omega\,\omega\,|\mathring{\phi}(\omega)|^2}\,\left(1+\mathcal{N}\,(G\hbar\mathring{\omega})^2\right). \tag{23}$$

where, as before, $\mathcal{N}$ is the expected number of 'photons' in the state $|\Psi_{\mathring{\phi}}\rangle$. Thus, the classical value of g_{RR} is recoverd *provided the frequency is low enough that* $\mathcal{N}\,G\hbar\mathring{\omega} \ll 1$. Interestingly, while in the matter sector the classical approximation becomes better as $\mathcal{N}$ increases, for geometry, it becomes worse as $\mathcal{N}$ increases even when $\mathring{\omega}$ is in the low-frequency regime. (I thank Don Marolf for pointing out that, in hindsight, this second condition can be understood by first noting the relation $g_{RR} = e^{GH_\circ}$ and then using the form (15) of fluctuations in $\hat{H}_\circ$. Although the result refers only to to the expectation value of $\hat{g}_{RR}$ and not to fluctuations, the argument makes the disagreement between classical expectations and exact quantum results plausible.)

In the high-frequency regime, $G\hbar\mathring{\omega} \gg 1$, there is *always a huge disagreement* between the expectation value and the classical expression:

$$\langle\Psi_{\mathring{\phi}}|\,\hat{g}_{RR}\,|\Psi_{\mathring{\phi}}\rangle \;\approx\; e^{\mathcal{N}(e^{G\hbar\mathring{\omega}})} \quad \text{so that} \quad \frac{\langle\Psi_{\mathring{\phi}}|\,\hat{g}_{RR}\,|\Psi_{\mathring{\phi}}\rangle}{g_{RR}} \approx e^{\mathcal{N}\,(e^{G\hbar\mathring{\omega}}\,-\,G\hbar\mathring{\omega})}\,. \tag{24}$$

In this regime, $\hbar$ does not disappear; in fact, terms involving $\hbar$ swamp the classical term since, in the numerator, $G\hbar\mathring{\omega}$ appears *exponentially* in the exponent itself. Further, again, the larger the expected number of 'photons' in the source, the larger the disagreement between the quantum and the classical results.

One can also calculate the quantum fluctuations. In the matter sector, these are very small provided $\mathcal{N} \gg 1$, irrespective of the frequency at which $\mathring{\phi}(\omega)$ is concentrated. The situation is again quite different for quantum geometry. Since $\hat{G}_{RR}^2 = e^{2G\hat{H}_\circ}$, the same calculation that was used to evaluate $\langle\Psi_{\mathring{\phi}}|\,\hat{g}_{RR}\,|\Psi_{\mathring{\phi}}\rangle$ yields

$$\langle\Psi_{\mathring{\phi}}|\,\hat{g}_{RR}^2\,|\Psi_{\mathring{\phi}}\rangle \;=\; e^{\frac{1}{\hbar}\int_0^\infty d\omega\,|\mathring{\phi}(\omega)|^2\,(e^{2G\hbar\omega}-1)}\,, \tag{25}$$

whence the relative uncertainty is given by

$$\left(\frac{\triangle\hat{g}_{RR}}{\langle\hat{g}_{RR}\rangle}\right)^2 \;=\; [e^{\frac{1}{\hbar}\int d\omega\,|\mathring{\phi}(\omega)|^2(1-e^{G\hbar\omega})^2}\;-\;1]\,. \tag{26}$$

Again, this exact expression can be simplified using $\mathring{\phi}(\omega)$, which is sharply peaked at some frequency $\mathring{\omega}$. Then, in the low-frequency regime with $G\hbar\mathring{\omega} \ll 1$, one has

$$\left(\frac{\triangle\hat{g}_{RR}}{\langle\hat{g}_{RR}\rangle}\right)^2 \approx e^{\mathcal{N}(G\hbar\mathring{\omega})^2} - 1\,. \tag{27}$$

Note, again that the uncertainties in geometry grow with the number $\mathcal{N}$ of 'photons'. Thus, there is an interesting tension: to reduce quantum fluctuations in the matter sector, one needs a large number $\mathcal{N}$ of 'photons' in the coherent state $|\Psi_{\mathring{\phi}}\rangle$. However, for a fixed $\mathring{\omega}$, the larger the value of $\mathcal{N}$—i.e., the more intense the beam—the larger its influence on gravity. In quantum theory, this gives rise to larger relative uncertainties. The qualitative nature of this effect is not surprising. Nonetheless, it is pleasing to see it appear in a sharp, precise, and quantitative form. This is possible because the model is exactly soluble.

In the high-frequency regime, as one would expect from the above results on expectation values, quantum fluctuations in geometry

$$\left(\frac{\triangle\hat{g}_{RR}}{\langle\hat{g}_{RR}\rangle}\right)^2 \approx e^{\mathcal{N}\,(e^{2G\hbar\omega_o})}\,, \tag{28}$$

are huge since $G\hbar\omega \gg 1$ and appears *exponentially* in the first exponent. Again, these effects refer to the asymptotic form of the metric in a region in which the classical profile $\mathring{\phi}$ can be vanishingly small. In the classical theory, the metric in this region is as tame as it can be since the curvature vanishes. Yet, quantum geometry exhibits very large, quintessentially quantum mechanical effects even though it has no degrees of freedom of its own and is completely determined by the matter sector.

Finally, for the total Hamiltonian $\hat{H}$ of the matter–gravity system, the analysis can be carried out using the same techniques. For completeness, let us list the final results. The expectation value of $\hat{H}$ is given by

$$\langle\Psi_{\mathring{\phi}}|\hat{H}|\Psi_{\mathring{\phi}}\rangle = \frac{1}{4G}\left[1 - e^{\frac{1}{\hbar}\int_0^\infty \mathrm{d}\omega\,|\mathring{\phi}(\omega)|^2(e^{-4G\hbar\omega}-1)}\right], \tag{29}$$

while the value of the classical Hamiltonian $H(\mathring{\phi})$ is

$$H(\mathring{\phi}) = \frac{1}{4G}\left(1 - e^{-4G\int_0^\infty \mathrm{d}\omega|\mathring{\phi}(\omega)|^2}\right). \tag{30}$$

In the low-frequency regime, $G\hbar\mathring{\omega} \ll 1$, one has

$$\langle\Psi_{\mathring{\phi}}|\hat{H}|\Psi_{\mathring{\phi}}\rangle \approx H(\mathring{\phi}) - \frac{4}{G}\mathcal{N}(G\hbar\mathring{\omega})^2 e^{-4GH_\circ(\mathring{\phi})}\,, \tag{31}$$

while in the high-frequency limit, $G\hbar\mathring{\omega} \gg 1$:

$$\langle\Psi_{\mathring{\phi}}|\hat{H}|\Psi_{\mathring{\phi}}\rangle \approx \frac{1}{4G}\left(1 - e^{-\frac{H_\circ(\mathring{\phi})}{\hbar\mathring{\omega}}}\right), \tag{32}$$

which is quite different from $H_\circ(\mathring{\phi})$. The uncertainties can also be calculated exactly. However, the significance of quantum effects is now overshadowed by the fact that the spectrum of $\hat{H}$ is bounded in the finite interval $[0, \frac{1}{4G}]$.

Let us summarize. One of the primary motivations here was to analyze the quantum effects on geometry induced by quantum matter. Therefore, one is naturally led to the following question: can matter induce interesting quantum effects on the geometrical/gravitational sectors, even when it is itself in 'tame' quantum states? This question led us to focus on standard coherent states in the matter sector because they provide canonical 'quantum representations of classical matter fields'. The analysis showed that the quantum effects induced on geometry/gravity can be very large even when quantum fluctuations in the matter sector are small. This unforeseen behavior arises because of the non-linear structure that is specific to Einstein dynamics: these non-linearities can magnify small fluctuations in the matter sector to huge quantum effects on geometry/gravity. Unsurprisingly, the effect is very pronounced when the classical configuration $\mathring{\phi}(\omega)$ is peaked at a trans-Planckian frequency $\mathring{\omega} \gg \omega_{\rm Pl} = \frac{1}{\hbar G}$. Then, the expectation value $\langle\Psi_{\mathring{\phi}}|\, \hat{g}_{RR}\, |\Psi_{\mathring{\phi}}\rangle$ is wildly different from the classical $g_{RR}(\mathring{\phi})$, even in the asymptotic region where $\mathring{\phi}$ vanishes. However, there are interesting effects also in the 'tame' sub-Planckian frequency regime $G\hbar\mathring{\omega} \ll 1$, because what enters in the explicit expressions of quantum effects in the gravitational sector is $\sqrt{\mathcal{N}} \times G\hbar\mathring{\omega}$—rather than just $G\hbar\mathring{\omega}$—where $\mathcal{N}$ is the expected number of 'photons' in the coherent state $|\Psi_{\mathring{\phi}}\rangle$. In the matter sector, classical theory provides better and better approximation to the full quantum theory as $\sqrt{\mathcal{N}}$ increases. Therefore, it is also natural to consider the limit in which $G\hbar\mathring{\omega} \ll 1$ but $\sqrt{\mathcal{N}}\,(G\hbar\mathring{\omega}) > 1$. In this regime, the matter sector is extremely well approximated by the classical theory. However, Equation (23) implies that $\langle\Psi_{\mathring{\phi}}|\, \hat{g}_{RR}\, |\Psi_{\mathring{\phi}}\rangle$ will not be well approximated by the g_{RR} of the classical solution. Furthermore, Equation (27) implies that in this regime quantum fluctuations will also be small in the geometrical sector. Thus, the state $|\Psi_{\mathring{\phi}}\rangle$ will be peaked on a classical geometry but different from the one determined by the classical source $\mathring{\phi}(\omega) \sim \mathring{\Phi}(R,T)$.

Finally, since $\hat{g}_{RR}$ is a (positive definite) self-adjoint operator on $\mathcal{F}$, one can use its spectral decomposition to construct states in $\mathcal{F}$ that are sharply peaked at a classical value $g_{RR}(\mathring{\phi})$. What happens to the surprising quantum effects in these states? This interesting issue is discussed in Ref. [13]. It turns out that the large quantum effects are then transferred to the matter sector. In this work the focus is on $|\Psi_{\mathring{\phi}}\rangle$ quantum effects on geometry *induced by matter*. Since matter observables have tame behavior in states $|\Psi_{\mathring{\phi}}\rangle$, these states are best suited to bring out the unforeseen effects induced on quantum geometry.

Remark

The point of departure of the current analysis was the standard description of general relativity in terms of the metric. Now, in the asymptotically flat context, there exists in the literature an alternate formulation of Einstein's equations where the basic variable is not a metric g_{ab} but a so-called 'cut-function', Z, that, at the end of the analysis, determines g_{ab}. In the final picture, Z specifies intersections of light cones, emanating from space-time points, with future null infinity, $\mathcal{I}^+$ [28]. The underlying idea was to pass to quantum general relativity via an appropriate quantization of Z, rather than g_{ab}, with the hope that in the final theory, space-time points themselves would become fuzzy [29,30]. In the 2 + 1 theory under consideration, the metric $g_{ab}(R,T)$ is replaced by a single function $Z(\xi;R,T)$, where (u,ξ) label points of $\mathcal{I}^+$ and, in the final picture, the 3-parameter family of 'cuts' defined by space-time points is given by $u = Z$. In the 3 + 1 theory, it has not been possible to implement this idea beyond linearized approximation. However, it could be carried out non-perturbatively in the 2 + 1 theory using the framework outlined in this Section as the point of departure [16]. Interestingly, it was found that, even though there are large quantum fluctuations in the metric, fluctuations in $\hat{Z}$ are strongly damped. In this framework, there are dynamical variables that correspond to space-time points and their fluctuations are similar to those of $\hat{g}_{RR}$. In this sense, the idea of 'fuzzy points' is realized in an interesting manner: space-time points can have large fluctuations but fluctuations of the intersections of their light cones with $\mathcal{I}^+$ are highly suppressed.

3.3. *Lessons from a Thought Experiment*

As explained in Section 1, exactly soluble models are often well suited to sharpen conceptual issues. One such issue is how does one's description of physical reality change when one passes from a less accurate theory to a more accurate one? General relativity, for example, opened entirely new classes of phenomena and possibilities that could not be envisaged in Newtonian gravity. These arise because, with both G and the velocity of light c at one's disposal, a new scale arises. One can now associate a length with a mass, namely the Schwarzschild radius, $R_{\rm Sch} = 2GM/c^2$, which is not available in the Newtonian theory which knows only about G. This new scale then unleashes an entirely new class of phenomena. Similarly, in the quantum theory, Planck's constant, $\hbar$, provides new scales, dramatically changing our understanding of the atomic world and leading to a plethora of unforeseen phenomena that have shaped the physics of the micro-world. In quantum gravity, one has access to all three of these fundamental constants and there have been speculations, dating back to Planck himself, on the nature of new physics that would arise from Planck length, Planck frequency, and Planck density. Exactly soluble models provide a clean-cut platform to discuss the nature of this new physics. In this Subsection, the idea is to leave c as in special relativity but switch on and off $\hbar$ and G, and examine how the nature of physical reality changes.

Consider a trivial thought experiment in a 3-dimensional space-time: Switch on a beam of laser light and then switch it off after some time. (The laser beam would be naturally axisymmetric. The focus will be just on the light beam and not on the details of the source that produced it. Therefore, the relevant source-free solution of Maxwell's equations is obtained as the retarded minus the advanced solution produced by the source.) The task is is to describe what 'really' happens in the space-time around us when this is done. Interestingly, there are four *quite different* answers to this question: (i) A description of a *classical physicist*, who ignores both general relativity (GR) and quantum field theory (QFT) and sees only a Maxwell field propagating in flat space-time, setting $\hbar = G = 0$; (ii) that of a *quantum field theorist*, who knows of $\hbar$ and describes the laser beam as a coherent state of photons but sets $G = 0$; (iii) that of a *general relativist*, who ignores $\hbar$ but knows that the beam curves space-time with surprising consequences in the non-perturbative regime; finally, (iv) that of a *quantum gravity theorist*, who knows about both G and $\hbar$. As the discussion of the last two Subsections shows, the fourth description of physical reality has both subtle and truly novel elements.

Let us begin with the first description. The physical system of interest for the classical physicists is the Maxwell field, $\mathring{F}_{ab} = \epsilon_{abc}\nabla^c\mathring{\Phi}$, that propagates in a 3-d Minkowski $(M,\, \overset{\circ}{g}_{ab})$. The Maxwell field has initial data of compact support on every Cauchy surface and, therefore, vanishes in a finite neighborhood of spatial infinity, i°. The total energy in the system is entirely from the Maxwell field and equals $H_\circ(\mathring{F})$, which can also be read-off at $\mathcal{I}$ (for detailed expressions, see Ref. [18]). This energy is unbounded from above and increases linearly with the frequency and intensity (i.e., $|\mathring{\phi}(\omega)|^2$) of the beam. With only the velocity of light, c, at one's disposal, there is no frequency (nor energy) scale in the theory. Therefore, one cannot speak of low or high frequency (nor of low or high energy) fields and the general description holds for any frequency or intensity of the laser beam. The second description is that of a quantum field theorists. With availability also of $\hbar$, the quantum theorists can regard the beam as consisting of photons. They would point out that what the classical physicist referred to as a classical Maxwell field $\mathring{F}_{ab} \sim \mathring{\Phi}$ is in fact a coherent state $|\Psi_{\mathring{\phi}}\rangle$ and calculate the expected number $\mathcal{N} = \frac{1}{\hbar}\int_0^\infty d\omega\, |\mathring{\phi}|^2$ of photons in it. They now point out that although the expectation value $\langle\Psi_{\mathring{\phi}}|\, \hat{\mathring{F}}_{ab}(R,T)\, |\Psi_{\mathring{\phi}}\rangle$ of the Maxwell field operator agrees with that of the classical physicist's $\mathring{F}_{ab}(R,T)$, there are quantum fluctuations that the classical physicist missed and these would become measurable if $\mathcal{N}$ were small, decaying as $1/\sqrt{\mathcal{N}}$. However, in this description, since $G = 0$, the physical space-time is again Minkowski space $(M, \overset{\circ}{g}_{ab})$.

The third description is that of the general relativists, who have access to G and Einstein's equations but not to $\hbar$. Therefore, the general relativists continue to describe the light beam using the classical Maxwell field. Their description of physical reality is quite different from that of the classical physicist. Now, the physical metric is g_{ab} rather than $\overset{\circ}{g}_{ab}$. This metric is dynamical within the support of the beam. More interestingly, although g_{ab} is flat outside the support, if one parallel transports a test spinning particle around a closed curve C lying *entirely* in this region, one would find that its state changes when it returns to the starting point. The change is encoded in the non-trivial holonomy $U^{(C)}$ of (9). Thus, by performing experiments with probes that never interact with the light beam, the general relativists can determine $H_\circ$—and hence, the non-trivial metric components $g_{RR}\, (= -g_{TT})$—and know that space-time is *not* described by the Minkowski metric $\overset{\circ}{g}_{ab}$

even in the asymptotic region. Perhaps the most startling discovery they would make is that the total energy H of the system, including the gravitational contributions, is *bounded from above*, irrespective of how high the frequency of the intensity of the laser beam is. This qualitatively new phenomenon occurs because G^{-1} has dimensions of mass that set the scale for the upper limit. Furthermore, the general relativist would discover that this interesting fact about nature cannot be captured using a perturbative expansion in G, no matter how high an order n in $(GM)^n$ is used in the truncation of the theory.

Finally, let us introduce the quantum gravity experts. With both G and $\hbar$ available, they can introduce a new notion of Planck frequency $\omega_{\rm Pl} = \frac{1}{G\hbar}$. With this new scale available, not only can they design new experiments to probe aspects of physical reality that have remained unexplored, but they can also solve some mysteries that the classic al physicists, quantum field theorist and general relativists were left with. The quantum gravity experts can design experiments to probe the quantum nature of geometry in the asymptotic region where there are no photons at all. Furthermore, since there are no gravitons—or radiative degrees of freedom—in the gravitational field, these are *'clean' tests* of the nature of quantum information in geometry induced entirely by the quantum properties of matter. If the quantum gravity experts could create laser beams peaked at a trans-Planckian frequency $\mathring{\omega} > 1/G\hbar$, they would discover that even in the asymptotic regions, $\langle\Psi_{\mathring{\phi}}|\, \hat{g}_{RR}\, |\Psi_{\mathring{\phi}}\rangle$ is *very* different from g_{RR} predicted by the classical theory. Much to the surprise of the general relativist, this can occur even when the amplitude of $\mathring{\phi}(\omega)$ is chosen to be sufficiently small for the space-time curvature computed using classical Einstein's equations to be quite small everywhere. A second regime would surprise the quantum field theorist. Consider the case when the intensity of the beam is so high that $\mathcal{N} \gg 1$. Then, the quantum fluctuations in the matter sector are small and the quantum field theorist would expect the induced quantum fluctuations in the geometry to be at least as small, if not much smaller. As shown in Section 3.2, this is not the case; as $\mathcal{N}$ increases, the relative quantum uncertainties in $\hat{g}_{RR}$ grow both in the low- and high-frequency regimes (see Equations (27) and (28), respectively).

Next, consider the regime where $G\hbar\mathring{\omega} \ll 1$ but $\mathcal{N} \times G\hbar\mathring{\omega} > 1$, discussed at the end of Section 3.2. Neither the quantum field theorist nor the general relativist would have a reason to suspect their descriptions because they would see nothing unusual in the matter sector. The quantum field theorist would only see a coherent state with a large number of photons, which is therefore well-approximated by a smooth, classical Maxwell field $\mathring{F}_{ab}$, and the general relativist would see the gravitational field $\mathring{F}_{ab}$ sources. Yet, both descriptions are inadequate. Had the general relativist carried out a careful measure of g_{RR}, they would have found that their theoretical prediction is significantly different from the observed value. This would have been regarded as an anomaly by the community, analogous to that in the perihelion shift of mercury before the discovery of general relativity. Had the quantum field theorist carried out a careful measurement of g_{RR} and constructed a statistical distribution, they would have been puzzled that an irreducible minimum persists in the relative uncertainty, no matter how accurately they measure this (to them) 'classical quantity'; this is somewhat like the puzzle experienced by Penzias and Wilson before they knew about the cosmic microwave background. Thus, the quantum gravity expert would be able to resolve the uncomfortable puzzles that the general relativity and quantum field theory communities had encountered by confronting their detailed theoretical calculations with careful measurements in specific regimes.

4. Discussion

Let us now explore the lessons offered by the discussion of the last two Sections of an exactly soluble sector of non-perturbative quantum gravity in 2 + 1 dimensions. Recall that there are a number of paradigms and conjectures in (3 + 1)-d quantum gravity. Are they realized in the soluble sector? Do results on the (2 + 1)-d model offer hints on the validity of these paradigms? Do they bring out subtleties that were missed in the (3 + 1)-d investigations?

Let us begin with the list of illustrative examples of proposals from Section 1. The first proposal pertains the possibility of violation of the local Lorentz invariance due to trans-Planckian quantum effects. Thanks to the underlying simplicity of the (2 + 1)-d sector, it was possible to carry out explicit calculations. They brought to the forefront the surprising role of trans-Planckian frequencies. If the matter profile, $\mathring{\phi}(\omega)$, at which the coherent state $|\Psi_{\mathring{\phi}}\rangle$ is peaked has even a small 'blip' at a frequency $\mathring{\omega} > \omega_{\rm Pl}$, the *induced* quantum effects on the space-time geometry become large; even in the asymptotic region away from the support of the matter source, the classical value g_{RR} is a poor approximation to the expectation value $\langle\Psi_{\mathring{\phi}}|\,\hat{G}_{RR}\,|\Psi_{\mathring{\phi}}\rangle$ and the relative quantum fluctuations in $\hat{g}_{RR}$ are large (recall that, starting from Section 3, the suffix 'asym' has been dropped in $\hat{g}_{RR}|_{\rm asym}$). Thus, the intuition that trans-Planckian frequencies may change the paradigm suggested by classical GR and standard QFT is indeed borne out in a precise sense. Do these effects, then, lead to the violation of local Lorentz invariance? The frequency $\mathring{\omega}$ does refer to the asymptotic time-translation Killing field $\partial/\partial T$. However, the *dynamics* do not break *local* Lorentz invariance. The field equation satisfied by $\hat{\Phi}(R,T)$, for example, is fully covariant. In particular, there is no quantum-gravity-induced vector field in dynamical equations that can be the source of such a breakdown. Next, let us consider the status of 'holography' in this model. As in 3 + 1 dimensions, neither general relativity nor quantum field theory have a built-in length scale to define an ultraviolet regime. Quantum gravity does and theory departs radically from classical GR and QFT in the ultraviolet regime $\omega > \omega_{\rm Pl} = \frac{1}{G\hbar}$. Do these effects, then, force one to adopt a holographic view in the non-perturbative quantum description? In the classical theory, in the (2+1)-theory the true degrees of freedom are encoded in the matter field that resides in the 3-manifold M. As the discussion of Section 3 showed, the situation is the same in non-perturbative quantum theory; the quantum scalar field continues to reside in the bulk as in any standard quantum field theory. Bulk degrees of freedom are not replaced by 'surface degrees of freedom'. Thus, even though the model exhibits novel features in the Planck regime, these features arise from the standard bulk degrees of freedom of quantum field theory. Now, one *can* construct the quantum Hilbert space $\mathcal{F}$ using just null infinity $\mathcal{I}^+$ [17,21] which serves as the future boundary of space-time. However, this is not what is meant by 'holography'. Indeed, one can also construct the standard Fock space of photons in 3 + 1 dimensions using just $\mathcal{I}^+$.

Next, recall that quantum geometry also has novel features already in the *low*-frequency regime. In particular, many 'seemingly tame' classical solutions turned out to be physically 'spurious'. Consider, for simplicity, solutions $(\mathring{\Phi},\, g_{ab})$, for which $\mathring{\phi}$ is sharply peaked at a frequency $\mathring{\omega} \ll \omega_{\rm Pl}$. It would seem natural to assume that such solutions are 'tame' and should arise in the classical limit of the full quantum theory. That is, one would have expected the quantum theory to admit states $|\Psi_{\mathring{\phi}}\rangle$ that are sharply peaked at such solutions. However, the discussion of Section 3 showed that, for this to occur, *two* inequalities have to be satisfied: $\mathcal{N} \gg 1$ and $\mathcal{N}(G\hbar\mathring{\omega})^2 \ll 1$. The first is required to ensure that $|\Psi_{\mathring{\phi}}\rangle$ is sharply peaked at $\mathring{\Phi}$ but the second inequality is surprising in that it requires that $\mathcal{N}$ cannot be 'too large'. Why is it needed? Suppose $\mathcal{N}$ is large as per the first inequality and $\mathcal{N}(G\hbar\mathring{\omega})^2 = K$. Then, $\langle\Psi_{\mathring{\phi}}|\hat{g}_{RR}|\Psi_{\mathring{\phi}}\rangle \approx e^{\frac{K}{2}}\, g_{RR}$. Thus, the classical answer would not be a good approximation even at the expectation value level unless $K \ll 1$. Consequently, a large class of apparently 'tame' classical solutions are simply not realized as approximations to the quantum theory; from the more complete perspective of quantum gravity, they are spurious. This surprising limitation seems to point in the same direction as Wheeler's 'space-time foam' idea that the quantum vacuum state would also be quite different from the perturbative vacuum that is peaked at Minkowski space. Is this expectation borne out? Recall that the perturbative vacuum $|0\rangle$ is the coherent state $|\Psi_{\mathring{\phi}}\rangle$ with $\mathring{\phi} = 0$, and is annihilated by the matter Hamiltonian $\hat{H}_\circ$. Now, the full Hamiltonian $\hat{H}$ and the asymptotic metric operator $\hat{g}_{RR}$ are both self-adjoint and well-defined functions of $\hat{H}_\circ$. Therefore, $|0\rangle$ is an eigenstate of both, with eigenvalue 0 for $\hat{H}$ and 1 for $\hat{g}_{RR}$. Thus, the perturbative vacuum $|0\rangle$ is in fact an eigenstate of the quantum metric operator with the Minkowski metric g°_{ab} as its eigenvalue. Contrary to one's first expectation, there is nothing unruly about the ground state. Finally, as noted in Section 2, there is a positive energy theorem in classical 2 + 1 dimensional general relativity [19]. It goes over to the quantum theory since $\hat{H}$ is a non-negative self-adjoint operator and the vacuum $|0\rangle$ is the unique state that is annihilated by it. Thus, again one finds that, while several aspects of proposals for novel physics in quantum gravity are realized, closely related aspects that were assumed to follow are not.

These considerations have implications also on semi-classical gravity, an approximation in which gravity is treated classically but matter quantum mechanically and the effect of quantum matter on the classical metric is computed using semi-classical Einstein's equations, $G_{ab} = 8\pi G \langle \hat{T}_{ab} \rangle$, in which the left side refers to a classical metric tensor g_{ab} and the right side to the expectation value of the (renormalized) stress-energy tensor of quantum matter that propagates on the classical geometry defined by g_{ab}. It is often assumed that this theory would serve as an excellent approximation to full quantum gravity away from the Planck regime. In 3 + 1 dimensions, exact solutions to semi-classical gravity are difficult to come by. Even for those available, it has not been possible to check if these solutions are good approximations because one does not have full quantum gravity. Let us examine this issue through the lens of the model discussed here, since one can work out predictions of full quantum gravity and compare them with the semi-classical theory. A solution to semi-classical gravity consists in specifying a classical metric g_{ab} and a quantum state of matter $|\Psi\rangle$ such that the quantum field $\hat{\Phi}$ satisfies its field equation with respect to g_{ab} and the pair $(g_{ab},\ |\Psi\rangle)$ satisfies the semi-classical Einstein's equation. Choose a classical solution (g_{ab}, Φ) in the (2+1)-d model. Then, it follows that in the asymptotic region the pair $(g_{ab},\ |\Psi_{\mathring{\phi}}\rangle)$ solves semi-classical equations exactly. Thus, now, there is an infinite class of solutions to semi-classical gravity. However, as results of Section 3 show, there are regimes—i.e., choices of $\mathring{\phi}(\omega)$—for which g_{ab} is a poor approximation to full quantum geometry. This occurs even in the asymptotic region where curvature vanishes identically. This suggests that the domain of validity of the semi-classical approximation can be quite subtle; it may not be dictated just by space-time curvature.

Interaction between quantum gravity and quantum information communities has increased over the last 5 years or so, giving rise to a healthy exchange of ideas. As mentioned in Section 1, this dialog has sparked interest in manifestation of quantum gravity effects that do not involve the 'radiative degrees of freedom' of gravity but are instead induced by the quantum nature of matter sources (see, e.g., Refs. [7–10]). These effects involve the 'Coulombic' part of the gravitational field. As discussed in Section 2, (2 + 1)-gravity is especially well suited to analyze these quantum effects because the gravitational field does not carry any radiative degrees of freedom—it is determined entirely by matter: while there is an infinite number of degrees of freedom, in our analysis they all reside in the matter sector. Therefore, *all* quantum properties of geometry/gravity are induced by the quantum properties of matter. Nonetheless, in quantum theory, the space-time metric has a number of quintessentially quantum properties. If the quantum state of matter $|\Psi_{\mathring{\phi}}\rangle$ is sharply peaked on a classical field, $\mathring{\Phi}(R,T)$, we are in a regime that is tame from the perspective of quantum matter. Yet, the expectation value of $\hat{g}_{RR}$ can differ significantly from that of the classical solution determined by $\mathring{\Phi}(R,T)$. The difference is entirely due to non-perturbative quantum effects, and of course goes to zero in the $\hbar \to 0$ limit. There is also a trans-Planckian regime in which matter continues to behaves in a tame manner—i.e., its quantum state $|\Psi_{\mathring{\phi}}\rangle$ is sharply peaked at a classical field $\mathring{\Phi}(R,T) \sim \mathring{\phi}(\omega)$ with small fluctuations—but the quantum metric experiences large quantum fluctuations. In both these regimes, one can carry out thought experiments. One can use spinning test particles in the asymptotic region where $\Phi(R,T)$ vanishes: there is no direct interaction between quantum matter that sources the gravitational field and test particles. One can parallel propagate the spinning particles around a closed loop C and measure the change in its spin. This results in a measurement of the holonomy $U^{(C)}$ of the gravitational spin connection. Although the curvature of the physical metric vanishes everywhere along C, the holonomy is non-trivial; this is the gravitational analog of the well-known Aharonov–Bohm effect in electromagnetism. Measurement of this holonomy determines the non-trivial metric component, $g_{RR} = -g_{TT}$, as well as the Hamiltonian/energy of the *total* matter plus gravity system. One can carry out repeated measurements and determine the expectation values and quantum uncertainties in these observables. Since these experiments are carried out entirely in the gravitational sector, they provide an unambiguous test of the quantum nature of geometry/gravity that is induced entirely by matter. As the discussion in Section 3 showed, the specific non-linear coupling between matter and geometry of general relativity can enhance the small quantum fluctuations in the matter sector to produce large quantum effects in geometry. These effects are *non-perturbative*, and not considered in current discussions where gravity is treated perturbatively and, furthermore, matter is generally modeled by non-relativistic particles rather than quantum fields. The frmework used in this paper is free of both these limitations. It provides a proof of principle that quantum information can be transferred to the gravitational sector from the matter sector through Coulombic interaction without any reference to gravitons or radiative modes.

It is natural to ask if the effects that were uncovered by our analysis are specific to gravity or if they would also arise in electromagnetism. (I thank Norbert Straumann for raising this question. The calculation summarized below resulted from that inquiry.) Are there interesting quantum effects on the non-radiative, Coulombic aspects of the Maxwell field by the charged quantum fields that source them? Now, as the discussion in Section 2 showed, in 2 + 1 dimensions, Maxwell fields are dual to scalar fields and, even with the axis-symmetric restriction, they carry radiative degrees of freedom. Therefore, (2 + 1)-d electromagnetic theory, the Maxwell field would not be determined entirely by charged sources; it would have its own radiative degree of freedom. A better candidate is provided by the spherically symmetric sector of (3 + 1)-d system consisting of Maxwell and charged Klein–Gordon fields. In this sector, there are no photons, just as in the current gravitational (2 + 1)-d model there are no gravitons. Therefore the quantum Maxwell field, $\hat{F}_{ab}$, is represented as an operator on the Hilbert space of $\mathring{\Phi}$, just as the quantum metric $\hat{g}_{ab}$ was an operator on the matter Fock space in Section 3. Thus, any quantum features one encounters in the Maxwell field $\hat{F}_{ab}$ are induced by the quantum Klein–Gordon field via Coulombic interaction. It turns out that this model is also exactly soluble in Minkowski space-time. Spherical symmetry singles out a time-translation Killing field t^a, which one can use to decompose F_{ab} into its electric and magnetic parts and introduce a canonical notion of frequency for the Klein–Gordon field. In the asymptotic region, one can express the only non-trivial component E^r of the electric field in terms of the total charge Q of the Klein–Gordon field, just as $g_{RR} = -g_{TT}$ could be expressed in terms of the total (Minkowski) energy $H_\circ$ of matter sources in the gravitational case. Again, one can introduce coherent states $|\Psi_{\mathring{\phi}}\rangle$ in the charged Klein–Gordon sector, where (for simplicity) $\mathring{\Phi}$ has compact support on every Cauchy slice, and calculate expectation values and fluctuations of $\hat{E}^r$ in the asymptotic region. These are again non-zero. However, whereas the sophisticated non-linearities of Einstein dynamics in the (2 + 1)-d theory imply the relation $g_{RR} = e^{GH_\circ}$, the (3 + 1)-d Maxwell theory implies that the relation between E^r and Q is linear: $E^r = Q/r^2$. Therefore, while quantum fluctuations in the Klein–Gordon sector do induce quantum effects in the Maxwell sector even in the absence of photons, they are not exponentially magnified as in the gravitational case. Indeed, the expressions one obtains in the Maxwell case resemble those in the weak field approximation of our (2 + 1)-d gravitational analysis. (In general relativity, the gravitational interaction 'dresses' the bare mass to yield surprising results also in 3 + 1 dimensions. These are genuinely non-perturbative effects that are lost if one expands the result in powers of Newton's constant; for a discussion of this phenomenon, see Chapter 1 of Ref. [31].)

These concrete insights on a number of conceptual issues could be reached because the model is exactly soluble. However, the very reasons that make it exactly soluble also limit its reach. As explained in Section 2, from a 4-d perspective, the model represents Einstein–Rosen waves, which have cylindrical symmetry. These symmetries make it totally unsuitable for the laboratory experiments being considered. Nonetheless, the analysis brought to light a number of unforeseen effects and conceptual subtleties in 3-d quantum gravity coupled to matter. To what extent can one take them over to four dimensions? An obvious strategy would be to consider general relativity coupled to scalar fields in 4-d but restrict oneself to the spherically symmetric sector. Then again, one would have an infinite dimensional midi-superspace where again there are no 'radiative degrees of freedom' in the gravitational field. The metric would be determined entirely by matter and one can investigate quantum gravity effects that are induced by quantum matter, just as in the present case. However, now, the analysis becomes significantly more complicated mathematically because one cannot decouple the scalar field dynamics from that of the metric. From a physical perspective, in 4-d is now the possibility of a gravitational collapse leading to a black hole. While this is a rich sector, it has its own deep puzzles associated with Hawking evaporation and, therefore, it is not directly useful to extract clean insights of the type that our 3-d model provided. However, one can focus just on 'sub-critical' initial data sets so that the scalar waves come in from infinity and scatter off to infinity. This restriction would prevent us from addressing interesting issues such as the Bekenstein bound [2] and related conjectures [3,4] that led to the idea of holography. However, one could still probe the status of other issues discussed in this section. That analysis will have to take into account another key difference between the 3-d and 4-d general relativity. In 2 + 1 dimensions, the metric is flat outside sources, while in the 4-d context, it only approaches the flat metric as $1/r$ as one recedes from the matter sources. Therefore, in 4-d asymptotically flat situations, the effects found here—such as those due to the presence of trans-Planckian frequencies—will decay as one moves away from sources and the fluctuations will be non-negligible only near the sources. However, this limitation is present also in the experiments that are currently contemplated. Thus, while the most striking predictions of the 3-d model will not carry over to 4-d situations, the subtleties illuminated by the analysis may provide useful guidance. At a conceptual level, its detailed analysis unearthed directions in which one can look for novel effects; at a mathematical level, it suggests strategies to represent matter sources by quantum fields, rather than non-relativistic particles, and to beyond perturbation theory in the gravitational sector.

Funding: This research was funded by the NSF grant PHY-1806356 and the Eberly Chair funds of Penn State.

Data Availability Statement: Not applicable

Acknowledgments: I would like to thank Fernando Barbero, Chris Beetle, Alex Corichi, Rodolfo Gambini, Badri Krishnan, Guillermo Mena, Monica Pierri, Jorge Pullin, Norbert Straumann, Thomas Thiemann and Madhavan Varadarajan for numerous discussions. Many of the results were presented at the Erwin Schrödinger Institute, Vienna; Schmidt-fest at the Albert Einstein Institute, Potsdam; Gravitational Waves Workshop at IIT Gandhinagar; and seminars elsewhere. I would like to thank the participants for their comments and questions that added clarity to the presentation of this work.

Conflicts of Interest: The author declares no conflict of interest.

References

1. Corley, S.; Jacobson, T. Hawking spectrum and high frequency dispersion. *Phys. Rev. D* **1996**, *54*, 1568–1586. [CrossRef]
2. Bekenstein, J. Entropy bounds and black hole remnants. *Phys. Rev. D* **1994**, *49*, 1912–1921. [CrossRef]
3. Susskind, L. The world as a hologram. *J. Math. Phys.* **1995**, *36*, 6377–6396. [CrossRef]
4. 't Hooft, G. Dimensional reduction in quantum gravity. *arXiv* **1993**, arXiv:gr-qc/9310026. [CrossRef]
5. Zaffaroni, A. AdS black holes, holography and localization. *Liv. Rev. Relativ.* **2020**, *23*, 2. [CrossRef]
6. Wheeler, J.A. Geometrodynamics and the issue of final state. In *Relativity, Groups and Topology*; DeWitt, C., DeWitt, B.S., Eds.; Gordon and Breach, Science Publishers, Inc.: New York, NY, USA, 1964; pp. 317–522. Available online: https://archive.org/details/relativitegroupe0000ecol/mode/2up (accessed on 5 November 2022).
7. Bose, S.; Mazumdar, A.; Morley, G.W.; Ulbricht, H.; Toroš, M.; Paternostro, M.; Geraci, A.A.; Barker, P.F.; Kim, M.S.; Milburn, G. Spin entanglement witness for quantum gravity. *Phys. Rev. Lett.* **2017**, *119*, 240401. [CrossRef]
8. Marletto, C.; Vedral, V. Gravitationally induced entanglement between two massive particles is sufficient evidence of quantum effects in gravity. *Phys. Rev. Lett.* **2017**, *119*, 240402. [CrossRef]
9. Carlesso, M.; Bassi, A.; Paternostro, M.; Ulbricht, H. Testing the gravitational field generated by a quantum superposition. *New J. Phys.* **2019**, *21*, 093052. [CrossRef]

10. Danielson, D.L.; Satischandran, G.; Wald, R.M. Gravitationally mediated entanglement: Newtonian field versus gravitons. *Phys. Rev. D* **2022**, *105*, 086001. [CrossRef]
11. Ashtekar, A. Large quantum gravity effects: Unforeseen limitations of the classical theory. *Phys. Rev. Lett.* **1996**, *77*, 4864–4867. [CrossRef]
12. Schmidt, B.G. Vacuum spacetimes with toroidal null infinities. *Class. Quant. Grav.* **1996**, *13*, 2811–2816. [CrossRef]
13. Gambini, R.; Pullin, J. Large quantum gravity effects: Backreaction on matter. *Mod. Phys. Lett. A* **1997**, *12*, 2407–2413. [CrossRef]
14. Marugan, G.A.M. Canonical quantization of the Gowdy model. *Phys. Rev. D* **1997**, *56*, 908–919. [CrossRef]
15. Beetle, C. Midi-superspace quantization of non-compact toroidally symmetric gravity. *Adv. Theor. Math. Phys.* **1998**, *2*, 471–495. [CrossRef]
16. Dominguez, A.E.; Tiglio, M.H. Large quantum gravity effects and nonlocal variables. *Phys. Rev. D* **1999**, *60*, 064001. [CrossRef]
17. Ashtekar, A.; Bičák, J.; Schmidt, B.G. Asymptotic structure of symmetry reduced general relativity. *Phys. Rev.* **1997**, *D55*, 669–686. [CrossRef]
18. Ashtekar, A.; Bičák, J.; Schmidt, B.G. Behavior of Einstein-Rosen waves at null infinity. *Phys. Rev. D* **1997**, *55*, 687–694. [CrossRef]
19. Ashtekar, A.; Varadarajan, M. Striking properties of the gravitational Hamiltonian. *Phys. Rev. D* **1994**, *50*, 4944–4956. [CrossRef]
20. Varadarajan, M. Gauge Fixing of one Killing field reductions of canonical gravity: The case of asymptotically flat induced two-geometry. *Phys. Rev. D* **1995**, *52*, 2020–2029. [CrossRef]
21. Ashtekar, A.; Pierri, M. Probing quantum gravity through exactly soluble midi-superspaces I. *J. Math. Phys.* **1996**, *37*, 6250–6270. [CrossRef]
22. Henneaux, M.; Energy-momentum, angular momentum, and supercharge in 2+1 supergravity. *Phys. Rev. D* **1983**, *29*, 2766–2768. [CrossRef]
23. Deser, S.; Jackiw, R.; 't Hooft, G. Three-dimensional Einstein gravity: Dynamics of flat space. *Ann. Phys.* **1984**, *152*, 220–235. [CrossRef]
24. Einstein, A.; Rosen, N. On gravitational waves. *J. Franklin Inst.* **1937**, *223*, 43–54. [CrossRef]
25. Stephani, H.; Kramer, D.; MacCallum, M.; Herlt, E. *Exact Solutions of Einstein's Field Equations*; Cambridge University Press: Cambridge, UK, 2003. [CrossRef]
26. Kuchař, K. Canonical quantization of cylindrical gravitational waves. *Phys. Rev.* **1971**, *D4*, 955–986. [CrossRef]
27. Allen, M. Canonical quantisation of a spherically symmetric, massless scalar field interacting with gravity in (2 + 1) dimensions. *Class. Quant. Grav.* **1987**, *4*, 149–169. [CrossRef]
28. Fritelli, S.; Kozameh, C.; Newman, E.T. GR via characteristic surfaces. *J. Math. Phys.* **1995**, *36*, 4984–5004. [CrossRef]
29. Fritelli, S.; Kozameh, C.N.; Newman, E.T.; Rovelli, C.; Tate, R.S. Fuzzy spacetime from a null-surface version of general relativity. *Class. Quant. Grav.* **1997**, *14*, A143–A154. [CrossRef]
30. Fritelli, S.; Kozameh, C.N.; Newman, E.T.; Rovelli, C.; Tate, R.S. Quantization of the null-surface formulation of general relativity. *Phys. Rev. D* **1997**, *56*, 889–907. [CrossRef]
31. Ashtekar, A. *Lectures on Non-Perturbative Canonical Gravity*; World-Scientific: Singapore, 1991. [CrossRef]

physics MDPI

Article

Parametrization of Deceleration Parameter in $f(Q)$ Gravity

Gaurav N. Gadbail, Sanjay Mandal and Pradyumn Kumar Sahoo *

Department of Mathematics, Birla Institute of Technology and Science-Pilani, Hyderabad Campus, Hyderabad 500078, India
* Correspondence: pksahoo@hyderabad.bits-pilani.ac.in

Abstract: In this paper, we investigate the modified symmetric teleparallel gravity or $f(Q)$ gravity, where Q is the nonmetricity, to study the evolutionary history of the universe by considering the functional form of $f(Q) = \alpha Q^n$, where α and n are constants. Here, we consider the parametrization form of the deceleration parameter as $q = q_0 + q_1 z/(1+z)^2$ (with the parameters q_0 (q at $z = 0$), q_1, and the redshift, z), which provides the desired property for a sign flip from a decelerating to an accelerating phase. We obtain the solution of the Hubble parameter by examining the mentioned parametric form of q, and then we impose the solution in Friedmann equations. Employing the Bayesian analysis for the Observational Hubble data (OHD), we estimated the constraints on the associated free parameters (H_0, q_0, q_1) with H_0 the current Hubble parameter to determine if this model may challenge the ΛCDM (Λ cold dark matter with the cosmological constant, Λ) limitations. Furthermore, the constrained current value of the deceleration parameter $q_0 = -0.832^{+0.091}_{-0.091}$ shows that the present universe is accelerating. We also investigate the evolutionary trajectory of the energy density, pressure, and EoS (equation-of-state) parameters to conclude the accelerating behavior of the universe. Finally, we try to demonstrate that the considered parametric form of the deceleration parameter is compatible with $f(Q)$ gravity.

Keywords: $f(Q)$ gravity; accelerated expansion; deceleration parameter; EoS (equation-of-state) parameter; cosmic chronometer dataset; observational constraint

Citation: Gadbail, G.N.; Mandal, S.; Sahoo, P.K. Parametrization of Deceleration Parameter in $f(Q)$ Gravity. *Physics* **2022**, *4*, 1403–1412. https://doi.org/10.3390/physics4040090

Received: 4 October 2022
Accepted: 1 December 2022
Published: 13 December 2022

1. Introduction

Recently, several cosmological observations [1–6] have supported the late-time cosmic acceleration expansion of the universe. However, based on the same cosmological observation, it is estimated that dark energy (DE) and dark matter (DM) cover up 95–96% of the universe's composition, comprising mysterious dark components, the so-called dark matter and dark energy, whereas baryonic matter covers up 4–5% of the content of the universe. Presently, general relativity (GR) is believed to be the most successful theory of gravitation, and its few gravitational tests have been discussed in Ref. [7]. However, it cannot provide a satisfactory explanation for the dark energy and dark matter problem; hence, it may not be regarded as the ultimate gravitational force theory for dealing with the current cosmological problems. Several alternative approaches have been proposed in the literature over the last several decades to overcome the current cosmological problems. Nowadays, the modified theory of gravity is the most admirable candidate to solve the current difficulties (the DE and DM problem) of the universe. One of the most prominent schemes to address the dark content issue of the universe is the modification of GR called the $f(R)$ theory of gravity, where R is the Ricci scalar [8]. Some other modified theories are also developed to solve this issue, such as the $f(\mathcal{T})$ theory, where $\mathcal{T}$ is the torsion [9,10]; the $f(R,T)$ theory [11,12]; the $f(R, L_m)$ theory, where L_m is the matter Lagrangian density [13,14]; the $f(R,G)$ theory, where G is the Gauss–Bonnet invariant [15,16]; and many more.

Jimenez et al. [17] recently proposed a novel proposal by considering a modification of the symmetric teleparallel equivalent to GR called $f(Q)$ gravity, where Q is a nonmetricity scalar. The nonmetricity, Q, of the metric geometrically characterizes the variation in the length of a vector in parallel transport, and it represents the primary geometric variable explaining the features of a gravitational interaction. Recently, several studies were conducted on $f(Q)$ gravity. Mandal et al. studied cosmography [18] and the energy condition [19] in nonmetric $f(Q)$ gravity. For the purpose of examining an accelerated expansion of the universe with the recent observations, Lazkoz et al. [20] examined several $f(Q)$ gravity models. Furthermore, Solanki et al. [21] studied the effect of bulk viscosity in the accelerating expansion of the universe in $f(Q)$ gravity. Esposito et al. [22] examined exact isotropic and anisotropic cosmological solutions using reconstruction techniques. Moreover, $f(Q)$ gravity easily overcomes the limits set by Big Bang Nucleosynthesis (BBN) [23]. Many other studies have been completed within the context of the $f(Q)$ gravity theory [24–29]. Although various theoretical approaches exist to explain the phenomenon of cosmic acceleration, none are definitively known as the appropriate one. The current model of late-time cosmic acceleration is known as reconstruction. This is the inverse method of locating a suitable cosmological model. There are two kinds of reconstruction: parametric reconstruction and non-parametric reconstruction. The parametric reconstruction relies on estimating the model parameters from various observational data. It is also known as the model-dependent approach. The main idea is to assume a specific evolution scenario and then determine the nature of the matter sector or the exotic component that is causing the alleged acceleration. Several authors have used this method to find a suitable solution [30–32].

In this paper, we consider the parametrization form of the deceleration parameter in terms of the redshift, z, as $q(z) = q_0 + q_1 z/(z+1)^2$ (with the parameters q_0 and q_1), which provides the desired property for the sign flip from a decelerating to an accelerating phase and investigate the Friedmann–Lemaître–Robertson–Walker (FLRW) universe in the framework of nonmetric $f(Q)$ gravity by using the functional form of $f(Q)$ as $f(Q) = \alpha Q^n$, where α and n are arbitrary constants. The present paper is arranged as follows. In Section 2, we start with the basic $f(Q)$ gravity formalism and develop the field equation for the FLRW line element. In Section 3, we adopt the parametric form of a deceleration parameter and then find the Hubble solution. In Section 4, we estimate the constraints on the associated free parameters (H_0, q_0, q_1) by employing the Bayesian analysis for the Observational Hubble data (OHD). Then, we check the evolutionary trajectory of the energy density, pressure, and the equation-of-state (EoS) parameters to conclude the accelerating behavior of the universe in Section 5. Lastly, we conclude our result in Section 6.

2. $f(Q)$ Gravity Formalism

The most generic action of nonmetric $f(Q)$ gravity is given by [17]

$$S = \int \left[\frac{1}{2} f(Q) + \mathcal{L}_{\mathrm{m}}\right] \sqrt{-g}\, d^4x, \tag{1}$$

where f is an arbitrary function of nonmetricity scalar Q, $\mathcal{L}_{\mathrm{m}}$ is the matter Lagrangian density, and g is a determinant of the metric tensor, $g_{\alpha\beta}$, where four-dimensional tensor indices are denoted by lower-case Greek letters and take the values 0 (time), 1, 2, 3 (space).

The definition of nonmetricity tensor in $f(Q)$ gravity is

$$Q_{\sigma\alpha\beta} = \nabla_\sigma\, g_{\alpha\beta} \tag{2}$$

and the corresponding traces are

$$Q_\sigma = Q_\sigma{}^{\alpha}{}_{\alpha}, \quad \tilde{Q}_\sigma = Q^{\alpha}{}_{\sigma\alpha}. \tag{3}$$

Moreover, the superpotential tensor $P^{\lambda}_{\mu\nu}$ is given by

$$4P^{\sigma}{}_{\alpha\beta} = -Q^{\sigma}{}_{\alpha\beta} + 2Q_{(\alpha}{}^{\sigma}{}_{\beta)} - Q^{\sigma} g_{\alpha\beta} - \tilde{Q}^{\sigma} g_{\alpha\beta} - \delta^{\sigma}_{(\alpha}\, Q_{\beta)}, \tag{4}$$

Hence, the nonmetricity scalar can be obtained as

$$Q = -Q_{\sigma\alpha\beta}P^{\sigma\alpha\beta}. \tag{5}$$

The gravitational field equation derived by varying the action (1) with regard to the metric tensor is presented below:

$$\frac{2}{\sqrt{-g}}\nabla_\sigma\left(f_Q\sqrt{-g}\,P^\sigma{}_{\alpha\beta}\right) + \frac{1}{2}f\,g_{\alpha\beta} + f_Q\left(P_{\alpha\sigma\lambda}Q_\beta{}^{\sigma\lambda} - 2Q_{\sigma\lambda\alpha}P^{\sigma\lambda}{}_\beta\right) = -T_{\alpha\beta}, \tag{6}$$

where $T_{\alpha\beta} \equiv -\frac{2}{\sqrt{-g}}\frac{\delta(\sqrt{-g})\mathcal{L}_m}{\delta g^{\alpha\beta}}$ and $f_Q = df/dQ$.

Similarly, by varying the action (1) with regard to the connection, the following result can be obtained:

$$\nabla_\alpha\nabla_\beta\left(f_Q\sqrt{-g}\,P^{\alpha\beta}{}_\sigma\right) = 0. \tag{7}$$

We shall consider a spatially flat FLRW universe throughout the investigation, whose metric is given by

$$ds^2 = -dt^2 + a^2(t)(dx^2 + dy^2 + dz^2). \tag{8}$$

Here, $a(t)$ is a cosmic scale factor. The nonmetricity scalar $Q = 6H^2$ obtained for the above FLRW metric, where $H = \dot{a}/a$ is the Hubble parameter, and the dot denotes the time derivative. In this case, the energy-momentum tensor of a perfect fluid, $T_{\alpha\beta} = (p+\rho)u_\alpha u_\beta + pg_{\alpha\beta}$, where p and ρ are pressure and energy density, respectively, and u_α denotes the four-velocity vector of the fluid.

For the metric (8), the corresponding Friedmann equations are [17]:

$$3H^2 = \frac{1}{2f_Q}\left(-\rho + \frac{1}{2}f\right), \tag{9}$$

$$\dot{H} + 3H^2 + \frac{\dot{f}_Q}{f_Q}H = \frac{1}{2f_Q}\left(p + \frac{1}{2}f\right). \tag{10}$$

Using the preceding Friedmann equations in the context of $f(Q)$ gravity, one may now study possible cosmological applications.

3. Parametrization of the Deceleration Parameter

The parametrization of the deceleration parameter q plays a significant role in determining the nature of the universe's expanding rate. In this regard, some research employed various parametric forms of deceleration parameters, while other research investigated non-parametric forms. These methods have been widely discussed in the literature to characterize the concerns with cosmological investigations, such as the initial singularity problem, the problem of all-time decelerating expansion, the horizon problem, Hubble tension, and so on [33–35]. Motivated by this fact, in this paper, we consider the simplest parametric form of the deceleration parameter q in terms of redshift z as [36]

$$q(z) = q_0 + \frac{q_1 z}{(z+1)^2}, \tag{11}$$

where $q_0 = q(z=0)$ indicates the present value of deceleration parameter, and q_1 depicts the variation in the deceleration parameter as a function of z. Certainly, one of the most well-liked parametrizations of the dark energy equation of state served as inspiration for this parametric form for $q(z)$ [37], and it seems to be versatile enough to fit the $q(z)$ behavior of a broad class of accelerating models.

The derivative of the Hubble parameter with respect to time t is $\dot{H} = -(1+q)H^2$. Then, there exists a relation between the Hubble parameter and the deceleration parameter in virtue of an integration:

$$H(z) = H_0 \exp\left[\int_0^z (1+q(x))d\ln(1+x)\right], \tag{12}$$

where x is a changing variable. By using Equation (11) in Equation (12), we obtained the Hubble parameter in terms of redshift z as

$$H(z) = H_0(z+1)^{q_0+1}e^{\frac{q_1 z^2}{2(z+1)^2}}, \tag{13}$$

where H_0 is the current Hubble constant (at $z = 0$). Furthermore, utilizing the relationship between redshift and the universe's scale factor $a(t) = \frac{1}{1+z}$, we may describe the relationship between cosmic time and redshift as

$$\frac{d}{dt} = \frac{dz}{dt}\frac{d}{dz} = -(1+z)H(z)\frac{d}{dz}. \tag{14}$$

Using Equations (13) and (14) in Fridemann equations, we obtained the energy density ρ, pressure p, and equation of state parameter ω in terms of redshift z as

$$\rho = \alpha\left(-2^{n-1}\right)3^n(2n-1)\left(H_0^2\,(z+1)^{2q_0+2}\,e^{\frac{q_1 z^2}{(z+1)^2}}\right)^n, \tag{15}$$

$$p = \alpha 6^{n-1}\left(H_0^2(z+1)^{2q_0+2}e^{\frac{q_1 z^2}{(z+1)^2}}\right)^n\left(-\frac{2n(q_0(z+1)^2+z(q_1+z+2)+1)}{(z+1)^2}\right.$$
$$\left.-\frac{4(n-1)n(z+1)^{-q_0-3}\,e^{-\frac{q_1 z^2}{2(z+1)^2}}\,(q_0(z+1)^2+z(q_1+z+2)+1)}{H_0}+6n-3\right), \tag{16}$$

$$w = -\frac{-\frac{4n(n-1)(z+1)^{(-q_0-3)}\,e^{-\frac{q_1 z^2}{2(z+1)^2}}\,(q_0(z+1)^2+z(q_1+z+2)+1)}{H_0} - \frac{2n\left(q_0(z+1)^2+z(q_1+z+2)+1\right)}{(z+1)^2}+6n-3}{3(2n-1)}, \tag{17}$$

respectively. The behavior and essential cosmological properties of the model described in Equation (11) are wholly dependent on the model parameters (q_0, q_1). In the next section, we constraint the model parameter (H_0, q_0, q_1) by using the recent observational datasets to investigate the behavior of the cosmological parameters.

4. Observational Constraints and Cosmological Applications

Now, one can deal with the various observational datasets to constraint the parameters H_0, q_0, q_1. In order to study the observational data, we use the standard Bayesian technique, and to obtain the posterior distributions of the parameters, we employ a Markov Chain Monte Carlo (MCMC) method. Moreover, we use the *emcee* package to perform the MCMC analysis. Here, in this study, we used the Hubble measurements (i.e., Hubble data) to complete the stimulation. The following *likelyhood* function is used to find the best fits of the parameters;

$$\mathcal{L} \propto exp(-\chi^2/2), \tag{18}$$

where χ^2 is the *pseudo chi-squared function* [38]. The χ^2 functions for various datasets are discussed below.

Cosmic Chronometer (CC) Sample

Recently, a list of Hubble measurements in the redshift range $0.07 \leq z \leq 1.965$ were compiled by Singirikonda and Desai [39]. This $H(z)$ dataset was measured from the differential ages Δt of galaxies [40–43]. The complete list of datasets is presented in Ref. [39]. To estimate the model parameters, we use the chi-squerd function which is given by

$$\chi^2_{CC}(p_s) = \sum_{i=1}^{31} \frac{[H_{th}(p_s, z_i) - H_{obs}]^2}{\sigma^2_{H(z_i)}}, \tag{19}$$

where $H_{th}(p_s, z_i)$, $H_{obs}(z_i)$ represents the Hubble parameter with the model parameters, observed Hubble parameter values, respectively. $\sigma^2_{H(z_i)}$ is the standard deviation obtained from observations. The marginalized constraining results are displayed in Figure 1. In Figure 2, the profile of our model against Hubble data is shown.

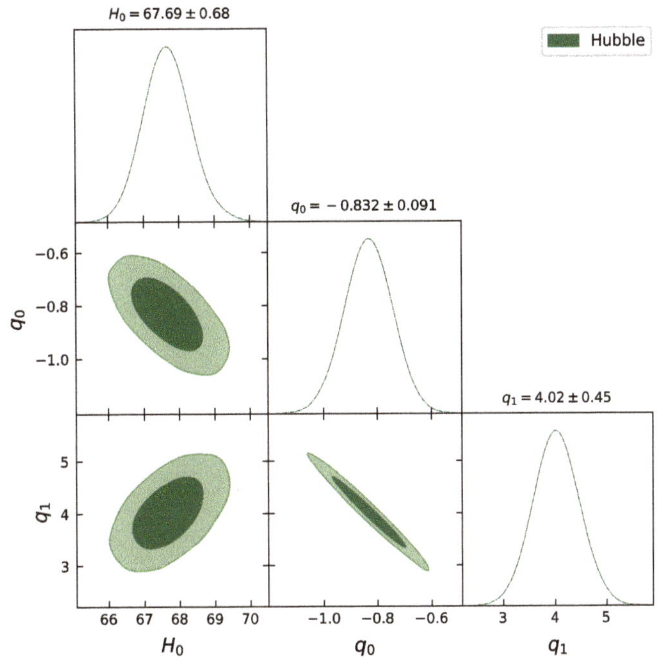

Figure 1. The marginalized constraints on the coefficients in the expression of Hubble parameter, $H(z)$, in Equation (13) are shown by using the Hubble sample [39].

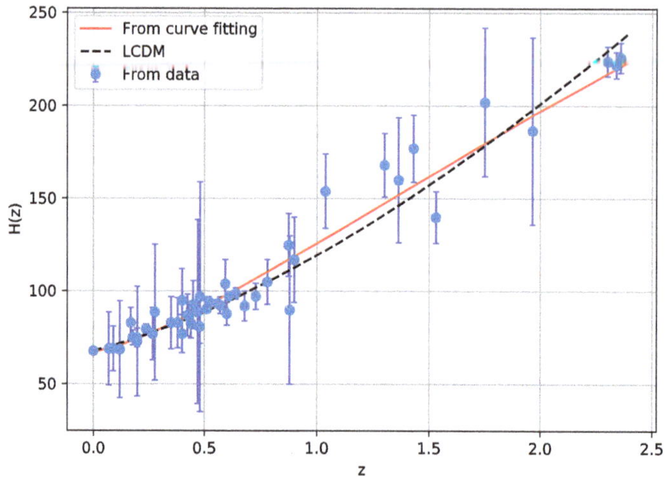

Figure 2. The evolution of Hubble parameter, $H(z)$, with respect to redshift z. The red line represents our model and dashed line indicates the ΛCDM model with the pressureless matter parameter, $\Omega_{m0} = 0.3$ [44] and the dark energy density parameter, $\Omega_{\Lambda 0} = 0.7$ [44]. The dots show the Hubble dataset with error bars [39].

5. Cosmological Parameters

One of the cosmological parameters that is significant in explaining the state of the expansion of our universe is the deceleration parameter q. When the value of the deceleration parameter is strictly less than zero, it shows the accelerating behavior of the universe, and when it is non-negative, the universe decelerates. Furthermore, the observational data employed in this study revealed that our current universe is in an accelerating phase, with the present value of the deceleration parameter becoming $q_0 = -0.832^{+0.091}_{-0.091}$. This type of result is seen in the existing literature [45,46].

Figure 3 indicates that the energy density of the universe increases with a redshift and still seems to as the universe expands, but Figure 4 demonstrates that the pressure decreases with the redshift and has large negative values throughout the cosmic evolution. The present cosmic acceleration induces this isotropic pressure behavior.

The EoS parameter w is also helpful in categorizing the decelerating and accelerating behavior of the universe, and it is defined as $w = p/\rho$. The EoS categorizes three possible states for the accelerating universe which are the quintessence $(-1 < w < -\frac{1}{3})$ era, phantom $(w < -1)$ era, and cosmological constant $(w = -1)$. Figure 5 shows the evolutionary trajectory of the EoS parameter, and it can be seen that the whole trajectory lies in the quintessence era. From Figure 5, One can see that $w < 0$ and the current value of the EoS parameter is $w_0 = -0.9^{+0.08}_{-0.12}$. Our result aligned with some of the studies [32,47], which indicates an accelerating phase.

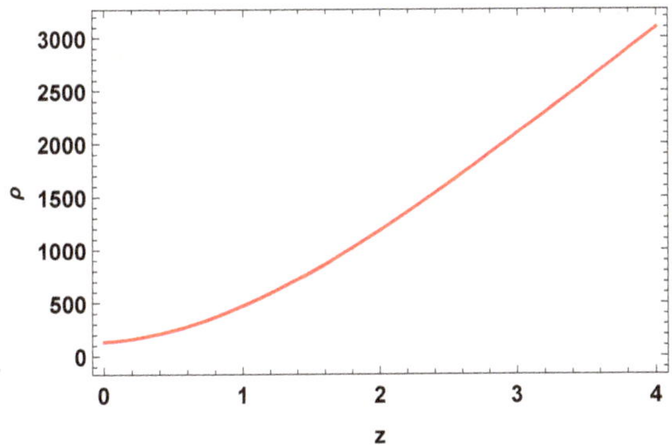

Figure 3. Evolution trajectory of the energy density, ρ (15), versus z with constraint values from the Hubble datasets [39] and $\alpha = -0.01, n = 1.2$.

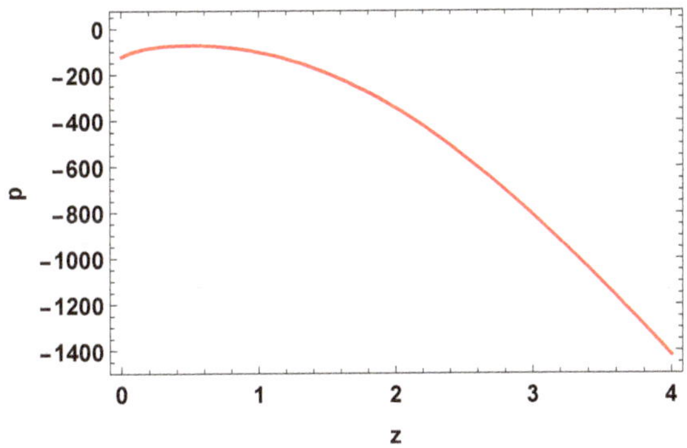

Figure 4. Evolution trajectory of the pressure, p (16), versus z with constraint values from the Hubble datasets [39] and $\alpha = -0.01, n = 1.2$.

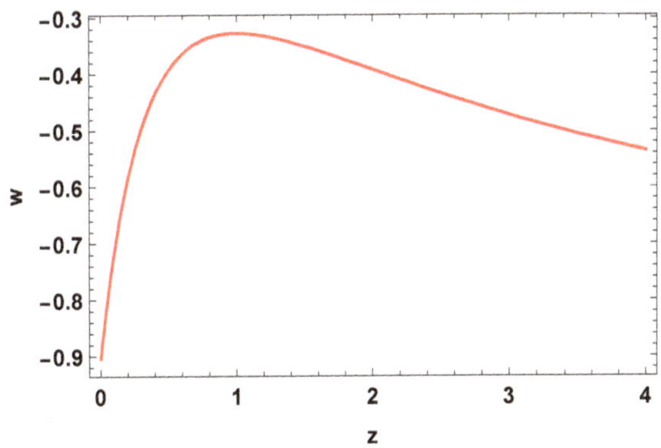

Figure 5. Evolution trajectory of the equstion-of-state parameter, w (17), versus z with constraint values from the Hubble datasets [39] and $\alpha = -0.01, n = 1.2$.

6. Conclusions

The current scenario of the accelerated expansion of the universe has grown increasingly fascinating over time. Numerous dynamical DE models and modified gravity theories have been employed in various ways to find a suitable description of the accelerating universe. In this paper, we explored the accelerated expansion of the universe by adopting the parametric form of the deceleration parameter in the framework of $f(Q)$ gravity, where Q is the nonmetricity scalar depicted in the gravitational interaction.

We have examined the functional form of $f(Q)$ as $f(Q) = \alpha Q^n$, where α and n are the arbitrary constants, and the parametrization form of the deceleration parameter as $q = q_0 + q_1 z/(1+z)^2$, where (q_0, q_1) are the model parameters. By utilizing the above parametric form, we find out the solution of the Hubble parameter as $H(z) = H_0(z+1)^{q_0+1} e^{\frac{q_1 z^2}{2(z+1)^2}}$. Furthermore, we used the Hubble datasets containing 31 data points to determine the best-fit values for the model parameters (H_0, q_0, q_1) as $H_0 = 67.69 \pm 0.68$, $q_0 = -0.832 \pm 0.091$, and $q_1 = 4.02 \pm 0.45$. Here, the q_0 shows the current value of the deceleration parameter, which depicts that the present expansion of the universe is accelerating. We analyzed the evolution of the various cosmological parameters corresponding to these best-fit values of the model parameters. The EoS parameter exhibits negative behavior and lies in the quintessence era, which depicts that the present universe is in an accelerating phase. Figure 3 indicates that the energy density of the universe increases with a redshift and still seems to as the universe expands, but Figure 4 demonstrates that the pressure decreases with the redshift and has large negative values throughout the cosmic evolution. Lastly, we conclude that the considered parametric form of the deceleration parameter in the framework of $f(Q)$ gravity theory plays an important role in driving the universe's accelerated expansion.

Author Contributions: Conceptualization, G.N.G.; data curation, S.M.; formal analysis, P.K.S.; investigation, G.N.G.; methodology, G.N.G.; project administration, P.K.S.; software, G.N.G. and S.M.; supervision, P.K.S.; validation, P.K.S.; visualization, S.M.; writing—original draft, G.N.G.; writing—review and editing, S.M. and P.K.S. All authors have read and agreed to the published version of the manuscript.

Funding: G.N.G. acknowledges the University Grants Commission (UGC), New Delhi, India, for awarding the Junior Research Fellowship (UGC-Ref. No.: 201610122060). S.M. acknowledges the Department of Science & Technology (DST), Govt. of India, New Delhi, for awarding the INSPIRE Fellowship (File No. DST/INSPIRE Fellowship/2018/IF180676). P.K.S. acknowledges the the Council of Scientific & Industrial Research (CSIR), New Delhi, India, for the financial support to carry out the Research project [No.03(1454)/19/EMR-II, Dt. 2 August 2019] and the Inter-University Centre for Astronomy and Astrophysics (IUCAA), Pune, India, for providing support through the visiting Associateship program.

Data Availability Statement: The data used can be found in the references cited.

Acknowledgments: Thanks to Department of Mathematics, Birla Institute of Technology and Science-Pilani, Hyderabad Campus, India, for providing DST-FIST lab, where the data simulation work is done.

Conflicts of Interest: The author declare no conflict of interest.

References

1. Perlmutter, S.; Aldering, G.; Goldhaber, G.; Knop, R.A.; Nugent, P.; Castro, P.G.; Deustua, S.; Fabbro, S.; Goobar, A.; Groom, D.E.; et al. [The Supernova Cosmology Project]. Measurements of Ω and Λ from 42 high-redshift supernovae. *Astrophys. J.* **1999**, *517*, 565–586. [CrossRef]
2. Riess, A.G.; Filippenko, A.V.; Challis, P.; Clocchiatti, A.; Diercks, A.; Garnavich, P.M.; Gilliland, R.L.; Hogan, J.C.; Jha, S.; Kirshner, P.R.; et al. Observational evidence from supernovae for an accelerating universe and a cosmological constant. *Astrophys. J.* **1998**, *116*, 1009–1038. [CrossRef]
3. Riess, A.G.; Strolger, L.G.; Tonry, J.; Casertano, S.; Ferguson, H.C.; Mobasher, B.; Challis, P.; Filippenko, A.V.; Jha, S.; Li, W.; et al. Type Ia supernova discoveries at $z > 1$ from the Hubble Space Telescope: Evidence for past deceleration and constraints on dark energy evolution. *Astrophys. J.* **2004**, *607*, 665–687. [CrossRef]
4. Spergel, D.N.; Verde, L.; Peiris, H.V.; Komatsu, E.; Nolta, M.R.; Bennett, C.L.; Halpern, M.; Hinshaw, G.; Jarosik, N.; Kogut, A.; et al. First-year Wilkinson Microwave Anisotropy Probe (WMAP) observations: Determination of cosmological parameters. *Astrophys. J. Suppl.* **2003**, *148*, 175–194. [CrossRef]
5. Koivisto, T.; Mota D.F. Dark energy anisotropic stress and large scale structure formation. *Phys. Rev. D* **2006**, *73*, 083502. [CrossRef]

6. Daniel, S.F.; Caldwell, R.R.; Cooray, A.; Melchiorri, A. Large scale structure as a probe of gravitational slip. *Phys. Rev. D* **2008**, *77*, 103513. [CrossRef]
7. Corda, C. Interferometric detection of gravitational waves: The definitive test for general relativity. *Int. J. Mod. Phys. D* **2009**, *18*, 2275–2282. [CrossRef]
8. Buchdahl, H.A. Non-linear Lagrangians and cosmological theory. *Mon. Not. R. Astron. Soc.* **1970**, *150*, 1–8. [CrossRef]
9. Ferraro, R.; Fiorini, F. Modified teleparallel gravity: Inflation without an inflaton. *Phys. Rev. D* **2007**, *75*, 084031. [CrossRef]
10. Myrzakulov, R. Accelerating universe from $F(T)$ gravity. *Eur. Phys. J. C* **2012**, *71*, 1752. [CrossRef]
11. Harko, T.; Lobo, F.S.; Nojiri, S.I.; Odintsov, S.D. $f(R,T)$ gravity. *Phys. Rev. D* **2011**, *84*, 024020. [CrossRef]
12. Sahoo, P.K.; Moraes, P.H.R.S.; Sahoo, P. Wormholes in R^2-gravity within the $f(R,T)$ formalism. *Eur. Phys. J. C* **2018**, *78*, 46. [CrossRef]
13. Harko, T.; Lobo, F.S.N. $f(R,L_m)$ gravity. *Eur. Phys. J. C* **2010**, *70*, 373–379. [CrossRef]
14. Jaybhaye, L.V.; Solanki, R.; Mandal, S.; Sahoo, P.K. Cosmology in $f(R,L_m)$ gravity. *Phys. Lett. B* **2022**, *831*, 137148. [CrossRef]
15. Elizalde, E.; Myrzakulov, R.; Obukhov, V.V.; Sáez-Gómez, D. ΛCDM epoch reconstruction from $F(R,G)$ and modified Gauss–Bonnet gravities. *Class. Quantum Grav.* **2010**, *27*, 095007. [CrossRef]
16. Bamba, K.; Odintsov, S.D.; Sebastiani, L.; Zerbini, S. Finite-time future singularities in modified Gauss-Bonnet and $\mathcal{F}(\mathcal{R},\mathcal{G})$ gravity and singularity avoidance. *Eur. Phys. J. C* **2010**, *67*, 295–310. [CrossRef]
17. Jimenez, J.B.; Heisenberg, L.; Koivisto, T. Coincident general relativity. *Phys. Rev. D* **2018**, *98*, 044048. [CrossRef]
18. Mandal, S.; Wang, D.; Sahoo, P.K. Cosmography in $f(Q)$ gravity. *Phys. Rev. D* **2020**, *102*, 124029. [CrossRef]
19. Mandal, S.; Sahoo, P.K.; Santos, J.R.L. Energy conditions in $f(Q)$ gravity. *Phys. Rev. D* **2020**, *102*, 024057. [CrossRef]
20. Lazkoz, R.; Lobo, F.S.N.; Ortiz-Baños, M.; Salzano, V. Observational constraints of $f(Q)$ gravity. *Phys. Rev. D* **2019**, *100*, 104027. [CrossRef]
21. Solanki, R.; Mandal, S.; Sahoo, P.K. Cosmic acceleration with bulk viscosity in modified $f(Q)$ gravity. *Phys. Dark Univ.* **2021**, *32*, 100820. [CrossRef]
22. Esposito, F.; Carloni, S.; Cianci, R.; Vignolo, S. Reconstructing isotropic and anisotropic $f(Q)$ cosmologies. *Phys. Rev. D* **2022**, *105*, 084061. [CrossRef]
23. Anagnostopoulos, F.K.; Gakis, V.; Saridakis, E.N.; Basilakos, S. New models and Big Bang Nucleosynthesis constraints in $f(Q)$ gravity. *arXiv* **2022**, arXiv:2205.11445. [CrossRef]
24. Harko, T.; Koivisto, T.S.; Lobo, F.S.; Olmo, G.J.; Rubiera-Garcia, D. Coupling matter Q in modified gravity. *Phys. Rev. D* **2018**, *98*, 084043. [CrossRef]
25. Frusciante, N. Signatures of $f(Q)$ gravity in cosmology. *Phys. Rev. D* **2021**, *103*, 044021. [CrossRef]
26. Khyllep, W.; Paliathanasis, A.; Dutta, J. Cosmological solutions and growth index of matter perturbations in $f(Q)$ gravity. *Phys. Rev. D* **2021**, *103*, 103521. [CrossRef]
27. Ayuso, I.; Lazkoz, R.; Salzano, V. Observational constraints on cosmological solutions of $f(Q)$ theories. *Phys. Rev. D* **2021**, *103*, 063505. [CrossRef]
28. Gadbail, G.; Mandal, S.; Sahoo, P.K. Reconstruction of ΛCDM universe in $f(Q)$ gravity. *Phys. Lett. B* **2022**, *835*, 137509. [CrossRef]
29. Capozziello, S.; D'Agostino, R. Model-independent reconstruction of $f(Q)$ non-metric gravity. *Phys. Lett. B* **2022**, *832*, 137229. [CrossRef]
30. Mukherjee, A. Acceleration of the universe: A reconstruction of the effective equation of state. *Mon. Not. R. Astron. Soc.* **2016**, *460*, 273–282. [CrossRef]
31. Arora, S.; Parida, A.; Sahoo, P.K. Constraining effective equation of state in $f(Q,T)$ gravity. *Eur. Phys. J. C* **2021**, *81*, 555. [CrossRef]
32. Gadbail, G.N.; Arora, S.; Kumar, P.; Sahoo, P.K. Interaction of divergence-free deceleration parameter in Weyl-type $f(Q,T)$ gravity. *Chin. J. Phys.* **2022**, *79*, 246–255. [CrossRef]
33. Banerjee, N.; Das, S. Acceleration of the universe with a simple trigonometric potential. *Gen. Relativ. Gravit.* **2005**, *37*, 1695–1703. [CrossRef]
34. Cunha, J.V.; Lima, J.A.S. Transition redshift: New kinematic constraints from supernovae. *Mon. Not. R. Astron. Soc.* **2008**, *390*, 210–217. [CrossRef]
35. Escamilla-Rivera, C.; Nájera, A. Dynamical dark energy models in the light of gravitational-wave transient catalogues. *J. Cosmol. Astropart. Phys.* **2022**, *03*, 060. [CrossRef]
36. Wang, Y.-T.; Xu, L.-X.; Lu, J.-B.; Gui, Y.-X. Reconstructing dark energy potentials from parameterized deceleration parameters. *Chin. Phys. B* **2010**, *19*, 019801. [CrossRef]
37. Gong, Y.; Wang, A. Observational constraints on the acceleration of the Universe. *Phys. Rev. D* **2006**, *73*, 083506. [CrossRef]
38. Hobson, M.P.; Jaffe, A.H.; Liddle, A.R.; Mukherjee, P.; Parkison, D. (Eds.) *Bayesian Methods in Cosmology*; Cambridge University Press: Cambridge, UK, 2009. [CrossRef]
39. Singirikonda, H.; Desai, S. Model comparison of ΛCDM vs $R_h = ct$ using cosmic chronometers. *Eur. Phys. J. C* **2020**, *80*, 694. [CrossRef]
40. Simon, J.; Verde, L.; Jimenez, R. Constraints on the redshift dependence of the dark energy potential. *Phys. Rev. D* **2005**, *71*, 123001. [CrossRef]
41. Stern, D.; Jimenez, R.; Verde, L.; Kamionkowski, M.; Stanford, S.A. Cosmic chronometers: Constraining the equation of state of dark energy. I: $H(z)$ measurements. *J. Cosmol. Astropart. Phys.* **2010**, *02*, 008. [CrossRef]

42. Moresco, M. Raising the bar: New constraints on the Hubble parameter with cosmic chronometers at z2. *Mon. Not. R. Astron. Soc.* **2015**, *450*, L16–L20. [CrossRef]
43. Ratsimbazafy, A.L.; Loubser, S.I.; Crawford, S.M.; Cress, C.M.; Bassett, B.A.; Nichol, R.C.; Väisänen, P. Age-dating luminous red galaxies observed with the Southern African Large Telescope. *Mon. Not. R. Astron. Soc.* **2017**, *467*, 3239–3254. [CrossRef]
44. Gao, L.-Y.; Zhao, Z.-W.; Xue, S.-S.; Zhang, X., Relieving the H_0 tension with a new interacting dark energy model. *J. Cosmol. Astropart. Phys.* **2021**, *07*, 005. [CrossRef]
45. Mamon, A.A.; Das, S. A divergence-free parametrization of deceleration parameter for scalar field dark energy. *Int. J. Mod. Phys. D* **2016**, *25*, 1650032. [CrossRef]
46. Hanafy, W.E.; Nashed, G.G.L. Phenomenological reconstruction of $f(T)$ teleparallel gravity. *Phys. Rev. D* **2019**, *100*, 083535. [CrossRef]
47. Gong, Y.; Wang, A. Reconstruction of the deceleration parameter and the equation of state of dark energy. *Phys. Rev. D* **2007**, *75*, 043520. [CrossRef]

physics

MDPI

Article

On Majorization Uncertainty Relations in the Presence of a Minimal Length

Alexey E. Rastegin

Department of Theoretical Physics, Irkutsk State University, K. Marx St. 1, Irkutsk 664003, Russia; alexrastegin@mail.ru

Abstract: The emergence of a minimal length at the Planck scale is consistent with modern developments in quantum gravity. This is taken into account by transforming the Heisenberg uncertainty principle into the generalized uncertainty principle. Here, the position-momentum commutator is modified accordingly. In this paper, majorization uncertainty relations within the generalized uncertainty principle are considered. Dealing with observables with continuous spectra, each of the axes of interest is divided into a set of non-intersecting bins. Such formulation is consistent with real experiments with a necessarily limited precision. On the other hand, the majorization approach is mainly indicative for high-resolution measurements with sufficiently small bins. Indeed, the effects of the uncertainty principle are brightly manifested just in this case. The current study aims to reveal how the generalized uncertainty principle affects the leading terms of the majorization bound for position and momentum measurements. Interrelations with entropic formulations of this principle are briefly discussed.

Keywords: generalized uncertainty principle; minimal observable length; majorization uncertainty relations

Citation: Rastegin, A.E. On Majorization Uncertainty Relations in the Presence of a Minimal Length. *Physics* **2022**, *4*, 1413–1425. https://doi.org/10.3390/physics4040091

Received: 22 September 2022
Accepted: 1 December 2022
Published: 14 December 2022

1. Introduction

One of the key problems of modern physics is to build a quantum theory of gravitation [1]. The existence of a minimal observable length has long been suggested due to these efforts. Let us refrain from listing them and refer to the papers [2,3]. There are proposals to investigate the testable effects of the minimal length, including astronomical observations [4,5] and experimental schemes seemingly feasible within current technology [6–8]. Papers [9–12] discussed measurements in which one may be able to probe the effects of quantum gravity. The implications of the deformed forms of the commutation relation have attracted large attention [13–18]. In particular, researchers analyzed the consequences for the harmonic oscillator [17,18], the free particle, and potentials with infinitely sharp boundaries [14]. Going beyond the linear regime in graphene, in Ref. [19], a generalized uncertainty framework compatible with quantum gravity scenarios with a minimal length was obtained.

The Heisenberg uncertainty principle [20] emphasizes fundamental limitations on the simultaneous knowledge of observables in the quantum world. Uncertainty relations in terms of the product of standard deviations were formally derived by Kennard [21] for position and momentum and later by Robertson [22] for any pair of observables. An alternative to this traditional approach is provided by entropic characterization. For the position-momentum pair, an entropic formulation was initiated by Hirschman [23] and later developed in Refs. [24,25]. With a primary focus on observables with discrete spectra, the use of entropies to characterize quantum uncertainties was explored in Refs. [26,27]. Being the subject of current research, entropic uncertainty relations are reviewed in Refs. [28–32]. The majorization approach provides another flexible way to pose uncertainty relations [33–37] with a natural transition to entropic characterization when required.

Heisenberg's uncertainty principle *per se* does not impose a restriction separately on the spreads of position or momentum. Below the scale linked to the Planck length, $\ell_{Pl} = \sqrt{G\hbar/c^3} \approx 1.616 \times 10^{-35}$ m, the very structure of space-time is an open question [38]. Here, G is the Newtonian constant of gravitation, $\hbar$ is the reduced Planck's constant, and c denotes the speed of light. The Heisenberg principle is replaced here with the generalized uncertainty principle, which declares a non-zero lower bound on the position spread [39–41]. The generalized uncertainty principle can be reinterpreted as an effective variation of the Planck constant [42], with a link to Dirac's large numbers hypothesis [43]. Using the preparation scenario, entropic uncertainty relations in the presence of a minimal length were examined in Refs. [44–46]. At each stage of the scenario of successive measurements, an actual pre-measurement state depends on the results of previous measurements [47,48]. This viewpoint is closer to Heisenberg's thought experiment with microscope [49]. The generalized uncertainty principle with successive measurements of position and momentum was analyzed in Ref. [50].

This paper is devoted to majorization uncertainty relations in the presence of a minimal length. To focus on changing the majorization bound for position and momentum measurements, a consideration is restricted here to the preparation scenario. In addition, the case of high-resolution measurements with small bins is most interesting from the physical viewpoints. Hence, one naturally obtains a small dimensionless parameter, with respect to which the quantities of interest can be expanded. For practical purposes, several leading terms in expansion of the majorization bound should be taken into account. It turns out that an effect of the generalized uncertainty relation is actually revealed in this way. The paper is organized as follows. Section 2 reviews the preliminary findings and fix the notation. The derivation of basic terms of the majorization bound is presented in Section 3. Section 4 concludes the paper with a summary of the results. In Appendix A, a perturbation theory is developed to solve an auxiliary eigenvalue problem.

2. Preliminaries

In this Section, the generalized uncertainty principle and related findings are recalled. Further, basic points of the majorization approach to quantum uncertainties are discussed.

2.1. The Generalized Uncertainty Principle

The generalized uncertainty principle declares the deformed commutation relation for the position and momentum operators [13]. Some different representations of the same algebra exist. However, the physical content is determined by the physical observables. Namely, these operators provide access to the explicit information on the position and momentum measurements [51]. For convenience, the wavenumber operator, $\hat{k}$, is used instead of the momentum operator, $\hbar\hat{k}$. Let us consider the commutation relation,

$$[\hat{x}, \hat{k}] = i\left(\mathbf{1} + \beta\hat{k}^2\right). \tag{1}$$

Here, the positive parameter β is rescaled by factor $\hbar^2$ from its known sense, and $\mathbf{1}$ is the identity operator. In the limit $\beta \to 0$, Equation (1) gives the known commutation relation of ordinary quantum mechanics. This is a most straightforward modification leading to the presence of a minimal length. Instead of Equation (1), more general forms of the additional term can be placed in the right-hand side [17]. Due to the results of [14], the used formulation allows us to study questions of interest with a more apparent analogy with the ordinary case. It is suitable at the first step in probing potential effects of the generalized uncertainty principle. In addition, the formulation (1) is asymmetric with respect to the role of position and momentum. It seems to be natural in topics concerning just the existence of a minimal length.

A quantum state is represented by a positive self-adjoint operator $\hat{\rho}$ with $\mathrm{Tr}(\hat{\rho}) = 1$ called the density matrix. Combining Equation (1) with the known Robertson formulation [22] leads to the inequality,

$$(\Delta\hat{x})_{\hat{\rho}}\,(\Delta\hat{\kappa})_{\hat{\rho}} \geq \frac{1}{2}\left(1 + \beta\langle\hat{\kappa}^2\rangle_{\hat{\rho}}\right) \geq \frac{1}{2}\left(1 + \beta(\Delta\hat{\kappa})^2_{\hat{\rho}}\right). \tag{2}$$

As in general, for any operator $\hat{Q}$ one has:

$$\langle\hat{Q}\rangle_{\hat{\rho}} = \mathrm{Tr}(\hat{Q}\hat{\rho}), \qquad (\Delta\hat{Q})^2_{\hat{\rho}} = \langle\hat{Q}^2\rangle_{\hat{\rho}} - \langle\hat{Q}\rangle^2_{\hat{\rho}}.$$

It further follows from Equation (2) that $(\Delta\hat{x})_{\hat{\rho}} \geq \sqrt{\beta}$ for every state $\hat{\rho}$. Thus, it is impossible to localize a particle below the scale corresponding to the square root of β.

It is helpful to introduce the auxiliary wavenumber operator $\hat{q}$ [14]. Let $\hat{x}$ and $\hat{q}$ be self-adjoint operators that obey $[\hat{x}, \hat{q}] = i\mathbf{1}$. In the q-space, the action of $\hat{q}$ results in multiplying a wave function $\varphi(q)$ by q, whereas $\hat{x}\,\varphi(q) = i\,\mathrm{d}\varphi/\mathrm{d}q$. Following [14], let us define

$$\hat{\kappa} = \frac{1}{\sqrt{\beta}}\tan\left(\sqrt{\beta}\,\hat{q}\right). \tag{3}$$

So, the auxiliary wavenumber obeys the ordinary commutation relation but ranges between $\pm\, q_{\max}(\beta) = \pm\,\pi/(2\sqrt{\beta})$. The function $q \mapsto \kappa = \tan(\sqrt{\beta}q)/\sqrt{\beta}$ provides a one-to-one correspondence between $q \in (-q_{\max}, +q_{\max})$ and $\kappa \in (-\infty, +\infty)$. So, the eigenvalues of $\hat{\kappa}$ fully cover the real axis. For a pure state, one actually has the three wave functions $\phi(\kappa)$, $\varphi(q)$, and $\psi(x)$. The auxiliary wave function $\varphi(q)$ is a convenient mathematical tool as connected with $\psi(x)$ via the Fourier transform. Let the eigenkets $|q\rangle$ of $\hat{q}$ be normalized through Dirac's delta function and satisfy the completeness relation,

$$\int_{-q_{\max}}^{+q_{\max}} \mathrm{d}q\,|q\rangle\langle q| = \mathbf{1}. \tag{4}$$

In the q-space, the eigenfunctions of $\hat{x}$ are expressed as $\langle q|x\rangle = \exp(-iqx)/\sqrt{2\pi}$. Combining this with Equation (4), any wave function in the coordinate space reads:

$$\psi(x) = \frac{1}{\sqrt{2\pi}}\int_{-q_{\max}}^{+q_{\max}} \exp(+iqx)\,\varphi(q)\,\mathrm{d}q\,. \tag{5}$$

Wave functions in the q- and x-spaces are connected by the Fourier transform [14], namely,

$$\varphi(q) = \frac{1}{\sqrt{2\pi}}\int_{-\infty}^{+\infty} \exp(-iqx)\,\psi(x)\,\mathrm{d}x\,. \tag{6}$$

The only distinction from ordinary quantum mechanics is that each wave function $\varphi(q)$ in the q-space should be treated as 0 for all $|q| > q_{\max}(\beta)$. However, a distribution of physical wavenumber values is determined by $\phi(\kappa)$. Let us consider the probability to find the momentum between two prescribed values. Due to the one-to-one correspondence between κ and q, there is a bijection between the intervals (κ_1, κ_2), and (q_1, q_2). Thus, the probability of interest is expressed as

$$\int_{\kappa_1}^{\kappa_2} |\phi(\kappa)|^2\,\mathrm{d}\kappa = \int_{q_1}^{q_2} |\varphi(q)|^2\,\mathrm{d}q\,, \tag{7}$$

whence the probability density functions are related via $|\phi(\kappa)|^2\,\mathrm{d}\kappa = |\varphi(q)|^2\,\mathrm{d}q$.

2.2. On Majorization Uncertainty Relations

Let us proceed to a general formulation of majorization uncertainty relations. Let $y = (y_1, \cdots, y_n)$ and $z = (z_1, \cdots, z_n)$ be two n-dimensional vectors with real components.

By adding zero components, one can always reach that the two vectors have the same number of elements. One says that y is majorized by z, in symbols $y \prec z$, if [52]

$$\sum_{j=1}^{m} y_j^{\downarrow} \leq \sum_{j=1}^{m} z_j^{\downarrow} \tag{8}$$

for all $m = 1, \dots, n$ and

$$\sum_{j=1}^{n} y_j^{\downarrow} = \sum_{j=1}^{n} z_j^{\downarrow} . \tag{9}$$

The arrows down mark that the components should be taken in non-increasing order.

To pose majorization uncertainty relations, the following notions will be used [33]. The infimum of a set of vectors is defined as the vector that is majorized by every element of the set and, in turn, majorizes any vector with that property [53]. The supremum is similarly defined as the vector that majorizes every element of the set and is, in turn, majorized by any vector with that property. The procedure to calculate the desired vectors are also discussed in Refs. [33,53] with a reference to the MATHEMATICA codes prepared for these purposes. Let us refrain from discussing some subtle points related to such calculations. Even if continuous observables are dealt with, one can nevertheless restrict a consideration to a finite set of large number of bins. This holds not only due to non-zero sizes of bins but also in view of boundness of values available to be measured in practice.

To formulate majorization uncertainty relations, one should fix operators that describe each measurement of interest. By $x_{\alpha 1} < x_{\alpha 2}$, one denotes the least points of α-th bin in the position measurement. The corresponding projection operator reads:

$$\hat{\Pi}_{\alpha} = \int_{x_{\alpha 1}}^{x_{\alpha 2}} \mathrm{d}x \, |x\rangle\langle x| \, . \tag{10}$$

To the momentum measurement, let us assign a set of projection operators of the form,

$$\hat{\Lambda}_{\gamma} = \int_{\kappa_{\gamma 1}}^{\kappa_{\gamma 2}} \mathrm{d}\kappa \, |\kappa\rangle\langle \kappa| \, , \tag{11}$$

where $\kappa_{\gamma 1} < \kappa_{\gamma 2}$ are the least points of γ-th bin. The above form of the operators corresponds to an orthogonal resolution of the identity in each case. Strictly speaking, the finiteness of the detector resolution is typically addressed in terms of acceptance functions [48]. Certainly, the projection operators are obtained with an acceptance function in the form of boxcar one. A consideration is restricted to boxcar acceptance functions since the aim here is to focus on the corollaries of the generalized uncertainty principle. At the same time, the use of Gaussian acceptance functions is apparently closer to practice [48]. On the other hand, the case of high-resolution measurements with small bins is of primary interest. Moreover, little changes of the form of acceptance functions have no actual bearing on the principal possibility to observe the effects of a minimal length.

One of the advantages of the majorization approach is that uncertainty relations are expressed directly in terms of probabilities. For the prepared pre-measurement state $\hat{\rho}$, one obtains the probabilities,

$$\mathrm{Tr}(\hat{\Pi}_{\alpha}\hat{\rho}) = \int_{x_{\alpha 1}}^{x_{\alpha 2}} \langle x|\hat{\rho}|x\rangle \, \mathrm{d}x \, , \qquad \mathrm{Tr}(\hat{\Lambda}_{\gamma}\hat{\rho}) = \int_{\kappa_{\gamma 1}}^{\kappa_{\gamma 2}} \langle \kappa|\hat{\rho}|\kappa\rangle \, \mathrm{d}\kappa \, ,$$

which respectively constitute the vectors $\mathrm{p}^{\hat{x}}(\hat{\rho})$ and $\mathrm{p}^{\hat{\kappa}}(\hat{\rho})$. The majorization uncertainty relation of the paper [33] is posed as follows. It was shown that

$$\mathrm{p}^{\hat{x}}(\hat{\rho}) \otimes \mathrm{p}^{\hat{\kappa}}(\hat{\rho}) \prec \sup\Big\{\mathrm{p}^{\hat{x}\oplus\hat{\kappa}}(\hat{\varrho}) : \, \hat{\varrho} \geq \mathbf{0}, \, \hat{\varrho}^{\dagger} = \hat{\varrho}, \, \mathrm{Tr}(\hat{\varrho}) = 1\Big\}. \tag{12}$$

In the case of discrete observables, one *a priori* has a unitary matrix connecting two orthonormal bases. Inspecting the norms of the submatices of this unitary matrix, the majorization uncertainty relations follow straight away [34,36]. Moreover, such relations are straightforwardly converted into inequalities for Rényi and Tsallis entropies. An application of these results to neutrino flavor and mass was studied in Ref. [54] since the Pontecorvo–Maki–Nakagawa–Sakata matrix is dealt with here. In a general case, however, finding the right-hand side of Equation (12) or some of its components can be more tractable than building a suitable unitary matrix. In addition, one will not necessarily be dealing with projective measurements.

In paper [33] it is described how to calculate the first term in the right-hand side of Equation (12). Overall, one seeks for the maximum value of $\mathrm{Tr}(\hat{\Pi}_\alpha\hat{\rho})\,\mathrm{Tr}(\hat{\Lambda}_\gamma\hat{\rho})$, where the projectors are fixed and $\hat{\rho}$ is varied. The desired extremal value is realized with a pure state, for example, $|\psi_*\rangle$. As was shown in Ref. [33], the task is reduced to the eigenvalue problem,

$$\hat{\Pi}_\alpha\,\hat{\Lambda}_\gamma\,\hat{\Pi}_\alpha|\eta_*\rangle=\mu^2\,|\eta_*\rangle \tag{13}$$

with $|\eta_*\rangle=\hat{\Pi}_\alpha|\psi_*\rangle$. One should find the maximal eigenvalue of the problem (13). The first component in the right-hand side of Equation (12) is then equal to $(1+\mu_{\max})^2/4$. Further terms are more difficult to calculate. For high-resolution measurements with small bins, however, many components differ little from the first, except for the tails of the distributions and the intermediate zones. In effect, the bins should be such that many of them are lying around distribution peaks. One can leave this assumption by replacing Equation (12) with the uncertainty relation in terms of the min-entropies. For the given probability distribution $\mathrm{p}=\{p_j\}$, its min-entropy is defined as

$$H_\infty(\mathrm{p})=-\ln\bigl(\max\,p_j\bigr)\,. \tag{14}$$

The latter is obtained when the order of Rényi's entropy [55] tends to infinity. It follows from Equations (12) and (14) that

$$H_\infty\bigl(\mathrm{p}^{\hat{x}}(\hat{\rho})\bigr)+H_\infty\bigl(\mathrm{p}^{\hat{\kappa}}(\hat{\rho})\bigr)\geq 2\ln 2-2\ln(1+\mu_{\max})\,. \tag{15}$$

The next Section examines how the above relations are affected by the generalized uncertainty principle.

3. Main Results

The previous Section provides a ground to study the question how the generalized uncertainty principle affects the majorization bound for position and momentum measurements. It is natural that the analysis here begins with the case $\beta=0$.

3.1. The Case of Ordinary Commutation Relation

In Ref. [33], the problem (13) was reformulated as

$$\frac{1}{\pi}\int\limits_{-\Delta x/2}^{+\Delta x/2}\frac{\sin\bigl(\Delta\kappa(x-x')/2\bigr)}{x-x'}\,\eta_*(x')\,\mathrm{d}x'=\mu^2\,\eta_*(x)\,, \tag{16}$$

where the variables x and x' are both restricted to the range $\bigl[-\Delta x/2,+\Delta x/2\bigr]$ and $\eta_*(x)=\langle x|\eta_*\rangle$. One uses Equation (16) under the assumption that the origins of both the x and κ axes are placed into centers of the two bins for which the optimality is reached. Surely, this holds for the ordinary commutation relation. To analyze consequences of the generalized uncertainty principle, one needs also to examine Equation (16) in more detail than it was made in the paper [33].

Substituting $x=\xi\Delta x$, $s=\Delta x\Delta\kappa/(2\pi)$ and $\mu^2=s\lambda$, one rewrites Equation (16) as

$$\int_{-1/2}^{+1/2}\frac{\sin[s\pi(\xi-\xi')]}{s\pi(\xi-\xi')}\,u(\xi')\,\mathrm{d}\xi'=\lambda\,u(\xi)\,, \tag{17}$$

where $u(\xi) = \eta_*(\xi \Delta x)$. Here, the kernel is expanded as

$$\frac{\sin[s\pi(\xi - \xi')]}{s\pi(\xi - \xi')} = 1 - \frac{s^2\pi^2}{3!}\,(\xi - \xi')^2 + \frac{s^4\pi^4}{5!}\,(\xi - \xi')^4 + \cdots . \tag{18}$$

It then follows from Equations (A16) and (A19) that

$$\mu^2_{\max}\Big|_{\beta=0} = s\left(1 - \frac{s^2\pi^2}{36} + O(s^4)\right), \tag{19}$$

$$\eta_*(x)\big|_{\beta=0} = 1 - \frac{s^2\pi^2}{6}\left(\frac{x^2}{\Delta x^2} - \frac{1}{12}\right) + O(s^4). \tag{20}$$

In the limit $s \to 0$, the term (19) tends to s as mentioned in Ref. [33]. The result (19) is useful also in the sense of characterizing a level of smallness for s. For example, for $s < 1/2$, one has:

$$\frac{s^2\pi^2}{36} < 0.069\,, \tag{21}$$

i.e., the eigenvalue correction turns out to be of several percents. Thus, the validity of perturbation expansions up to the first order does not require exceptionally high resolution. Within the given scheme, one can further derive second and higher-order perturbations, though complexity of expressions grows quickly.

3.2. *The Case of Modified Commutation Relation*

For $\beta \neq 0$, one keeps kernels of the Fourier transform by means of the auxiliary wavenumber. The eigenvalue problem (13) straightforwardly leads to

$$\begin{aligned}\mu^2\langle x|\eta_*\rangle &= \int_{\kappa_{\gamma 1}}^{\kappa_{\gamma 2}} \mathrm{d}\kappa \int_{x_{\alpha 1}}^{x_{\alpha 2}} \mathrm{d}x'\,\langle x|\kappa\rangle\langle\kappa|x'\rangle\,\eta_*(x') \\ &= \frac{1}{2\pi}\int_{x_{\alpha 1}}^{x_{\alpha 2}} \mathrm{d}x' \int_{q_{\gamma 1}}^{q_{\gamma 2}} \frac{2\mathrm{d}q}{1+\cos(2\sqrt{\beta}q)}\,\exp[i(x-x')q]\,\eta_*(x')\,,\end{aligned} \tag{22}$$

where one used

$$\frac{\mathrm{d}\kappa}{\mathrm{d}q} = \frac{2}{1+\cos(2\sqrt{\beta}q)}\,. \tag{23}$$

Let us now seek a possibility to translate the wavenumber axis as it was made to obtain Equation (16). Strictly speaking, this step can be used with the generalized uncertainty principle only approximately. Indeed, the standard kernels of the Fourier transform stand in Equations (5) and (6) due to the auxiliary wavenumber that ranges between $\pm\, q_{\max}(\beta)$. The latter indeed prevents translational invariance with respect to the q axis. On the other hand, the value $q_{\max}(\beta)$ corresponds to extremely high energies that are completely beyond the capabilities of modern experiments. Actually, reaching such high energies is inevitably coupled with approaching the Planck scale *per se*. Therefore, one deals with wavepackets supported in the momentum space far away from values of the mentioned order. In addition, such bins are advisably inclined to use that are small in comparison with characteristic spreading of typical wavepackets. Under these circumstances one is able to use shifts along the q and κ axes.

The eigenvalue problem (16) is then replaced with

$$\frac{1}{2\pi}\int\limits_{-\Delta x/2}^{+\Delta x/2} \eta_*(x')\,\mathrm{d}x' \int\limits_{-\Delta q/2}^{+\Delta q/2} \frac{2\exp[i(x-x')q]\,\mathrm{d}q}{1+\cos(2\sqrt{\beta}q)} = \mu^2\,\eta_*(x)\,. \tag{24}$$

Separating explicitly the term assigned to $\beta = 0$, one obtains:

$$\frac{1}{\pi}\int\limits_{-\Delta x/2}^{+\Delta x/2} \eta_*(x')\,\mathrm{d}x' \Bigg\{ \frac{\sin\big(\Delta q(x-x')/2\big)}{x-x'} + \frac{1}{2}\int\limits_{-\Delta q/2}^{+\Delta q/2}\left(\frac{2}{1+\cos(2\sqrt{\beta}q)} - 1\right)\exp[i(x-x')q]\,\mathrm{d}q \Bigg\} = \mu^2\,\eta_*(x)\,. \tag{25}$$

Let us take $q = y\Delta q/2$, $s = \Delta x \Delta q/(2\pi)$ and

$$\varkappa = (x-x')\,\frac{\Delta q}{2} = s\pi(\xi - \xi')\,, \tag{26}$$

then

$$\frac{1}{2}\int\limits_{-\Delta q/2}^{+\Delta q/2}\left(\frac{2}{1+\cos(2\sqrt{\beta}q)} - 1\right)\exp[i(x-x')q]\,\mathrm{d}q = \frac{\beta\Delta q^3}{8}\,I_2(\varkappa) + O(\beta^2)\,, \tag{27}$$

with $I_n(\varkappa) = (1/2)\int_{-1}^{+1} y^n \exp(i\varkappa y)\,\mathrm{d}y$ for $n = 1, 2, \ldots$. One can see from Equations (25) and (27) that

$$\int\limits_{-\Delta x/2}^{+\Delta x/2}\left(\frac{\sin\varkappa}{\pi(x-x')} + \frac{\beta\Delta q^3}{8\pi}\,I_2(\varkappa) + O(\beta^2)\right)\eta_*(x')\,\mathrm{d}x' = \mu^2\,\eta_*(x)\,.$$

By changing the variable to $\xi' = x'/\Delta x$, one finally obtains:

$$\int_{-1/2}^{+1/2}\left(\frac{s\sin\varkappa}{\varkappa} + \frac{s\beta\Delta q^2}{4}\,I_2(\varkappa) + O(\beta^2)\right)u(\xi')\,\mathrm{d}\xi' = \mu^2\,u(\xi)\,. \tag{28}$$

Substituting $\mu^2 = s\lambda$ again, Equation (28) is reduced to the form,

$$(\mathsf{K}^{(0)} + \varepsilon\,\mathsf{K}^{(1)} + \cdots)u(\xi) = \lambda\,u(\xi)\,, \tag{29}$$

with intent to use the formulas of Appendix A for $\varepsilon = \beta\Delta q^2/4$,

$$\mathsf{K}^{(0)} f(\xi) = \int_{-1/2}^{+1/2} \frac{\sin\varkappa}{\varkappa}\, f(\xi')\,\mathrm{d}\xi'\,,$$

$$\mathsf{K}^{(1)} f(\xi) = \int_{-1/2}^{+1/2} \frac{\varkappa^2\sin\varkappa + 2\varkappa\cos\varkappa - 2\sin\varkappa}{\varkappa^3}\, f(\xi')\,\mathrm{d}\xi'\,.$$

It follows from Equation (A15) that the factor of $\beta\Delta q^2/4$ in the first-order correction reads as

$$\langle \widetilde{w}_0, \mathsf{K}^{(1)}\widetilde{w}_0\rangle = \int_{-1/2}^{+1/2}\mathrm{d}\xi\int_{-1/2}^{+1/2}\mathrm{d}\xi'\,\frac{\varkappa^2\sin\varkappa + 2\varkappa\cos\varkappa - 2\sin\varkappa}{\varkappa^3}\,\widetilde{w}_0(\xi)\,\widetilde{w}_0(\xi')\,, \tag{30}$$

where

$$\widetilde{w}_0(\xi) = 1 - \frac{s^2\pi^2}{6}\left(\xi^2 - \frac{1}{12}\right) + O(s^4)$$

due to Equation (A19). Again, a consideration is aimed to be restricted to the case of sufficiently high resolution. The calculations show that

$$\frac{\varkappa^2\sin\varkappa + 2\varkappa\cos\varkappa - 2\sin\varkappa}{\varkappa^3} = \frac{1}{3} - \frac{s^2\pi^2}{10}\,(\xi - \xi')^2 + \cdots\,, \tag{31}$$

and further,

$$\langle \varpi_0, \mathsf{K}^{(1)} \varpi_0 \rangle = \int_{-1/2}^{+1/2} d\xi \int_{-1/2}^{+1/2} d\xi' \left\{ \frac{1}{3} - \frac{s^2\pi^2}{10} (\xi - \xi')^2 - \frac{s^2\pi^2}{18} \left(\xi^2 - \frac{1}{12} \right) \right.$$
$$\left. - \frac{s^2\pi^2}{18} \left(\xi'^2 - \frac{1}{12} \right) + \cdots \right\} = \frac{1}{3} - \frac{s^2\pi^2}{60} + O(s^4) \,. \tag{32}$$

Summing up, the result is obtained in the form,

$$\frac{\mu_{\max}^2}{s} = \lambda = 1 + \frac{\beta \Delta q^2}{12} - \frac{s^2\pi^2}{12} \left(\frac{1}{3} + \frac{\beta \Delta q^2}{20} \right) + \cdots \,. \tag{33}$$

The latter allows us to probe how the generalized uncertainty principle affects majorization uncertainty relations for position and momentum.

3.3. Discussion

Thus, expressions have been obtained for the leading term in majorization uncertainty relations in the presence of a minimal length. As is seen from the right-hand side of Equation (33), for measurements with sufficiently high resolution the maximal eigenvalue grows with an increase in β. Hence, the lower bound of the uncertainty relation (15) will decrease. This tendency is interesting in comparison with other uncertainty relations. It also follows from Equation (33) that the effects of changing Δx and Δq on the actual level of uncertainty differ. The position bin governs $\mu_{\max}$ only via s, whereas the auxiliary-wavenumber bin Δq does also through the terms involving β. At the fixed s, changes of a typical size of momentum bins are more influential on the amount of uncertainties. Certainly, these findings are stipulated by the initial choice of the deformed commutation relation.

Substituting Equation (33) into Equation (15) leads straight to

$$H_\infty(\mathrm{p}^{\hat{x}}(\hat{\rho})) + H_\infty(\mathrm{p}^{\hat{k}}(\hat{\rho}))$$
$$\geq 2\ln 2 - 2\ln\left\{ 1 + \sqrt{s}\left(1 + \frac{\beta \Delta q^2}{24} - \frac{s^2\pi^2}{24} \left(\frac{1}{3} + \frac{\beta \Delta q^2}{20} \right) + \cdots \right) \right\}. \tag{34}$$

It is instructive to compare Equation (34) with the uncertainty relation,

$$H_1(\mathrm{p}^{\hat{x}}(\hat{\rho})) + H_1(\mathrm{p}^{\hat{k}}(\hat{\rho})) \geq \ln\left(\frac{e\pi}{\Delta x \Delta \kappa} \right) + \langle \ln(\mathbf{1} + \beta \hat{k}^2) \rangle_{\hat{\rho}} \,, \tag{35}$$

proved in Ref. [45]. By $H_1(\mathrm{p})$, the Shannon entropy of the corresponding probability distribution is meant. Since the left-hand side of Equation (34) includes the two min-entropies, it differs from entropic uncertainty relations of Refs. [45,50]. Therein, the entropic parameters are connected due to the use of inequalities between the corresponding norms of a function and its Fourier transform. Hence, the obtained inequalities cannot involve min-entropies for both the observables. In this regard, the discussion here completed the consideration of the paper [45]. Another difference is that the right-hand side of Equation (34) decreases with β, at least for high-resolution measurements. In contrast, the correction term in the right-hand side of Equation (35) increases. This distinction reflects that the min-entropies depend only on the maximal probabilities. Naturally, the scope of Equation (34) is restricted to measurements with sufficiently high resolution. In the meantime, only such measurements are recognized as capable to verify uncertainty relations of various forms.

A natural question arises about the right-hand side of Equation (33). What are, in order of magnitude, the terms depending on β? To answer the question, let us make some plausible assumptions about the typical values of Δx and Δq. Surely, they are mainly determined by the capabilities of modern experimental techniques. Any detailed discussion is beyond the scope of this paper. Instead, one can refer to the concrete experimental results of the verification of the Heisenberg uncertainty principle [56]. On average, typical bins can be estimated as $\Delta x \sim 100$ nm and $\Delta\kappa \sim 10^7$ 1/m. The latter also holds for Δq in view of $\kappa = \tan(\sqrt{\beta}q)/\sqrt{\beta}$ and given that

$$\sqrt{\beta} \sim \ell_{\mathrm{Pl}} = 1.616 \times 10^{-35}\,\mathrm{m}\,.$$

The calculations then give $s \sim 1/(2\pi) \approx 0.159$ and

$$\frac{\beta\Delta q^2}{s^2\pi^2} \sim 10^{-55}\,. \tag{36}$$

This value characterizes a ratio of the second term to the third one in the right-hand side of Equation (33). It is not surprising that the effects of the generalized uncertainty principle are estimated as extremely small.

Direct observational evidence for a foamed structure of space-time at the Planck scale seem to be currently unfeasible with an elementary particle as probe. In this way, the theoretical results of the form of Equation (34) are also unable to assist the presence of a minimal length in testing. Instead, paper [7] considered the use of a macroscopic probe for exploring space-time "roughness" at the relevant scale. It was found that, within the given level of ultrahigh vacuum and cryogenic technology, the proposed tabletop experiment could already be sensitive sufficiently. A witness for space-time "roughness" is provided the frequency of a certain event with a single photon turns out to be significantly less than the expected level. Applications of uncertainty relations to experiments of such a kind deserve to be studied in a separate investigation.

4. Conclusions

We have considered majorization uncertainty relations for position and momentum measurements in the presence of a minimal length. In particular, the uncertainty relation in terms of min-entropies was also formulated. In general, the proposed approach develops the treatment of Ref. [33] in combination with the generalized uncertainty principle. It was advisable from the physical viewpoint to focus on position and momentum measurements with sufficiently high resolution. In this way, one has derived corrections to leading terms of the majorization bound for the corresponding observables discretized into bins. Naturally, the changes of interest are determined by the parameter β that controls the modified commutation relation (1). The presented results allow us to reveal typical behavior and to estimate an order of corrections induced by the presence of a minimal length.

The obtained expressions with β are closely related to the structure of the modified commutation relation. These terms reveal some features that one could expect from the physical viewpoint. It is natural enough that lower bounds on uncertainty quantifiers will rather increase with the growth of β. The commutation relation (1) is indeed asymmetric in handling with position and momentum. The correction terms reflect this property, and the shortening of the momentum bins has a greater effect than the shortening of the position ones. At the same time, all of the mentioned changes lie in a so narrow range that they can hardly be probed within the capabilities of the modern experiment. A relative weight of the correction terms was estimated at a level undetectable in practice. It is apparent that a real successful experiment to show the presence a minimal length would be an exceptional advance.

Funding: This research received no external funding.

Data Availability Statement: Not applicable.

Conflicts of Interest: The author declares no conflict of interest.

Appendix A. Solution of the Eigenvalue Problem

It is useful to consider the eigenvalue problem in a form posed as

$$\int_{-1/2}^{+1/2} k(\xi,\xi')\,u(\xi')\,\mathrm{d}\xi' = \lambda\,u(\xi)\,, \tag{A1}$$

where the kernel is expanded as

$$k(\xi,\xi') = k^{(0)}(\xi,\xi') + \varepsilon\,k^{(1)}(\xi,\xi') + \varepsilon^2 k^{(2)}(\xi,\xi') + \cdots\,. \tag{A2}$$

The case (18) is gained for $\varepsilon = s^2$, $k^{(0)}(\xi,\xi') = 1$, $k^{(1)}(\xi,\xi') = A(\xi-\xi')^2$ with $A = -\pi^2/6$, and so on. The operator in the left-hand side of Equation (A1) is a Hilbert–Schmidt one, whenever

$$\int_{-1/2}^{+1/2} \mathrm{d}\xi \int_{-1/2}^{+1/2} \mathrm{d}\xi'\,\left|k(\xi,\xi')\right|^2 < \infty\,. \tag{A3}$$

It is known that Hilbert–Schmidt integral operators are both continuous and compact. The current study deals with symmetric real kernels, i.e., $k(\xi,\xi') = k(\xi',\xi)$ and $k(\xi,\xi') = k(\xi,\xi')^*$. Then, the integral operators of interest are self-adjoint.

For $\varepsilon = 0$, one obtains from Equation (17) the eigenvalue problem,

$$\int_{-1/2}^{+1/2} w(\xi')\,\mathrm{d}\xi' = \nu\,w(\xi)\,, \tag{A4}$$

with the eigenvalues $\nu_0 = 1$ and $\nu_1 = 0$. The function $w_0(\xi) = 1$ corresponding to $\nu_0 = 1$ is normalized as

$$\int_{-1/2}^{+1/2} w_0(\xi)^2 \mathrm{d}\xi = 1\,. \tag{A5}$$

The eigenvalue $\nu_1 = 0$ is degenerate, and there is a countably infinite set $\{w_n(\xi)\}_{n=1}^{\infty}$ of eigenfunctions such that

$$\int_{-1/2}^{+1/2} w_n(\xi)\,\mathrm{d}\xi = 0\,, \qquad \int_{-1/2}^{+1/2} w_m(\xi)\,w_n(\xi)\,\mathrm{d}\xi = \delta_{mn}\,, \tag{A6}$$

where δ_{mn} is the Kronecker delta. The first of these formulas implies $\langle w_0, w_n\rangle = 0$ for $n = 1,2,\ldots$ as well. One can recall here the Legendre polynomials $P_n(y)$ with the generating function,

$$\frac{1}{\sqrt{1-2yt+t^2}} = \sum_{n=0}^{\infty} P_n(y)\,t^n\,. \tag{A7}$$

Taking $y = 2\xi$, one can herewith write

$$w_n(\xi) = \sqrt{2n+1}\,P_n(2\xi) \qquad (n = 0,1,2,\ldots)\,. \tag{A8}$$

Apparently, the choice (A8) is not unique, but it is sufficient for the purposes here. In general, the quantities of interest are represented by expansions,

$$\lambda_n = \lambda_n^{(0)} + \varepsilon\,\lambda_n^{(1)} + \varepsilon^2\lambda_n^{(2)} + \cdots\,, \tag{A9}$$

$$u_n(\xi) = u_n^{(0)}(\xi) + \varepsilon\,u_n^{(1)}(\xi) + \varepsilon^2 u_n^{(2)}(\xi) + \cdots\,. \tag{A10}$$

For the problem (17), one has $\lambda_n^{(0)} = \delta_{n0}$ and $u_n^{(0)}(\xi) = w_n(\xi)$. One also deals with the Hilbert–Schmidt operators of the form,

$$\mathsf{K}^{(n)} f(\xi) = \int_{-1/2}^{+1/2} k^{(n)}(\xi, \xi') \, f(\xi') \, \mathrm{d}\xi' \,, \tag{A11}$$

$$\mathsf{K} = \mathsf{K}^{(0)} + \varepsilon \, \mathsf{K}^{(1)} + \varepsilon^2 \mathsf{K}^{(2)} + \cdots . \tag{A12}$$

Let us write the total equation as

$$\begin{aligned} &\big(\mathsf{K}^{(0)} + \varepsilon \, \mathsf{K}^{(1)} + \cdots\big)\big(u_n^{(0)}(\xi) + \varepsilon \, u_n^{(1)}(\xi) + \cdots\big) \\ &= \big(\lambda_n^{(0)} + \varepsilon \, \lambda_n^{(1)} + \cdots\big)\big(u_n^{(0)}(\xi) + \varepsilon \, u_n^{(1)}(\xi) + \cdots\big) \,, \end{aligned} \tag{A13}$$

whence

$$\mathsf{K}^{(1)} u_n^{(0)}(\xi) + \mathsf{K}^{(0)} u_n^{(1)}(\xi) = \lambda_n^{(1)} u_n^{(0)}(\xi) + \lambda_n^{(0)} \, u_n^{(1)}(\xi) \tag{A14}$$

in the first order. Due to self-adjointness, one finally obtains:

$$\lambda_n^{(1)} = \langle u_n^{(0)}, \mathsf{K}^{(1)} u_n^{(0)} \rangle \,. \tag{A15}$$

For the problem (17), the first-order correction to the eigenvalue reads as

$$\lambda_n^{(1)} = (2n+1) \int_{-1/2}^{+1/2} \mathrm{d}\xi \int_{-1/2}^{+1/2} \mathrm{d}\xi' \, k^{(1)}(\xi, \xi') \, P_n(2\xi) \, P_n(2\xi') \,.$$

For $n = 0$, this formula gives

$$\lambda_0^{(1)} = \int_{-1/2}^{+1/2} \mathrm{d}\xi \int_{-1/2}^{+1/2} \mathrm{d}\xi' \, k^{(1)}(\xi, \xi') = \frac{A}{6} \,, \tag{A16}$$

provided that $k^{(1)}(\xi, \xi') = A(\xi - \xi')^2$.

In order to examine Equation (29), one needs to know the corresponding eigenfunction of the problem (17). For $n = 0$, the first-order correction to the eigenfunction is obtained from $\langle w_0, u_0^{(1)} \rangle = 0$. The latter can be rewritten as

$$\mathsf{K}^{(0)} u_0^{(1)}(\xi) = \int_{-1/2}^{+1/2} u_0^{(1)}(\xi') \, \mathrm{d}\xi' = 0 \,, \tag{A17}$$

since $w_0(\xi) = 1$ and $k^{(0)}(\xi, \xi') = 1$. Combining Equation (A14) with Equation (A17) finally gives

$$u_0^{(1)}(\xi) = \mathsf{K}^{(1)} w_0(\xi) - \lambda_0^{(1)} w_0(\xi) \,. \tag{A18}$$

By $k^{(1)}(\xi, \xi') = A(\xi - \xi')^2$, one has:

$$u_0^{(1)}(\xi) = A\left(\xi^2 + \frac{1}{12}\right) - \frac{A}{6} = A\left(\xi^2 - \frac{1}{12}\right) . \tag{A19}$$

References

1. Rovelli, C. *Quantum Gravity*; Cambridge University Press: Cambridge, UK, 2004. [CrossRef]
2. Garay, L.J. Quantum gravity and minimum length. *Int. J. Mod. Phys. A* **1995**, *10*, 145–165. [CrossRef]
3. Hossenfelder, S. Minimal length scale scenarios for quantum gravity. *Living Rev. Relativ.* **2013**, *16*, 2. [CrossRef] [PubMed]
4. Amelino-Camelia, G.; Ellis, J.; Mavromatos, N.E.; Nanopoulos, D.V.; Sarkar, S. Tests of quantum gravity from observations of γ-ray bursts. *Nature* **1998**, *393*, 763–765. [CrossRef]
5. Jacob, U.; Piran, T. Neutrinos from gamma-ray bursts as a tool to explore quantum-gravity-induced Lorentz violation. *Nature Phys.* **2007**, *3*, 87–90. [CrossRef]
6. Pikovski, I.; Vanner, M.R.; Aspelmeyer, M.; Kim, M.; Brukner, Č. Probing Planck-scale physics with quantum optics. *Nature Phys.* **2012**, *8*, 393–397. [CrossRef]

7. Bekenstein, J.D. Is a tabletop search for Planck scale signals feasible? *Phys. Rev. D* **2012**, *86*, 124040. [CrossRef]
8. Marin, F.; Marino, F.; Bonaldi, M.; Cerdonio, M.; Conti, L.; Falferi, P.; Mezzena, R.; Ortolan, A.; Prodi, G.A.; Taffarello, L.; et al. Gravitational bar detectors set limits to Planck-scale physics on macroscopic variables. *Nature Phys.* **2013**, *9*, 71–73. [CrossRef]
9. Das, S.; Vagenas, E. Universality of quantum gravity corrections. *Phys. Rev. Lett.* **2008**, *101*, 221301. [CrossRef]
10. Ali, A.F.; Khalil, M.M.; Vagenas, E.C. Minimal length in quantum gravity and gravitational measurements. *EPL (Europhys. Lett.)* **2015**, *112*, 20005. [CrossRef]
11. Bousso, K.; Halpern, I.; Koeller, J. Mass bounds for compact spherically symmetric objects in generalized gravity theories. *Phys. Rev. D* **2016**, *94*, 064047. [CrossRef]
12. Howl, R.; Hackermüller, L.; Bruschi, D.E.; Fuentes, I. Gravity in the quantum lab. *Adv. Phys. X* **2018**, *3*, 1383184. [CrossRef]
13. Kempf, A.; Mangano, G.; Mann, R.B. Hilbert space representation of the minimal length uncertainty relation. *Phys. Rev. D* **1995**, *52*, 1108–1118. [CrossRef] [PubMed]
14. Pedram, P. New approach to nonperturbative quantum mechanics with minimal length uncertainty. *Phys. Rev. D* **2012**, *85*, 024016. [CrossRef]
15. Bosso, P. Rigorous Hamiltonian and Lagrangian analysis of classical and quantum theories with minimal length. *Phys. Rev. D* **2018**, *97*, 126010. [CrossRef]
16. Chung, W.S.; Hassanabadi, H. A new higher order GUP: One dimensional quantum system. *Eur. Phys. J. C* **2019**, *79*, 213. [CrossRef]
17. Bosso, P.; Luciano, G.G. Generalized uncertainty principle: From the harmonic oscillator to a QFT toy model. *Eur. Phys. J. C* **2021**, *81*, 982. [CrossRef]
18. Kim, M.S.; Hwang, M.-R.; Jung, E.; Park, D.K. Rényi and von Neumann entropies of thermal state in generalized uncertainty principle-corrected harmonic oscillator. *Mod. Phys. Lett. A* **2021**, *36*, 2150250. [CrossRef]
19. Iorio, A.; Pais, P.; Elmashad, I.A.; Ali, A.F.; Faizal, M.; Abou-Salem, L.I. Generalized Dirac structure beyond the linear regime in graphene. *Int. J. Mod. Phys. D* **2018**, *27*, 1850080. [CrossRef]
20. Heisenberg, W. Über den anschaulichen inhalt der quanten theoretischen kinematik und mechanik. *Z. Phys.* **1927**, *43*, 172–198. [CrossRef]
21. Kennard, E.H. Zur quantenmechanik einfacher bewegungstypen. *Z. Phys.* **1927**, *44*, 326–352. [CrossRef]
22. Robertson, H.P. The uncertainty principle. *Phys. Rev.* **1929**, *34*, 163–164. [CrossRef]
23. Hirschman, I.I. A note on entropy. *Amer. J. Math.* **1957**, *79*, 152–156. [CrossRef]
24. Beckner, W. Inequalities in Fourier analysis. *Ann. Math.* **1975**, *102*, 159–182. [CrossRef]
25. Białynicki-Birula, I.; Mycielski, J. Uncertainty relations for information entropy in wave mechanics. *Commun. Math. Phys.* **1975**, *44*, 129–132. [CrossRef]
26. Deutsch, D. Uncertainty in quantum measurements. *Phys. Rev. Lett.* **1983**, *50*, 631–633. [CrossRef]
27. Maassen, H.; Uffink, J.B.M. Generalized entropic uncertainty relations. *Phys. Rev. Lett.* **1988**, *60*, 1103–1106. [CrossRef] [PubMed]
28. Wehner, S.; Winter, A. Entropic uncertainty relations—A survey. *New J. Phys.* **2010**, *12*, 025009. [CrossRef]
29. Białynicki-Birula, I.; Rudnicki, Ł. Entropic uncertainty relations in quantum physics. In *Statistical Complexity*; Sen, K.D., Ed.; Springer: Dordrecht, The Netherlands, 2011; pp. 1–34. [CrossRef]
30. Coles, P.J.; Berta, M.; Tomamichel, M.; Wehner, S. Entropic uncertainty relations and their applications. *Rev. Mod. Phys.* **2017**, *89*, 015002. [CrossRef]
31. Toscano, F.; Tasca, D.S.; Rudnicki, Ł.; Walborn, S.P. Uncertainty relations for coarse-grained measurements: An overview. *Entropy* **2018**, *20*, 454. [CrossRef]
32. Hertz, A.; Cerf, N.J. Continuous-variable entropic uncertainty relations. *J. Phys. A Math. Theor.* **2019**, *52*, 173001. [CrossRef]
33. Partovi, M.H. Majorization formulation of uncertainty in quantum mechanics. *Phys. Rev. A* **2011**, *84*, 052117. [CrossRef]
34. Puchała, Z.; Rudnicki, Ł.; Życzkowski, K. Majorization entropic uncertainty relations. *J. Phys. A Math. Theor.* **2013**, *46*, 272002. [CrossRef]
35. Friedland, S.; Gheorghiu, V.; Gour, G. Universal uncertainty relations. *Phys. Rev. Lett.* **2013**, *111*, 230401. [CrossRef]
36. Rudnicki, Ł.; Puchała, Z.; Życzkowski, K. Strong majorization entropic uncertainty relations. *Phys. Rev. A* **2014**, *89*, 052115. [CrossRef]
37. Rastegin, A.E.; Życzkowski, K. Majorization entropic uncertainty relations for quantum operations. *J. Phys. A Math. Theor.* **2016**, *49*, 355301. [CrossRef]
38. Amati, D.; Ciafaloni, M.; Veneziano, G. Can spacetime be probed below the string size? *Phys. Lett. B* **1989**, *216*, 41–47. [CrossRef]
39. Scardigli, F. Generalized uncertainty principle in quantum gravity from micro-black hole gedanken experiment. *Phys. Lett. B* **1999**, *452*, 39–44. [CrossRef]
40. Bambi, C. A revision of the generalized uncertainty principle. *Class. Quantum Grav.* **2008**, *25*, 105003. [CrossRef]
41. Tawfik, A.N.; Diab, A.M. A review of the generalized uncertainty principle. *Rep. Prog. Phys.* **2015**, *78*, 126001. [CrossRef]
42. Ali, A.F.; Elmashad, I.; Mureika, J. Universality of minimal length. *Phys. Lett. B* **2022**, *831*, 137182. [CrossRef]
43. Dirac, P.A.M. Cosmological models and the large numbers hypothesis. *Proc. R. Soc. Lond. A* **1974**, *338*, 439–446.
44. Abdelkhalek, K.; Chemissany, W.; Fiedler, L.; Mangano, G.; Schwonnek, R. Optimal uncertainty relations in a modified Heisenberg algebra. *Phys. Rev. D* **2016**, *94*, 123505. [CrossRef]
45. Rastegin, A.E. On entropic uncertainty relations in the presence of a minimal length. *Ann. Phys.* **2017**, *382*, 170–180. [CrossRef]

46. Hsu, L.-Y.; Kawamoto, S.; Wen, W.-Y. Entropic uncertainty relation based on generalized uncertainty principle. *Mod. Phys. Lett. A* **2017**, *32*, 1750145. [CrossRef]
47. Srinivas, M.D. Optimal entropic uncertainty relation for successive measurements in quantum information theory. *Pramana J. Phys.* **2003**, *60*, 1137–1152. [CrossRef]
48. Distler, J.; Paban, S. Uncertainties in successive measurements. *Phys. Rev. A* **2013**, *87*, 062112. [CrossRef]
49. Rastegin, A.E. Entropic uncertainty relations for successive measurements of canonically conjugate observables. *Ann. Phys.* **2016**, *528*, 835–844. [CrossRef]
50. Rastegin, A.E. Entropic uncertainty relations for successive measurements in the presence of a minimal length. *Entropy* **2018**, *20*, 354. [CrossRef]
51. Bosso, P.; Petruzziello, L.; Wagner, F. The minimal length is physical. *Phys. Lett. B* **2022**, *834*, 137415. [CrossRef]
52. Bhatia, R. *Matrix Analysis*; Springer: New York, NY, USA, 1997. [CrossRef]
53. Partovi, M.H. Correlative capacity of composite quantum states. *Phys. Rev. Lett.* **2009**, *103*, 230502. [CrossRef]
54. Rastegin, A.E.; Shemet, A.M. Flavor-mass majorization uncertainty relations and their links to the mixing matrix. *Mod. Phys. Lett. A* **2021**, *36*, 2150211. [CrossRef]
55. Rényi, A. On Measures of Entropy and Information. In *Proceedings of the Fourth Berkeley Symposium on Mathematical Statistics and Probability (Berkeley/Los Angeles, CA, USA, June 20 – July 30, 1960). Volume 1: Contributions to the Theory of Statistics*; Neyman, J., Ed.; University of California Press: Berkeley/Los Angeles, CA, USA, 1961; pp. 547–561. Available online: https://projecteuclid.org/proceedings/berkeley-symposium-on-mathematical-statistics-and-probability/proceedings-of-the-fourth-berkeley-symposium-on-mathematical-statistics-and-probability-volume-1-contributions-to-the-theory-of-statistics/toc/bsmsp/1200512153 (accessed on 25 November 2022).
56. Nairz, O.; Arndt, M.; Zeilinger, A. Experimental verification of the Heisenberg uncertainty principle for fullerene molecules. *Phys. Rev. A* **2002**, *65*, 032109. [CrossRef]

physics

MDPI

Article

On Momentum Operators Given by Killing Vectors Whose Integral Curves Are Geodesics

Thomas Schürmann

Jülich Supercomputing Centre, Jülich Research Centre, D-52425 Jülich, Germany; t.schurmann@icloud.com

Abstract: The paper considers momentum operators on intrinsically curved manifolds. Given that momentum operators are Killing vector fields whose integral curves are geodesics, the corresponding manifold is flat or of the compact type with positive constant sectional curvature and dimensions equal to 1, 3, or 7. Explicit representations of momentum operators and the associated Casimir element are discussed for the 3-sphere S^3. It is verified that the structural constants of the underlying Lie algebra are proportional to $2\,\hbar/R$, where R is the curvature radius of S^3 and $\hbar$ is the reduced Planck's constant. This results in a countable energy and momentum spectrum of freely moving particles in S^3. The maximal resolution of the possible momenta is given by the de Broglie wave length, $\lambda_R = \pi R$, which is identical to the diameter of the manifold. The corresponding covariant position operators are defined in terms of geodesic normal coordinates, and the associated commutator relations of position and momentum are established.

Keywords: generalized uncertainty principle; curved spacetime; extended uncertainty principle; quantization in curved space

Citation: Schürmann, T. On Momentum Operators Given by Killing Vectors Whose Integral Curves Are Geodesics. *Physics* **2022**, 4, 1440–1452. https://doi.org/10.3390/physics4040093

Received: 23 September 2022
Accepted: 17 November 2022
Published: 15 December 2022

1. Introduction

Every generalization of the ordinary momentum operator in quantum mechanics to intrinsically curved manifolds strongly depends on the assumptions that are supposed to be established. Those assumptions are mostly based on the rules of quantum mechanics in Cartesian coordinates of the flat Euclidean space.

Let M be an n-dimensional smooth Riemannian manifold with metric g (occasionally denoted by $\langle\cdot,\cdot\rangle$). At every point $p \in M$, smooth manifolds admit a tangent space, T_pM, which is an n-dimensional real vector space. For every smooth function f on M, consider the action of differential form, df. Since df is a map on tangential space T_pM at point $p \in M$, the gradient of f is defined, such that

$$df(v) = \langle v, \mathrm{grad} f\rangle \tag{1}$$

for every smooth vector field $v \in T_pM$. From this definition, one can obtain the formal expression of the gradient vector field by

$$\mathrm{grad} f = (\mathrm{grad} f)^k \partial_k \tag{2}$$

with contravariant components $(\mathrm{grad} f)^i = g^{ik}\partial_k f$, where g^{ik} are components of the inverse metric g, $\{\partial_k\}$ is the natural basis of T_pM, and the Latin letters denote the space indexes, $k = 1, 2, \cdots, n$. With this notation, the traditional momentum operator of Euclidean space in Cartesian coordinates is given by covariant components, $\hat{p}_k f = -i\hbar(\mathrm{grad} f)_k$, which are commonly written as

$$\hat{p}_k = -i\hbar\partial_k \tag{3}$$

in physics textbooks; here, $\hbar$ is the reduced Planck's constant. Following De Witt [1,2], for intrinsically curved manifolds, one can obtain more general momentum operators that also satisfy the canonical commutation relations:

$$\left[\hat{x}^j, \hat{p}_k\right] = i\hbar\delta^j_k, \tag{4}$$

$$\left[\hat{p}_j, \hat{p}_k\right] = \left[\hat{x}_j, \hat{x}_k\right] = 0. \tag{5}$$

This yields an enhanced quantization rule

$$\hat{p}_k = -i\hbar\left(\partial_k + \Gamma^j_{jk}(x)\right), \tag{6}$$

where the curvature of the manifold is reflected by the contracted Christoffel symbols $\Gamma^j_{jk}(x)$. Compared to the Cartesian quantization rule in Euclidean space (3), the ordinary partial derivative is replaced by the definition (6). An advantage of this quantization rule is its applicability to a wide range of curved spaces and the validity of the canonical commutator relations. However, in order to set up a quantum Hamiltonian, one has to keep in mind that there is no unique prescription to quantize the classical curved space Hamilton function. This is because of operator orderings of the kinetic energy term result in different, inequivalent quantum corrections, such that the correct Hamiltonian can only be confirmed empirically.

On the other hand, a particular property that is, in a sense, self-evident for the Cartesian case is that the partial derivatives ∂_k of the momentum representation in Equation (3) can be understood as *orthonormal* Killing vectors whose isometries are "translations" in Euclidean space. Strictly speaking, the integral curves of this Killing vectors (called Killing trajectories) are geodesics, especially for a particle moving in a force-free surrounding. If this idea is generalized, for instance, to a freely moving particle on the 3-sphere of radius R, then it is already known that the (geodesic) Killing trajectories are the greater circles. Hence, the structural coefficients of the underlying Lie algebra are proportional to the fraction $\hbar/R$ and thereby different from zero; one is compelled to relax the form of canonical commutator relations (4) and (5).

One of the first attempts in this direction was proposed by Segal [3], and later developed by Śniatycki [4], Doebner, Tolar, and Nattermann [5,6]. Without going into any detail (a review can be found in Ref. [7]), a generalized momentum operator is obtained by projection onto a given smooth vector field X on M according to

$$P_X = -i\hbar\left(\nabla_X + \frac{1}{2}\mathrm{div}X\right), \tag{7}$$

where $\mathrm{div}X$ is the covariant divergence of vector field X. A straightforward computation (that is here omitted) yields the general commutation relation,

$$[P_X, P_Y] = -i\hbar P_{[X,Y]}, \tag{8}$$

where $[X, Y]$ denotes the commutator of two vector fields X and Y in the usual sense of the theory of manifolds [3]. In the case of a linear manifold, this vanishes for two infinitesimal translations, and Equation (8) specializes to the commutativity of the conventional linear momenta.

Apparently, operator (6) of De Witt can be recovered for vector fields X of the special form $X_k = \partial_k$, with $\Gamma^j_{jk} = \partial_k \log\sqrt{g} \equiv \Gamma_k$. However, that means $\mathrm{div}X_k = \Gamma_k \neq 0$, such that the vector fields X_k cannot be Killing vectors if the underlying manifold is intrinsically curved.

The paper is organized as follows. In Section 2, the question of the hermiticity of momentum operators (7) is discussed. This draws attention to the special importance of Killing frame fields. A classification of manifolds with such a structure and the corresponding Lie algebra are discussed in Section 3. Possible examples S^1, S^3, and S^7 are considered in Section 4. In Section 5, a covariant position operator on S^3 is defined in terms of geodesic normal coordinates and the associated commutator relations of position and momentum are established. Lastly, a summary and outlook are given in Section 6.

2. Hermitian Momentum Operators

Let us consider the Hilbert space of square integrable complex functions $L^2(U,\mu)$ on a compact subset, $U \subseteq M$, with smooth boundary ∂U endowed by the inner product,

$$(f,g) = \int_U d\mu f^* g, \quad f,g \in L^2(U,\mu), \tag{9}$$

where μ is the standard volume measure on $U \subseteq M$. The statement that momentum operators are Hermitian with respect to an inner product is typically based on the assumption that the boundary terms vanish after partial integration. Indeed, one cannot define a momentum operator on a bounded domain without specifying boundary conditions. In mathematical terms, choosing the boundary conditions amounts to choosing an appropriate domain for the operator. If one uses no boundary conditions, too many functions are eigenvectors, so the spectrum of P_X is the whole complex plane. On the other hand, if Dirichlet boundary conditions are imposed, the situation is too restrictive, and one cannot find an orthonormal basis. Thus, if functions $f \in L^2(U,\mu)$ are smooth on U, but constant functions at ∂U, finding a domain such that P_X is self-adjoint is a compromise to obtain an orthonormal basis of countable spectrum. In what follows, the focus is on the Hilbert spaces $\mathcal{H}_U \subset L^2(M,\mu)$, with

$$\mathcal{H}_U = \left\{ f \in C^1(U) : f|_{\partial U} = \text{const.} \right\} \tag{10}$$

At this point, one must check whether momentum operator (7) still remains Hermitian because the elements in $\mathcal{H}_U$ are not supposed to vanish at the boundary.

Proposition 1. *Let X be a smooth and divergenceless vector field on $U \subseteq M$. Then, P_X is Hermitian on $\mathcal{H}_U$.*

Proof. Let $f, h \in \mathcal{H}_U$. The divergence of product fX can be written as

$$\mathrm{div}(fX) = f\mathrm{div}(X) + \nabla_X f. \tag{11}$$

On the other hand, one has the decomposition

$$\begin{aligned} h^* \nabla_X f &= \langle h^* X, \mathrm{grad} f \rangle \\ &= \mathrm{div}(f h^* X) - f \mathrm{div}(h^* X). \end{aligned} \tag{12}$$

This equation can be integrated with respect to the volume form $d\mu$ on U as follows:

$$\begin{aligned} (h, \nabla_X f) &= -\int_U d\mu f \mathrm{div}(h^* X) \\ &\quad + \int_{\partial U} d\mu_\partial f h^* \langle X, \nu \rangle. \end{aligned} \tag{13}$$

Here, Stokes' theorem is applied, where $d\mu_\partial$ is the volume measure with respect to the boundary ∂U and ν is the non-negative outward normal on ∂U. Now, since f and h were assumed to be constant at the boundary of U, they could be taken out of the integration in (13), and one can apply Stokes' theorem once more, such that the remaining boundary integral on the right-hand side in (13) becomes

$$\int_{\partial U} d\mu_\partial \langle X, \nu \rangle = \int_U d\mu \mathrm{div} X. \tag{14}$$

With the assumption $\text{div}X = 0$, it follows

$$(h, \nabla_X f) = -\int_U d\mu f \text{div}(h^* X). \tag{15}$$

Finally, Equation (11) is applied and one gets

$$(h, \nabla_X f) + (\nabla_X h, f) = -\int_U d\mu h^* f \text{div}(X). \tag{16}$$

The term on the right-hand side can be absorbed into each term on the left-hand side with a prefactor 1/2. By multiplication of the equation with $-i\hbar$, and after applying definition (7), one lastly obtains

$$(h, P_X f) = (P_X h, f) \tag{17}$$

for all $f, h \in \mathcal{H}_U$. □

At this point, to emphasise is that the boundary term in Equation (14) is not necessarily zero under more general conditions. This renders the divergence criterion of the vector field X necessary to ensure hermiticity. The possible manifolds that are available under these circumstances are discussed in the next Section.

3. Geodesic Momentum Operators

The classification of manifolds that are compatible to the conditions of Proposition 1 can be described with the following definition of Killing frames:

Definition 1. *(Killing frame) [8]. A Riemannian manifold M has the Killing property if, in some neighborhood of each point of M, there exists an orthonormal frame, $X_1, ..., X_n$, such that each X_i, $i = 1, ..., n$ is a Killing vector field (local infinitesimal isometry). Such a frame is called a Killing frame.*

Since any linear combination with constant coefficients of Killing vector fields is again a Killing vector field, a manifold has the Killing property if and only if it is always possible to find frames consisting of Killing vector fields, such that $\langle X_i, X_j \rangle = \text{const}$ for each choice of i and j. The normality condition of the definition implies that the integral curves of the isometries are geodesics, since a necessary and sufficient condition for this is that the Killing vector fields have constant length ([9], p. 349; [10], p. 50).

For instance, let x, y and r, φ denote Cartesian and polar coordinates on the Euclidean plane $\mathbb{R}^2$ endowed with Euclidean metric. Then, the Killing vector fields corresponding to translations and rotations are $X_1 = \partial_x$, $X_2 = \partial_y$ and $X_3 = \partial_\varphi$. Their squared vector norms are $X_1^2 = 1, X_2^2 = 1$ and $X_3^2 = r^2$. Killing vector fields X_1 and X_2 had a constant length on the whole plane. Their trajectories are straight lines, which are geodesics. The Killing trajectories corresponding to rotations X_3 are concentric circles around the origin. The length of X_3 is constant along the circles, but nonconstant on the whole plane. The corresponding Killing trajectories are circles that are not geodesics [11].

In the context given so far, one comes to the following.

Definition 2. *(Momentum operator).*

Let M be an n-dimensional Riemannian manifold with Killing frame $X_1, ..., X_n$ on M. The set of operators defined by

$$P_{X_k} = -i\hbar X_k, \tag{18}$$

$k = 1, ..., n$, are called (geodesic) momentum operators in the direction X_k on M.

This definition is compatible with Equation (7), since every Killing vector field X_i is a priori divergenceless, i.e., $\text{div}X_i = 0, i = 1, ..., n$. Moreover, the Lie bracket of two Killing fields is still a Killing field. The momentum operators (18) thus form a Lie subalgebra of vector fields on M. If M is a complete manifold, this is the Lie algebra of the translation group. In this case, the commutation relations of the Killing vector fields are given by

$$[X_i, X_j] = c_{ij}^k X_k, \tag{19}$$

where structural coefficients, c_{ij}^k, express the multiplication of pairs of vectors as a linear combination. The corresponding commutator relations of the momenta are obtained by multiplication with the physical units $(-i\hbar)^2$ on both sides of Equation (19) and subsequently applying the definition (18), i.e.,

$$[P_{X_i}, P_{X_j}] = -i\hbar c_{ij}^k P_{X_k}, \tag{20}$$

which is compatible with the general expression (8). The associated Casimir element of this Lie algebra is given by [8]:

Proposition 2. *Let M be an n-dimensional manifold and $X_1, ..., X_n$ be a Killing frame on M. There is a decomposition of the Laplace–Beltrami operator, such that*

$$\sum_{j=1}^{n} P_{X_j}^2 = -\hbar^2 \Delta. \tag{21}$$

Proof. Vector fields, X_j, can be expressed as a linear combination of the coordinate vector fields, $\partial_\alpha = \partial/\partial x^\alpha$, with the Greek letters denoting indices of the local chart, such that

$$X_i = \xi_i^\alpha \partial_\alpha, \tag{22}$$

where each ξ_i^α is a function. For every smooth f on M, one can write

$$\sum_j X_j^2 f = \partial_\alpha (g^{\alpha\beta} \partial_\beta f) - \delta^{ij} (\partial_\alpha \xi_i^\alpha) \xi_j^\beta \partial_\beta f, \tag{23}$$

which was obtained by the product rule of differentiation. On the other hand, the Laplace–Beltrami operator in the natural frame is

$$\Delta f = \partial_\alpha (g^{\alpha\beta} \partial_\beta f) + \frac{1}{\sqrt{g}} (\sqrt{g})_{,\alpha}\, g^{\alpha\beta} \partial_\beta f. \tag{24}$$

Now, it follows that expression (23) is equal to Equation (24) for every f if it can be shown that

$$\frac{1}{\sqrt{g}} (\sqrt{g})_{,\alpha}\, g^{\alpha\beta} + \delta^{ij} (\partial_\alpha \xi_i^\alpha) \xi_j^\beta = 0. \tag{25}$$

Using the basic property, $\nabla g = 0$, of the Levi-Civita connection together with $\delta^{ij} \xi_i^\alpha \xi_j^\beta = g^{\alpha\beta}$, one obtains the following condition:

$$\delta^{ij} (\nabla_{X_i} \xi_j^\alpha) = 0. \tag{26}$$

This identity can be confirmed as follows:

$$\begin{aligned}0 &= \nabla_{\partial_l} g^{lk} \\ &= \delta^{ij} \nabla_{\partial_l} (\xi_i^l \xi_j^k) \\ &= \delta^{ij} \left(\mathrm{div} X_i \xi_j^k + \nabla_{X_i} \xi_j^k \right) \\ &= \delta^{ij} \nabla_{X_i} \xi_j^k \end{aligned} \tag{27}$$

and thus,

$$\sum_j X_j^2 f = \Delta f. \tag{28}$$

With Definition (18), in physical units, one obtains statement (21). □

According to this decomposition of the covariant Laplacian, there is no ambiguity concerning operator orderings of the kinetic energy term. Decomposition (21) is independent of the particular choice of the orthonormal basis [12]. Moreover, the commutator of Δ with the elements X_j of Lie algebra (19) is given by

$$[X_j, \Delta] = 0. \tag{29}$$

From a mathematical point of view, the Casimir element has a meaning only for the theory of representations, but not as an element of the Lie algebra, since the product in Equation (21) is not defined for the algebra itself. However, from linear algebra we know that the eigenvectors of a linear operator always form a basis for the vector space in question. In addition, for any Lie group, one or more of the generators can be simultaneously diagonalized using similarity transformations. The set of generators that can be diagonalized simultaneously are called Cartan generators. Thus, a suggestive and particularly easy basis for the vector space of each representation is given by the eigenvectors of the Cartan generators (see below).

The scope of the concept given so far asks for a mathematical classification of manifolds with Killing property. A Riemannian manifold having the Killing property must be locally symmetric [8]. Thus, each point of a connected Riemannian manifold having the Killing property has an open neighbourhood that is isometric to an open neighbourhood in a simply connected Riemannian symmetric space M. Then, M also has the Killing property and global Killing frames. Actually, a local Killing frame exists on M because of the given local isometry and can be extended uniquely to give a global Killing frame. The extension of each Killing vector field to a global Killing vector field is possible since the symmetry implies completeness. This extension remains orthonormal, since the Riemannian structure on M is subordinate to a real analytic Riemannian structure (cf. [13], p. 240; [14], p. 187). A simply connected Riemannian symmetric space is said to be irreducible if it is not the product of two or more Riemannian symmetric spaces. It can then be shown that any simply connected Riemannian symmetric space is a Riemannian product of irreducible ones.

Therefore, the calculations furthermore are restricted to the irreducible, simply connected Riemannian symmetric spaces. Any simply connected Riemannian symmetric space M is of one of the following three types: (i) Euclidean type: M has a vanishing curvature and is thereby isometric to a Euclidean space. (ii) Compact type: M has a non-negative (but not identically zero) sectional curvature. (iii) Noncompact type: M has a nonpositive (but not identically zero) sectional curvature. Strictly negatively curved manifolds imply that there are no nontrivial (real valued) orthonormal Killing fields.

Manifolds of constant positive curvature [8] have the Killing property only if the dimension of M is equal to 1, 3, or 7. For the spheres S^1, S^3 and S^7, actually, there is a global Killing frame. The construction depends essentially on the existence of a multiplication in $\mathbb{R}^2$ (complex numbers), $\mathbb{R}^4$ (quaternions), and $\mathbb{R}^8$ (Cayley numbers).

4. Applications

From the discussion of the previous Section and the particular role of Killing frames, it is straightforward to consider the designated cases of constant curvature manifolds S^1, S^3 and S^7 in more detail. Let us begin with the "trivial" case S^1.

4.1. Circle

For the circle, one can take X_1 to be the unit tangent vector field pointing, say, in the counterclockwise direction. More precisely, consider the situation of a compact subset $M \subset S^1$ embedded in $\mathbb{R}^2$. The general solution of the Killing equation $\mathcal{L}_{X_1} g = 0$ on S^1, with metric $ds^2 = \rho^2 d\varphi^2$ is given by $X_1 = \xi^\varphi \partial_\varphi$, for $\xi^\varphi \in \mathbb{R}$. Let $\xi^\varphi = 1/\rho$, where ρ is the constant (hyper-)radius of the circle; then, we have $X_1^2 = 1$, and the Killing trajectory is a geodesic. The associated momentum operator is

$$P_\varphi = -i\hbar \frac{1}{\rho} \frac{\partial}{\partial \varphi}. \tag{30}$$

This operator is symmetric on any compact set $M \subset S^1$, with $f = \text{const.}$ on the boundary. In Ref. [15], it is reported that, in quantum mechanics on a circle with standard commutation relation for φ and p_φ, the uncertainty relation cannot be stronger than $\sigma_p \sigma_\varphi \geq 0$, where σ_φ and σ_p are the standard deviations of position and momentum. Indeed, this inequality is not informative at all, since a product of two nonnegative values cannot be negative. Alternatively, one is referred to the approach in Ref. [16], which is not affected by difficulties arising in defining a proper measure of position uncertainty on manifolds mentioned in Ref. [17]. By applying the substitution $r = \rho\varphi$ that corresponds to the arc-length on S^1, the uncertainty principle of Ref. [16] is given by

$$\sigma_p \Delta r \geq \pi\hbar, \tag{31}$$

where Δr is the measure (length) of a compact domain on S^1.

Before turning over to the case of S^3, let us briefly mention that indeed each single component, L_1, L_2, and L_3 of the ordinary textbook angular momentum operator $\mathbf{L}$ is a Killing vector on S^2 and moreover, $\mathbf{L}^2 = L_1^2 + L_2^2 + L_3^2$ actually corresponds to the Laplace–Beltrami operator on S^2. Although this seems quite promising, these vector fields are not normalizable. All vector fields on the 2-sphere are actually inappropriate for this purpose because of the hairy-ball theorem of differential topology that states that there is generally no nonvanishing continuous tangent vector field on even-dimensional n-spheres. This hinders thinking about what kind of vector fields should be appropriate for an adequate description of momentum operators on S^2. The discussion in the literature regarding which momentum operators on S^2 might be considered to be appropriate extends up to the present day.

4.2. 3-Sphere

The 3-sphere of radius $R > 0$ can be understood as the three-dimensional hypersurface in the four-dimensional Euclidean space. This can be naturally described by the standard spherical coordinates of $\mathbb{R}^4$ given by [18]

$$\begin{aligned} x^1 &= R\cos\chi, \\ x^2 &= R\sin\chi\cos\theta, \\ x^3 &= R\sin\chi\sin\theta\cos\varphi, \\ x^4 &= R\sin\chi\sin\theta\sin\varphi. \end{aligned}$$

In order to cover all points of the 3-sphere with both positive and negative values of the coordinates x^i, it is necessary that $0 \leq \chi, \theta \leq \pi$, $0 \leq \varphi < 2\pi$. In these coordinates, the metric of S^3 takes the form:

$$ds^2 = R^2(d\chi^2 + \sin^2\chi(d\theta^2 + \sin^2\theta d\varphi^2)). \tag{32}$$

The corresponding Killing equation is solved for the unit sphere ($R = 1$) and the following orthonormal Killing frame is selected :

$$\begin{aligned} X_1 &= \sin\theta\cos\varphi\partial_\chi \\ &+ (\cot\chi\cos\theta\cos\varphi - \sin\varphi)\partial_\theta \\ &- (\cot\chi\csc\theta\sin\varphi + \cot\theta\cos\varphi)\partial_\varphi, \end{aligned} \tag{33}$$

$$\begin{aligned} X_2 &= \sin\theta\sin\varphi\partial_\chi \\ &+ (\cot\chi\cos\theta\sin\varphi + \cos\varphi)\partial_\theta \\ &+ (\cot\chi\csc\theta\cos\varphi - \cot\theta\sin\varphi)\partial_\varphi, \end{aligned} \tag{34}$$

$$X_3 = \cos\theta\partial_\chi - \cot\chi\sin\theta\partial_\theta + \partial_\varphi. \tag{35}$$

Case $R \neq 1$ can be obtained by dividing the right-hand side by R. The orthonormality relation $g(X_i, X_j) = \delta_{ij}$ is straightforwardly verified. The corresponding representation in Cartesian coordinates, $p = (x^1, x^2, x^3, x^4)$, of the Euclidean embedding space $\mathbb{R}^4$ is also determined and given by

$$X_1(p) = (-x^4, -x^3, x^2, x^1), \tag{36}$$
$$X_2(p) = (x^3, -x^4, -x^1, x^2), \tag{37}$$
$$X_3(p) = (-x^2, x^1, -x^4, x^3), \tag{38}$$

which satisfies $X_i(p) \cdot X_j(p) = \delta_{ij}$ with respect to the Euclidean scalar product. One can also check that the Lie algebra generated by $\{X_1, X_2, X_3\}$ is given by the following commutation relations,

$$[X_i, X_j] = -\frac{2}{R}\epsilon_{ijk}X_k, \tag{39}$$

where ϵ_{ijk} is the Levi-Civita symbol in three dimensions. In physical units, this can be rewritten as

$$[P_{X_i}, P_{X_j}] = \frac{2i\hbar}{R}\epsilon_{ijk}P_{X_k}. \tag{40}$$

The corresponding Hamilton operator, H, of a free particle is given by

$$H = \frac{1}{2m}\sum_{i=1}^{3} P_{X_i}^2 = -\frac{\hbar^2}{2m}\Delta, \tag{41}$$

which is equal to the Casimir element of Proposition 2 in three dimensions. Thus, it follows that

$$[H, P_{X_i}] = 0 \tag{42}$$

for $i = 1, 2$, and 3.

Alternative decompositions of the Laplacian in Equation (41) by using six (nonorthonormal) Killing vector fields instead of three were proposed in Santander et al. [19]. One essential point of the approach in Ref. [19] is that the structural coefficients of the associated commutator relations are not constants, such that the Hamiltonian cannot be considered as a Casimir element of the operator algebra. This makes the analysis of the eigenvalues and the corresponding eigenspaces quite complicated. There seems to be no reason why one should regard a decomposition of the free Hamiltonian in terms of additional angular momenta whose integral curves are not geodesics.

Another interesting approach is the momentum-space quantization of a particle moving on the $SU(2)$ group manifold by Guerrero et al. [20]. Their algorithm also exhibits a proper and unambiguous realization of the basic operators and of the Hamiltonian, which also turns out to be the Laplace–Beltrami operator on S^3. Although the right-invariant generators (62) in Ref. [20] are different from the Killing frame fields X_i introduced above, they are compatible with the algebra given in (39). However, the question whether the generators in Ref. [20] also form a Killing frame was not explicitly discussed.

The eigenvalues of H can be obtained by the hyperspherical harmonics on S^3, which were discussed as part of investigations of a variety of gravitational physics problems in spaces with the topology of the 3-sphere [21,22]. According to Ref. [22], these hyperspherical harmonics on S^3 are denoted by Y^{nlm}. The integers n, l and m with $n \geq l \geq 0$ and $l \geq m \geq -l$ indicate the order of the harmonic. These harmonics are eigenfunctions of the covariant Laplacian according to

$$\Delta Y^{nlm} = -\frac{n(n+2)}{R^2} Y^{nlm}. \tag{43}$$

The corresponding energy eigenvalues E_n of H are given by

$$E_n = \frac{\hbar^2}{2m} \frac{n(n+2)}{R^2}. \tag{44}$$

for $n = 0, 1, ...$

Now let us consider the corresponding eigenvalue spectrum of the momentum operators. Although P_{X_i} and H are commuting Hermitian operators, it is not necessarily given that each eigenbasis of H is also an eigenbasis of P_{X_i}. Indeed, most of the functions Y^{nlm} given in Equation (43) are not eigenfunctions of the momentum operator P_{X_3}. A simultaneous eigenbasis can be obtained by applying the standard textbook formalism, but on the basis of the specific algebra given by Equation (40). Rather then working with the operators P_{X_1} and P_{X_2}, it is convenient to work with the non-Hermitian linear combinations:

$$P_{\pm} = P_{X_1} \pm i P_{X_2}, \tag{45}$$

where by definition $(P_{-})^{\dagger} = P_{+}$. Using (40) and (42), it is straightforward to show that

$$[P_{+}, P_{-}] = 4\frac{\hbar}{R} P_{X_3}, \tag{46}$$

$$[P_{X_3}, P_{\pm}^k] = \pm 2k \frac{\hbar}{R} P_{\pm}^k \tag{47}$$

for $k = 0, 1, 2, \ldots$ Certainly, one also has

$$[H, P_{\pm}^k] = 0. \tag{48}$$

In order to obtain a simultaneous eigenbasis of H and P_{X_3} for every fixed $n \in \mathbb{N}$, let us consider the set of orthogonal states $\{\psi_{\pm}^{nk}\}_{k=0}^{n}$ given by applying the "ladder" operators $P_{\pm}$ according to

$$\psi_{\pm}^{nk} = P_{\pm}^k Y^{nn(\mp n)}, \tag{49}$$

where $k = 0, 1, ..., n$. From this definition it follows that $P_{\mp}\psi_{\pm}^{n0} = 0$. By applying the general commutator rule (47), all of these states are eigenstates of the momentum operator, such that

$$P_{X_3}\psi_{\pm}^{nk} = \pm p_{nk}\psi_{\pm}^{nk}, \tag{50}$$

$$p_{nk} = (k - \frac{n}{2})\frac{2\hbar}{R} \tag{51}$$

for $k = 0, 1, ..., n$. The physical interpretation becomes straightforward by recalling that the diameter of the 3-sphere is, by definition, the maximal possible geodesic distance (πR) between two points on S^3. If one applies the original definition of Planck's constant $\hbar = h/2\pi$, the maximal resolution $\Delta p_{nk} = p_{nk+1} - p_{nk}$ of the possible momenta in Equation (51) is given by

$$\Delta p_{nk} = \frac{h}{\pi R}. \tag{52}$$

This unit of momentum corresponds to the de Broglie wavelength,

$$\lambda_R = \pi R, \tag{53}$$

which is identical to the diameter of the manifold. It is hard to think of higher resolutions than this.

On the other hand, the energy eigenvalues can be verified by applying commutator (48), such that we obtain the following eigenvalue equations:

$$H\psi_{\pm}^{nk} = E_n\psi_{\pm}^{nk}, \tag{54}$$

where E_n is given in Equation (44). For numerical purposes, it is helpful to know the explicit form of the initial functions $\psi_{\pm}^{n0}$, which are given by

$$\psi_{\pm}^{n0} = C_n \sin^n\chi \sin^n\theta e^{\mp in\varphi}, \tag{55}$$

and the normalization constant is

$$C_n = \frac{\sqrt{2^{2n-1}(n+1)}}{\pi}\frac{n!(2n-1)!!}{(2n)!}. \tag{56}$$

The representation of operators P_{X_1} and P_{X_2} can be obtained by inverting relation (45) in order to express them in terms of the ladder operators. This completes the brief analysis of momentum operators on S^3.

4.3. 7-Sphere

Physical applications involving higher-dimensional spheres can be found almost exclusively in the context of $N = 1$ supergravity in 11 dimensions, which is beyond the scope of this study. Therefore, let us discuss (for information only) some aspects regarding the approach given so far. The sphere S^7, considered to be a Riemannian manifold embedded in $\mathbb{R}^8$ in the usual way, is also designated to have the Killing property. Explicitly, writing points p in $\mathbb{R}^8$ as column vectors and identifying the tangent spaces to S^7 with hyperplanes, the vector fields $X_i(p)$, for $i = 1, ..., 7$, are expressed in Table 1 below [8]. Since $p \cdot X_i(p) = 0$ and $X_i(p) \cdot X_j(p) = \delta_{ij}$, this gives a global orthonormal frame on S^7, which is also a Killing frame. If $\mathbb{R}^4$ is embedded in $\mathbb{R}^8$ as a subset $x^5 = x^6 = x^7 = x^8 = 0$, the restrictions of X_1, X_2, X_3 yield a Killing frame on S^3 corresponding to the previous Section (up to sign conventions).

Table 1. Orthonormal Killing vector representation at point p on S^7 embedded in $\mathbb{R}^8$.

p	$X_1(p)$	$X_2(p)$	$X_3(p)$	$X_4(p)$	$X_5(p)$	$X_6(p)$	$X_7(p)$
x^1	x^2	x^3	x^4	x^5	x^6	x^7	x^8
x^2	$-x^1$	$-x^4$	x^3	$-x^6$	x^5	$-x^8$	x^7
x^3	x^4	$-x^1$	$-x^2$	$-x^7$	x^8	x^5	$-x^6$
x^4	$-x^3$	x^2	$-x^1$	x^8	x^7	$-x^6$	$-x^5$
x^5	x^6	x^7	$-x^8$	$-x^1$	$-x^2$	$-x^3$	x^4
x^6	$-x^5$	$-x^8$	$-x^7$	x^2	$-x^1$	x^4	x^3
x^7	x^8	$-x^5$	x^6	x^3	$-x^4$	$-x^1$	$-x^2$
x^8	$-x^7$	x^6	x^5	$-x^4$	$-x^3$	x^2	$-x^1$

According to this representation, the Killing vector fields X_i of the frame bundle and their associated momentum operators P_{X_i} on S^7 can be well expressed in terms of hyperspherical coordinates. Moreover, the calculation of eigenvectors and eigenvalues of the corresponding Laplace–Beltrami operator on S^7 is straightforward and can be obtained by projecting harmonic fields in Euclidean $\mathbb{R}^8$ onto the unit sphere.

5. Covariant Position Operator on S^3

Lastly, some remarks concerning the notion of position operators. Segal defines [3] the position operator Q as follows: if f is a general function on M, then Q_f is defined as the operation of multiplication by f. For real f, g, operators Q_f and Q_g are Hermitian, such that there is no difficulty in verifying the commutation relations:

$$\left[Q_f, P_X\right] = i\hbar Q_{Xf}, \tag{57}$$

$$\left[Q_f, Q_g\right] = 0. \tag{58}$$

In order to define a covariant position operator on S^3, it is straight to consider the notion of geodesic distance. Since no point of S^3 is particularly distinguished, without loss of generality, the origin in hyperspherical coordinates is chosen by the "North Pole" $p = (R, 0, 0, 0) \in \mathbb{R}^4$. Geodesic normal coordinates, $q = (q^1, q^2, q^3)$, around this origin are such that every element of S^3 can be reached by the exponential map

$$\exp_p : \quad T_pS^3 \longrightarrow S^3; \tag{59}$$

$$\exp_p(X) = p\cos\left(\frac{s}{R}\right) + R\sin\left(\frac{s}{R}\right)\frac{X(p)}{||X(p)||}, \tag{60}$$

with $X = q^i X_i \in T_pS^3$ and the geodesic distance function,

$$s : q \longrightarrow s = ||q||, \tag{61}$$

where $||\cdot||$ is the Euclidean norm in T_pS^3. Let

$$X_q = \hat{q}^k X_k \tag{62}$$

be the corresponding tangent vector field in the unit direction, $\hat{q} = q/s$. Then, applying the smooth transition map from the hyperspherical chart to geodesic coordinates,

$$\begin{aligned} q^1 &= R\chi \sin\theta \cos\varphi, \\ q^2 &= R\chi \sin\theta \sin\varphi, \\ q^3 &= R\chi \cos\theta, \end{aligned}$$

it follows via straightforward computation that X_q can be expressed by

$$X_q = \frac{\partial}{\partial s}. \tag{63}$$

Accordingly, a geodesic position operator on S^3 is defined by Q_s, and commutator (57) can be expressed by

$$[Q_s, P_{X_q}] = i\hbar. \tag{64}$$

A covariant uncertainty relation of position and momentum that is compatible with this approach and was applied to S^3 can be found in Refs. [23,24]. A further generalization to the case of S^7 is also possible, but it is left to further considerations.

6. Summary and Outlook

The description of momentum operators by Killing vector fields is a long-established concept in momentum-space quantization on differentiable manifolds. On the other hand, from classical general relativity, the fundamental importance of geodesic trajectories as a key concept of the theory is also known.

If one wishes to unify these two concepts together in the approach of momentum-space quantization, this leads to the notion of Killing frames on manifolds. These special frames are implicitly part of the classical Cartan formalism. However, the orthonormal frame fields in the Cartan approach are usually not provided as Killing vector fields. A fundamental principle in the Cartanian approach is to chose the moving frames most suitable to the particular problem. Some consequences for the possible manifolds arising from the additional Killing frame property were discussed.

The present study is mainly focused on irreducible, simply connected Riemannian manifolds. However, many product manifolds can be constructed out of these irreducible components which also possess the Killing frame property. A straightforward example would be the temporally infinite but spatially finite case $\mathbb{R} \times S^3$. A further generalization is the case of 11-dimensional supergravity with $\mathbb{R} \times S^3 \times S^7$. Such possibilities and the extended analysis of the associated spin connections are the subject of further investigations.

Funding: This research received no external funding.

Data Availability Statement: Not applicable.

Conflicts of Interest: The author declares no conflict of interest.

References

1. DeWitt, B.S. Point transformations in quantum mechanics. *Phys. Rev.* **1952**, *85*, 653–661. [CrossRef]
2. DeWitt, B.S. Dynamical theory in curved spaces. I. A review of the classical and quantum action principles. *Rev. Mod. Phys.* **1957**, *29*, 377–397. [CrossRef]
3. Segal, I.E. Quantization of nonlinear systems. *J. Math. Phys.* **1960**, *1*, 468–488. [CrossRef]
4. Śniatycki, J. *Geometric Quantization and Quantum Mechanics*; Springer: New York, NY, USA, 1980. [CrossRef]
5. Doebner, H.-D.; Nattermann, P. Borel quantization: kinematics and dynamics. *Acta Phys. Polon.* **1996**, *27*, 2327–2339.
6. Doebner, H.-D.; Šťovíček, P.; Tolar, J. Quantization of kinematics on configuration manifolds. *Rev. Math. Phys.* **2001**, *13*, 1–47. [CrossRef]
7. Ali, S.T.; Engliš, M. Quantization Methods: A guide for physicists and analysts. *Rev. Math. Phys.* **2005**, *17*, 391–490. [CrossRef]
8. D'Atri, J.E.; Nickerson, H.K. The existence of special orthonormal frames. *J. Differ. Geom.* **1968**, *2*, 393–409. [CrossRef]
9. Schouten, J.A. *Ricci-Calculus*; Springer: Berlin/Heidelberg, Germany, 1954. [CrossRef]
10. Yano, K. *The Theory of Lie Derivatives and Its Applications*; North-Holland Publishing Co.: Amsterdam, The Netherlands, 1955. Available online: https://archive.org/details/theoryofliederiv029601mbp/ (accessed on 15 November 2022).
11. Katanaev, M.O. Killing vector fields and a homogeneous isotropic universe. *Phys.-Usp.* **2016**, *59*, 689–700. . UFNe.2016.05.037808 [CrossRef]
12. Hall, B.C. *Lie Groups, Lie Algebras, and Representations*; Springer International Publishing Switzerland: Cham, Switzerland, 2015. [CrossRef]

13. Wolf, J.A. *Spaces of Constant Curvature*; AMS Chelsea Publishing: Providence, RI, USA, 2011. Available online: https://bookstore.ams.org/chel-372-h (accessed on 15 November 2022).
14. Helgason, S. *Differential Geometry and Symmetric Spaces*; Academic Press: New York, NY, USA, 1978. Available online: https://www.sciencedirect.com/bookseries/pure-and-applied-mathematics/vol/80/suppl/C (accessed on 15 November 2022).
15. Golovnev, A.V.; Prokhorov, L.V. Uncertainty relations in curved spaces. *J. Phys. A Math. Gen.* **2004**, *37*, 2765–2775. [CrossRef]
16. Schürmann, T.; Hoffmann, I. A closer look at the uncertainty relation of position and momentum. *Found. Phys.* **2009**, *39*, 958. [CrossRef]
17. Trifonov, D.A. Position uncertainty measures on the sphere. In *Proceedings of the Fifth International Conference on Geometry, Integrability and Quantization, Varna, Bulgaria, 5–12 June 2003);* Mladenov, I.M., Hirshfeld, A.S.; Eds.; SOFTEX: Sofia, Bulgaria, 2003; pp. 211–224. [CrossRef]
18. Griffiths, J.B.; Podolský, J. *Exact Space-Times in Einstein's General Relativity*; Cambridge University Press: New York, NY, USA, 2009. [CrossRef]
19. Cariñena, J.F.; Rañada, M.F.; Santander, M. The quantum free particle on spherical and hyperbolic spaces: A curvature dependent approach. II. *J. Math. Phys.* **2012**, *53*, 102109. [CrossRef]
20. Guerrero, J.; López-Ruiz, F.F.; Aldaya, V. $SU(2)$-particle sigma model: momentum-space quantization of a particle on the sphere S^3. *J. Phys. A Math. Theor.* **2020**, *53*, 145301. [CrossRef]
21. Lachièze-Rey, M.; Caillerie, S. Laplacian eigenmodes for spherical spaces. *Class. Quant. Grav.* **2005**, *22*, 695–708. [CrossRef]
22. Lindblom, L.; Taylor, N.W.; Zhang, F. Scalar, vector and tensor harmonics on the three-sphere. *Gen. Rel. Grav.* **2017**, *49*, 140. [CrossRef]
23. Schürmann, T. Uncertainty principle on 3-dimensional manifolds of constant curvature. *Found. Phys.* **2018**, *48*, 716–725 [CrossRef]
24. Schürmann, T. On the uncertainty principle in Rindler and Friedmann spacetimes. *Eur. Phys. J. C* **2020**, *80*, 141. [CrossRef]

physics

MDPI

Article

BFSS Matrix Model Cosmology: Progress and Challenges

Suddhasattwa Brahma [1,*], Robert Brandenberger [2] and Samuel Laliberte [2]

[1] Higgs Centre for Theoretical Physics, School of Physics & Astronomy, University of Edinburgh, Edinburgh EH9 3FD, UK
[2] Department of Physics, McGill University, Montreal, QC H3A 2T8, Canada
* Correspondence: suddhasattwa.brahma@gmail.com

Abstract: We review a proposal to obtain an emergent metric space-time and an emergent early universe cosmology from the Banks–Fischler–Shenker–Susskind (BFSS) matrix model. Some challenges and directions for future research are outlined.

Keywords: early universe; superstring cosmology; emergent cosmology

1. Introduction

The current models of the very early universe are generally based on effective field theory (EFT), i.e., on considering matter fields minimally coupled to Einstein gravity. This is true both for the inflationary scenario [1–4] and for most alternatives to it (see, e.g., [5–7] for a comparative review of alternatives to inflation). However, there are conceptual challenges that effective-field-theory descriptions of cosmology face. For example, in the case of scalar field matter—ubiquitously used in early universe models—the range of such fields [8] and the shape of [9] of their potentials are constrained if we demand that the effective fields descend from some fundamental physical theory (the best candidate of which is superstring theory). Independent of superstring theory, the Trans-Planckian Censorship Conjecture (TCC) [10] severely constrains inflationary models that are based on an EFT analysis [11] (see, e.g., [12,13] for discussions of some arguments in support of the TCC; note that scenarios of inflation that are not based on EFT analyses (see [14–19] for examples) are not necessarily subject to these constraints).

An EFT analysis of matter also leads to the famous cosmological constant problem: considering matter fields coupled to Einstein gravity, and canonically quantizing the fields by expanding them in plane wave modes and quantizing each mode like a harmonic oscillator, then the vacuum energy of all of these modes yields a contribution to the cosmological constant that is many orders of magnitude (120 orders in a non-supersymmetric model) larger than what is compatible with current observations (to avoid the Planck catastrophe, an ultraviolet (UV) cutoff scale must be imposed, the value of which determines the predicted cosmological constant).

The abovementioned problems indicate that one needs to go beyond point-particle-based EFT to obtain a consistent picture of cosmology. Here, we review an approach [20,21] to obtain a model of the very early universe that is not based on EFT. Our starting point is the Banks–Fischler–Shenker–Susskind (BFSS) matrix model [22], a proposed non-perturbative definition of superstring theory (see also [23,24] for early paper on the quantum theory of membranes, and [25–33] for other matrix models related to superstring theory). We outline an avenue of how to obtain an emergent space-time, an emergent metric, and an emergent early universe cosmology. We consider the matrix model to be in a thermal state. This leads to density fluctuations and gravitational waves on the emergent space-time, and we indicate how the resulting power spectra of these fluctuations are scale-invariant and consistent with current observations.

In this context, we remind the reader that the inflationary scenario [1–4] is not the only early universe scenario that is consistent with current cosmological observations (see,

Citation: Brahma, S.; Brandenberger, R.; Laliberte, S. BFSS Matrix Model Cosmology: Progress and Challenges. *Physics* **2023**, *5*, 1–10. https://doi.org/10.3390/physics5010001

Received: 17 October 2022
Revised: 20 November 2022
Accepted: 23 November 2022
Published: 22 December 2022

e.g., [5–7] for comparative reviews of inflation and alternatives). An alternate scenario is the *emergent scenario*, which is based on the assumption that there is a primordial phase that can be modelled as quasi-static (from the point of view of the Einstein frame) and that then undergoes a phase transition to the Standard Big Bang phase of expansion.

A model for an emergent scenario is *string gas cosmology* (SGC) [34] (see also [35] for earlier study and [36] for a review), where it is assumed that the universe begins in a quasi-static *Hagedorn* phase, a phase in which a gas of closed superstrings is in thermal equilibrium at a high temperature close to the limiting Hagedorn value [37]. The initial size of all of the spatial dimensions is taken to be the same. However, a dimension of space can only expand if the winding modes in that direction disappear, and since strings are two-dimensional world sheets, the intersection of winding strings (which is necessary for winding modes to disappear) is not possible in more than four large space–time dimensions (in this argument, it is assumed that there are no long-range attractive forces between winding strings. Since the local curvature perpendicular to gauge strings vanishes [38], it is reasonable to make this assumption). This helps to clarify why only three of the nine spatial dimensions that superstring theory predicts can become large. It can then be shown [39,40] that thermal fluctuations of this string gas lead to scale-invariant spectra of cosmological perturbations and gravitational waves, with a slight red tilt for the curvature fluctuations and a slight blue tilt for the gravity wave spectrum.

What was missing in SGC is a consistent dynamical framework for analyzing the cosmological evolution. To obtain such a framework, one needs to go beyond an EFT analysis, and this is where our study comes in (see also [41] for a different approach to obtaining an emergent cosmology from superstring theory that matches well with the SGC model).

Note that, as we review below, the computation of cosmological perturbations and gravitational waves in our BFSS model is based on the same formalism used in SGC, with very similar results.

In the next Section, we present a brief review of the BFSS matrix model. In Section 3, we then present our proposal for emergent time, space, and metric from the matrix model. On this basis, in Section 4 we develop a new picture of an emergent universe and compute the curvature fluctuations and gravitational waves that are predicted if one starts with a high temperature state of the BFSS matrix model. We use natural units in which the speed of light and Planck's constant are set to one.

2. BFSS Matrix Model

Starting point is the BFSS matrix model [22], a quantum theory involving ten bosonic time-dependent $N \times N$ Hermitean matrices, $A_0(\tau)$ and $X_i(\tau), i = 1, \ldots, 9$, and their sixteen Fermionic superpartners $\theta_{a,b}(\tau)$ (also $N \times N$ matrices), which transform as spinors under the $SO(9)$. The Lagrangian is

$$
\begin{aligned}
L &= \frac{1}{2g^2}\Big[\mathrm{Tr}\Big\{\frac{1}{2}(D_\tau X_i)^2 - \frac{1}{4}[X_i, X_j]^2\Big\} \\
&\quad -\theta^T D_\tau \theta - \theta^T \gamma_i \left[\theta, X^i\right]\Big], \qquad (1)
\end{aligned}
$$

where g is a coupling constant and the covariant derivate involves the temporal bosonic matrix A_0:

$$D_\tau := \partial_\tau - i[A_0(\tau), \cdot], \qquad (2)$$

with $\partial_\tau \equiv \partial/\partial\tau$.

In the above, τ is the BFSS time, and γ_i are Clifford algebra matrices (see, e.g., [42,43] for reviews of the BFSS matrix model).

This Lagrangian has both a $U(N)$ symmetry and a $SO(9)$ symmetry under rotations of the spatial matrices (and the corresponding transformation of the spinors). The partition function of the theory is given by the functional integral,

$$Z = \int dAd\theta e^{iS_{\text{BFSS}}}, \tag{3}$$

where S_{BFSS} is the action obtained by integrating the Lagrangian over time, and the integration measures are the Haar measure for the bosonic variables and the spinorial counterpart.

This matrix model was introduced [22] as a candidate for a non-perturbative definition of superstring theory (more precisely of M-theory). String theory emerges in the limit $N \to \infty$ with $\lambda \equiv g^2 N$ held fixed.

Here, we consider this matrix model in a high-temperture state, with the temperature being denoted by T. By this assumption, there is no evolution of the system in BFSS time τ. However, below we extract an emergent time, t (which is different from the time τ) in terms of which there are non-trivial dynamics.

In the high-temperature state considered here, the spatial matrices, X_i, are periodic in Euclidean BFSS time τ_E, and one can expand them in Matsubara modes:

$$X_i(\tau_E) = \sum_n X_i^n e^{2\pi n T \tau_E}, \tag{4}$$

where n runs over the integers (note that this integer n is unrelated to the integers n_i used in later Sections to introduce comoving coordinates). At high temperatures, the BFSS action is dominated by the $n = 0$ modes,

$$S_{\text{BFSS}} = S_{\text{IKKT}} + \mathcal{O}(1/T), \tag{5}$$

where S_{IKKT} is the bosonic part of the action of the Ishibashi–Kawai–Kitazawa–Tsuchiya (IKKT) matrix model [44], which, in terms of rescaled matrices, A_i,

$$A_i \equiv T^{-1/4} X_i^0, \tag{6}$$

is given by

$$S = -\frac{1}{g^2}\text{Tr}\left(\frac{1}{4}\left[A^a, A^b\right][A_a, A_b]\right). \tag{7}$$

In what follows, the matrices A_0 and A_i (i.e., the zero modes of the BFSS matrices A_0 and X_i) are used to extract an emergent space-time. The $n \neq 0$ modes of the matrices play an important role when obtaining cosmological fluctuations and gravitational waves.

3. Emergent Metric Space-Time

Without loss of generality, one can work in a basis in which the temporal matrix A_0 is diagonal. The expection values of the diagonal elements are identified with the *emergent time*. We order the diagonal elements in increasing value,

$$A_0 = \text{diag}(t_1, ..., t_N), \tag{8}$$

with $t_i \leq t_j$ for $i < j$.

Numerical simulations of the IKKT model [45] indicate that

$$\frac{1}{N}\langle \text{Tr} A_0^2 \rangle \sim \kappa N, \tag{9}$$

where κ is a (small) constant. Assuming that the eigenvalues are separated equidistantly, it then follows that the maximal eigenvalue (the maximal value of the emergent time) scales as

$$t_{\max} \sim \sqrt{N}, \tag{10}$$

and that hence the difference between the discrete values of time scales as

$$\Delta t \sim \frac{1}{\sqrt{N}}. \tag{11}$$

Thus, in the $N \to \infty$ limit, one obtains emergent continuous and infinite time. Since the eigenvalues are symmetric about $t = 0$, the time that emerges is infinite both in the past and in the future. Hence, there is neither a Big Bang nor a Big Crunch singularity.

Turning to the spatial matrices, we follow the proposal of [46] and consider, for each integer m, $n_i \times n_i$ (where n_i is an integer sufficiently smaller than N) spatial submatrices $\bar{A}_i(t(m))$ centered a distance m down the diagonal (see Figure 1):

$$(\bar{A}_i)_{I,J}(t(m)) \equiv (A_i)_{m+I,m+J}, \tag{12}$$

where $t(m)$ is the m'th temporal eigenvalue. Following [46], one can define the *extent of space* parameters $x_i(t)$ in the i'th spatial direction at the time t via

$$x_i(t)^2 \equiv \left\langle \frac{1}{n}\mathrm{Tr}(\bar{A}_i)(t))^2 \right\rangle. \tag{13}$$

With this definition of *emergent space*, there is non-trivial time evolution in terms of the emergent time.

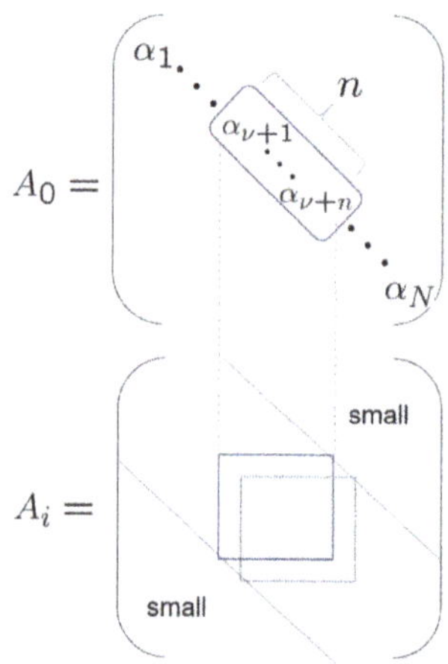

Figure 1. In the basis in which the A_0 matrix is diagonal with the diagonal elements defining emergent time, we can use spatial $n_i \times n_i$ submatrices $\bar{A}_i(t)$ located a distance t down the diagonal to define the i'th spatial direction at time t. The figure is taken from [42] with permission.

In the IKKT model, it has been observed via numerical simulations that, in the presence of the fermionic terms in the action, the state that minimizes the free energy breaks the $SO(9)$ symmetry into $SO(6) \times SO(3)$. Of the nine extent of space parameters, six remain small and three increase [47–52]. This result can also be confirmed by means of Gaussian expansion calculations [53,54]. The emergence of three macroscopic spatial dimensions with the other six remaining microscopic dimensions reminds one of the same phenomenon that was discovered in the context of string gas cosmology [34], where string winding modes prevent spatial dimensions from expanding. In order to obtain large dimensions, the string winding modes near these dimensions must be annihilated, and this is not possible in more than three large spatial dimensions since string winding modes will have vanishing intersection probability (see [55,56] for a more detailed analysis).

Very recently, we have been able to show [57] that in the presence of the fermionic terms, the $SO(9)$ symmetry will also be spontaneously broken in the BFSS matrix model. This result was derived using a Gaussian expansion calculation. With this tool, it is not yet possible to determine the symmetry of the state that minimizes the free energy, but the intuition gained from string gas cosmology leads us to expect that the symmetry breaking pattern will be the same as what was observed in the IKKT model.

Making use of the Riemann–Lebesgue Lemma, it can be shown from the bosonic IKKT action that the off-diagonal elements of the spatial matrices decay once one departs beyond a critical distance, n_c, from the diagonal:

$$\sum_i \langle |A_i|^2_{ab} \rangle \;\rightarrow\; 0 \text{ for } |a-b| > n_c\,, \tag{14}$$

where

$$n_c \;\sim\; \sqrt{N} \tag{15}$$

This result can be confirmed by numerical simulations (see [58–61] for recent investigations of the IKKT matrix model), which also indicate that

$$\sum_i \langle |A_i|^2_{ab} \rangle \;\sim\; \text{constant when } |a-b| < n_c\,, \tag{16}$$

a result, for which we at present do not have any analytical evidence.

Our proposal [21] is to view n_i as the i'th comoving spatial coordinate and

$$l^2_{\text{phys},i}(n_i,t) \;\equiv\; \left\langle \text{Tr}(\bar{A}_i)(t))^2 \right\rangle, \tag{17}$$

as the physical length square of the curve from $n_i = 0$ to n_i along the i'th coordinate direction. It then follows from Equation (16) that

$$l_{\text{phys},i}(n_i) \;\sim\; n_i \text{ for } n_i < n_c\,. \tag{18}$$

Given the definition of length and comoving coordinates, we can extract the spatial metric component g_{ii} in the usual way:

$$g_{ii}(n_i)^{1/2} \;=\; \frac{d}{dn_i} l_{\text{phys},i}(n_i) \tag{19}$$

and we obtain

$$g_{ii}(n_i,t) \;=\; \mathcal{A}(t)\delta_{ii} \;\; i = 1,2,3\,, \tag{20}$$

where $\mathcal{A}(t)$ is an increasing function of t at late times and δ_{ii} is the Kronecker delta. Making use of the remnant $SO(3)$ symmetry, one obtains:

$$g_{ij}(\mathbf{n},t) \;=\; \mathcal{A}(t)\delta_{ij} \;\; i,j = 1,2,3\,, \tag{21}$$

where $\mathbf{n}$ stands for the comoving 3-vector corresponding to the three large dimensions.

We have thus indicated how one may obtain emergent time, space, and spatial metrics from the zero modes of the BFSS matrix model. The emergent space-time is infinite both in temporal and spatial directions, and from Equation (21) it follows that the metric that emerges on the emergent large three-dimensional space is spatially flat. Hence, in the scenario that we are proposing the famous horizon and flatness problems of Standard Big Bang cosmology are automatically solved without the need for any phase of inflation [1–4] or superslow contraction [62,63]. In the next Section we show that the thermal fluctuations in the BFSS state automatically lead to approximately scale-invariant spectra for cosmological perturbations and gravitational waves.

4. Emergent Early Universe Cosmology

The cosmology that emerges from the BFSS matrix model is an *emergent cosmology*. The classical notion of space-time only makes sense at late emergent time. Solving the classical bosonic matrix equations for the three large spatial dimensions at late times yields [64–66],

$$\mathcal{A}(t) \sim t^{1/2}, \tag{22}$$

i.e., expansion like in a radiation-dominated universe. We expect that the inclusion of the fermionic sector will change this scaling and produce a matter component that scales as pressureless dust. Thus, in our cosmology there is a direct transition from the emergent phase to that of Standard Big Bang cosmology, like in string gas cosmology. Note that there is no sign of a cosmological constant. Thus, it appears that the cosmology that emerges from the BFSS matrix model will be free of the infamous cosmological constant problem. On the other hand, it is at the present time unclear how dark energy will emerge. Based on the swampland criteria [8,9] and the TCC [10], it follows that dark energy cannot be a cosmological constant. It may be a quintessence field [67–69] that emerges from the matter sector (see, e.g., [70] for an attempt to obtain a unified dark sector (dark matter and dark energy) from string theory).

Since we are considering the BFSS matrix model in a high-temperature state, there will be thermal fluctuations, and these will lead to curvature fluctuations and gravitational waves like in SGC (see [39,40] for analyses of the generation of cosmological perturbations and gravitational waves in SGC).

At late times, the classical metric of the $3+1$ dimensional space-time can be expressed at the level of linearized fluctuations as (see, e.g., [71,72] for reviews of the theory of cosmological perturbations)

$$ds^2 = a^2(\eta)\big((1+2\Phi)d\eta^2 - [(1-2\Phi)\delta_{ij} + h_{ij}]dx^i dx^j\big), \tag{23}$$

where Φ (which depends on space and time) represents the curvature fluctuations, η denotes the conformal time, and the transverse tracelss tensor, h_{ij}, stands for the gravitational waves (in writing, the perturbed metric in this form we have chosen to work in longitudinal gauge).

According to the linearized Einstein equations, the curvature fluctuations of the geometry are determined in terms of the energy density perturbations via

$$\langle|\Phi(k)|^2\rangle = 16\pi^2 G^2 k^{-4} \langle \delta T^0{}_0(k) \delta T^0{}_0(k)\rangle, \tag{24}$$

and the gravitational waves are given by the off-diagonal pressure fluctuations,

$$\langle|h(k)|^2\rangle = 16\pi^2 G^2 k^{-4} \langle \delta T^i{}_j(k) \delta T^i{}_j(k)\rangle \ \ (i \neq j), \tag{25}$$

where G is the Newtonian constant of gravitation.

Since the energy density perturbations in a region of radius R are determined by the specific heat capacity, $C_V(R)$, on that scale, the power spectrum of curvature fluctuations is given by

$$P(k) = k^3(\delta\Phi(k))^2 = 16\pi^2 G^2 k^2 T^2 C_V(R), \tag{26}$$

where k is the momentum scale associated with R.

For thermal fluctuations, the matter correlation functions are given by taking partial derivatives of the finite temperature partition function. Since the partition function of the BFSS matrix model is our starting point, we are able to compute the observables. Specifically, the specific heat capacity is related to the internal energy, $E(R)$, via

$$C_V(R) = \frac{\partial}{\partial T} E(R), \tag{27}$$

and the internal energy is given by the derivative of the partition function with respect to the inverse temperature, β:

$$E = -\frac{\partial}{\partial \beta} \ln Z(\beta) . \tag{28}$$

The computation of the internal energy $E(R)$ was carried out in Ref. [20], based on the study [73] in the context of the IKKT matrix model. We find:

$$E^2 = N^2 \langle \mathcal{E} \rangle_{\rm BFSS} \tag{29}$$

with

$$\mathcal{E} = -\frac{3}{4N\beta} \int_0^\beta dt \mathrm{Tr}([X_i, X_j]^2) . \tag{30}$$

Evaluating $\mathcal{E}$, we obtain:

$$E^2 = \frac{3}{4} N^2 \chi_2 T - \frac{3}{4} N^4 \alpha \chi_1 T^{-1/2} , \tag{31}$$

where χ_2 is a constant while χ_1 depends on R via

$$\chi_1 = \langle R^2 \rangle_{\rm BFSS} T^{-1/2} . \tag{32}$$

The first term in Equation (31) yields a Poisson spectrum that dominates on ultraviolet scales, while the second term yields a scale-invariant contribution to the curvature power spectrum that dominates on the infrared scales relevant to current cosmological observations.

5. Challenges and Future Directions

To summarize, we have presented a proposal to obtain time, space, and a metric as emergent phenomena from the BFSS matrix model, taken in a high-temperature state. In the $N \to \infty$ limit, the time that emerges is infinite in both directions and continuous. Similarly, the emergent space is of infinite extent. Given our proposal, the background metric that emerges is spatially flat and isotropic. Since we are considering a thermal state of the BFSS model, there are thermal fluctuations, and we have reviewed here how these perturbations yield scale-invariant spectra of curvature fluctuations and gravitational waves. Hence, our cosmological model solves the famous horizon and flatness problems of standard cosmology and provides a causal mechanism for the origin of scale-invariant spectra of curvature fluctuations and gravitational waves, all without the need to postulate a phase of cosmological inflation. The cosmology is non-singular, and the cosmological constant problem is absent.

The $SO(9)$ symmetry of the BFSS action is spontaneously broken. In the case of the IKKT model, the state that minimizes the energy has $SO(6) \times SO(3)$ symmetry, with the extent of space increasing only in three spatial dimensions. Based on solving the classical matrix equations, it appears that the cosmological expansion of the three large spatial dimensions is as in a radiation-dominated universe. The presence of fermions plays a key role in obtaining this symmetry breaking. On the other hand, the fermionic sector has not been included in studying the late time dynamics of the three large spatial dimensions. An important open problem is to include the fermionic sector. A careful study of the full BFSS matrix model may also shed some light on the nature of dark energy.

An important open problem is to study the dynamics of this phase transition and to determine the key physical reason for the distinguished symmetry breaking pattern. We speculate that the reason may be related to the reason why three dimensions become large in SGC. Once the dynamics of the phase transition are understood, one may be able to make predictions for the tilts n_s and n_t of the spectra of cosmological perturbations and gravitational waves, and determine consistency relations between the tilts. One may be able to recover the same consistency relation $n_t = 1 - n_s$ as in SGC (and as in the recently proposed ekpyrotic scenario mediated by an S-brane [74]).

While our scenario yields emergent space and time, and an emergent metric, it is important to investigate what effective action describes the evolution of localized matrix excitations. At low energies, we need to be able to recover Einstein gravity coupled to regular matter and radiation.

In conclusion, although we have indicated a promising approach to the emergence of space and time, and of early universe cosmology, a lot of work needs to be done before one will be able to make successful contact with testable low-energy physics.

Author Contributions: Conceptualization, S.B., R.B. and S.L.; Methodology, S.B., R.B. and S.L.; Investigation, S.B., R.B. and S.L.; Writing—original draft, S.B., R.B. and S.L. All authors have read and agreed to the published version of the manuscript.

Funding: S.B. is supported in part by the Higgs Fellowship. S.L. is supported in part by the Fonds de recherche du Québec - Nature et technologies (FRQNT), Canada. The research at McGill is supported in part by funds from the Natural Sciences and Engineering Research Council of Canada (NSERC) and from the Canada Research Chair program.

Data Availability Statement: Not applicable.

Acknowledgments: We thank B. Ydri for permission to use Figure 1, which is taken from [42].

Conflicts of Interest: The authors declare no conflict of interest.

References

1. Guth, A.H. The inflationary universe: A possible solution to the horizon and flatness problems. *Phys. Rev. D* **1981**, *23*, 347–356. Reprinted: *Adv. Ser. Astrophys. Cosmol.* **1987**, *3*, 139–148. [CrossRef]
2. Brout, R.; Englert, F.; Gunzig, E. The creation of the universe as a quantum phenomenon. *Ann. Phys.* **1978**, *115*, 78–106. [CrossRef]
3. Starobinsky, A.A. A New type of isotropic cosmological models without singularity. *Phys. Lett. B* **1980**, *91*, 99–102. [CrossRef]
4. Sato, K. First order phase transition of a vacuum and expansion of the universe. *Mon. Not. R. Astron. Soc.* **1981**, *195*, 467–479. [CrossRef]
5. Brandenberger, R.H. Alternatives to the inflationary paradigm of structure formation. *Int. J. Mod. Phys. Conf. Ser.* **2011**, *1*, 67–79. [CrossRef]
6. Brandenberger, R.H. Cosmology of the very early universe. *AIP Conf. Proc.* **2010**, *1268*, 3–70.
7. Brandenberger, R.; Peter, P. Bouncing cosmologies: Progress and problems. *Found. Phys.* **2017**, *47*, 797–850. [CrossRef]
8. Ooguri, H.; Vafa, C. On the geometry of the string landscape and the swampland. *Nucl. Phys. B* **2007**, *766*, 21–33. [CrossRef]
9. Obied, G.; Ooguri, H.; Spodyneiko, L.; Vafa, C. De Sitter space and the swampland. *arXiv* **2018**, arXiv:1806.08362. [CrossRef]
10. Bedroya, A.; Vafa, C. Trans-Planckian censorship and the swampland. *J. High Energy Phys.* **2020**, *2009*, 123. [CrossRef]
11. Bedroya, A.; Brandenberger, R.; Loverde, M.; Vafa, C. Trans-Planckian censorship and inflationary cosmology. *Phys. Rev. D* **2020**, *101*, 103502. [CrossRef]
12. Brandenberger, R. String cosmology and the breakdown of local effective field theory. *arXiv* **2021**, arXiv:2112.04082. [CrossRef]
13. Brandenberger, R.; Kamali, V. Unitarity problems for an effective field theory description of early universe cosmology. *Eur. Phys. J. C* **2022**, *82*, 818. [CrossRef] [PubMed]
14. Dvali, G.; Gomez, C.; Zell, S. Quantum break-time of de Sitter. *J. Cosmol. Astrpart. Phys.* **2017**, *6*, 028. [CrossRef]
15. Dvali, G.; Gomez, C. On exclusion of positive cosmological constant. *Fortsch. Phys.* **2019**, *67*, 1800092. [CrossRef]
16. Dvali, G.; Gomez, C.; Zell, S. Quantum breaking bound on de Sitter and swampland. *Fortsch. Phys.* **2019**, *67*, 1800094. [CrossRef]
17. Brahma, S.; Dasgupta, K.; Tatar, R. Four-dimensional de Sitter space is a Glauber-Sudarshan state in string theory. *J. High Energy Phys.* **2021**, *2021*, 114. [CrossRef]
18. Brahma, S.; Dasgupta, K.; Tatar, R. de Sitter space as a Glauber-Sudarshan state. *J. High Energy Phys.* **2021**, *2021*, 104. [CrossRef]
19. Bernardo, H.; Brahma, S.; Dasgupta, K.; Tatar, R. Crisis on infinite earths: Short-lived de Sitter vacua in the string theory landscape. *J. High Energy Phys.* **2021**, *2021*, 37. [CrossRef]
20. Brahma, S.; Brandenberger, R.; Laliberte, S. Emergent cosmology from matrix theory. *J. High Energy Phys.* **2022**, *3*, 67. [CrossRef]
21. Brahma, S.; Brandenberger, R.; Laliberte, S. Emergent metric space-time from matrix theory. *J. High Energy Phys.* **2022**, *9*, 31. [CrossRef]
22. Banks, T.; Fischler, W.; Shenker, S.H.; Susskind, L. M theory as a matrix model: A Conjecture. *Phys. Rev. D* **1997**, *55*, 5112–5128. [CrossRef]
23. Hoppe, J. Quantum Theory of a Massless Relativistic Surface. Ph.D. Thesis, Massachusetts Institute of Technology (MIT), Cambridge, MA, USA, 1982. Available online: http://dspace.mit.edu/handle/1721.1/15717 (accessed on 5 December 2022).
24. de Witt, B.; Hoppe, J.; Nicolai, H. On the quantum mechanics of supermembranes. *Nucl. Phys. B* **1988**, *305*, 545–581. [CrossRef]
25. Das, S.R.; Jevicki, A. String field theory and physical interpretation of $D = 1$ strings. *Mod. Phys. Lett. A* **1990**, *5*, 1639–1650. [CrossRef]

26. Dijkgraaf, R.; Verlinde, E.P.; Verlinde, H.L. Matrix string theory. *Nucl. Phys. B* **1997**, *500*, 43–61. [CrossRef]
27. Craps, B.; Sethi, S.; Verlinde, E.P. A matrix big bang. *J. High Energy Phys.* **2005**, *10*, 5. [CrossRef]
28. Blau, M.; O'Loughlin, M. DLCQ and plane wave matrix big bang models. *J. High Energy Phys.* **2008**, *9*, 97. [CrossRef]
29. Berenstein, D.E.; Maldacena, J.M.; Nastase, H.S. Strings in flat space and pp waves from $N = 4$ superYang-Mills. *J. High Energy Phys.* **2002**, *4*, 13. [CrossRef]
30. Maldacena, J.M.; Nunez, C. Towards the large N limit of pure $N = 1$ superYang-Mills. *Phys. Rev. Lett.* **2001**, *86*, 588–591. [CrossRef]
31. Craps, B.; Evnin, O. Light-like big bang singularities in string and matrix theories. *Class. Quant. Grav.* **2011**, *28*, 204006. [CrossRef]
32. Smolin, L. M theory as a matrix extension of Chern-Simons theory. *Nucl. Phys. B* **2000**, *591*, 227–242. [CrossRef]
33. Yargic, Y.; Lanier, J.; Smolin, L.; Wecker, D. A cubic matrix action for the Standard Model and beyond. *arXiv* **2022**, arXiv:2201.04183. [CrossRef]
34. Brandenberger, R.H.; Vafa, C. Superstrings in the early universe. *Nucl. Phys. B* **1989**, *316*, 391–410. [CrossRef]
35. Kripfganz, J.; Perlt, H. Cosmological impact of winding strings. *Class. Quant. Grav.* **1988**, *5*, 453. [CrossRef]
36. Brandenberger, R.H. String gas cosmology; In *String Cosmology*; Erdmenger, J., Ed.; Wiley-VCH Verlag GmbH & Co. KGaA: Weinheim, Germany, 2009; pp. 193–230. Available online: https://www.lehmanns.ch/shop/technik/24731252-9783527628070-string-cosmology (accessed on 5 December 2022).
37. Hagedorn, R. Statistical thermodynamics of strong interactions at high-energies. *Nuovo Cim. Suppl.* **1965**, *3*, 147–186. Reprinted: *Quark–Gluon Plasma: Theoretical Foundations*; Kapusta, J.I., Müller, B., Rafelski, J., Eds.; North-Holland: Amsterdam, The Netherlands; 2003; pp. 24–63. Available online: http://cds.cern.ch/record/346206 (accessed on 5 December 2022).
38. Vilenkin, A. Gravitational field of vacuum domain walls and strings. *Phys. Rev. D* **1981**, *23*, 852–857. [CrossRef]
39. Nayeri, A.; Brandenberger, R.H.; Vafa, C. Producing a scale-invariant spectrum of perturbations in a Hagedorn phase of string cosmology. *Phys. Rev. Lett.* **2006**, *97*, 021302. [CrossRef]
40. Brandenberger, R.H.; Nayeri, A.; Patil, S.P.; Vafa, C. Tensor modes from a primordial Hagedorn phase of string cosmology. *Phys. Rev. Lett.* **2007**, *98*, 231302. [CrossRef]
41. Agrawal, P.; Gukov, S.; Obied, G.; Vafa, C. Topological gravity as the early phase of our universe. *arXiv* **2020**, arXiv:2009.10077. [CrossRef]
42. Ydri, B. Review of M(atrix)-theory, type IIB matrix model and Matrix String Theory. *arXiv* **2017**, arXiv:1708.00734. [CrossRef]
43. Taylor, W. M(atrix) theory: Matrix quantum mechanics as a fundamental theory. *Rev. Mod. Phys.* **2001**, *73*, 419–462. [CrossRef]
44. Ishibashi, N.; Kawai, H.; Kitazawa, Y.; Tsuchiya, A. A large-N reduced model as superstring. *Nucl. Phys. B* **1997**, *498*, 467–491. [CrossRef]
45. Aoki, H.; Iso, S.; Kawai, H.; Kitazawa, Y.; Tada, T. Space-time structures from IIB matrix model. *Prog. Theor. Phys.* **1998**, *99*, 713–746. [CrossRef]
46. Kim, S.W.; Nishimura, J.; Tsuchiya, A. Expanding (3+1)-dimensional universe from a Lorentzian matrix model for superstring theory in (9+1)-dimensions. *Phys. Rev. Lett.* **2012**, *108*, 011601. [CrossRef] [PubMed]
47. Nishimura, J.; Vernizzi, G. Spontaneous breakdown of Lorentz invariance in IIB matrix model. *J. High Energy Phys.* **2000**, *4*, 15. [CrossRef]
48. Ambjorn, J.; Anagnostopoulos, K.N.; Bietenholz, W.; Hotta, T.; Nishimura, J. Monte Carlo studies of the IIB matrix model at large *N*. *J. High Energy Phys.* **2000**, *7*, 11. [CrossRef]
49. Nishimura, J.; Sugino, F. Dynamical generation of four-dimensional space-time in the IIB matrix model. *J. High Energy Phys.* **2002**, *5*, 1. [CrossRef]
50. Kawai, H.; Kawamoto, S.; Kuroki, T.; Matsuo, T.; Shinohara, S. Mean field approximation of IIB matrix model and emergence of four-dimensional space-time. *Nucl. Phys. B* **2002**, *647*, 153–189. [CrossRef]
51. Aoyama, T.; Nishimura, J.; Okubo, T. Spontaneous breaking of the rotational symmetry in dimensionally reduced super Yang-Mills models. *Prog. Theor. Phys.* **2011**, *125*, 537–563. [CrossRef]
52. Nishimura, J.; Okubo, T.; Sugino, F. Systematic study of the SO(10) symmetry breaking vacua in the matrix model for type IIB superstrings. *J. High Energy Phys.* **2011**, *10*, 135. [CrossRef]
53. Kabat, D.N.; Lifschytz, G.; Lowe, D.A. Black hole thermodynamics from calculations in strongly coupled gauge theory. *Int. J. Mod. Phys. A* **2001**, *16*, 856–865. [CrossRef]
54. Kabat, D.N.; Lifschytz, G.; Lowe, D.A. Black hole entropy from nonperturbative gauge theory. *Phys. Rev. D* **2001**, *64*, 124015. [CrossRef]
55. Watson, S.; Brandenberger, R. Stabilization of extra dimensions at tree level. *J. Cosmol. Astrpart. Phys.* **2003**, *11*, 8. [CrossRef]
56. Patil, S.P.; Brandenberger, R.H. The cosmology of massless string modes. *J. Cosmol. Astrpart. Phys.* **2006**, *1*, 5. [CrossRef]
57. Brahma, S.; Brandenberger, R.; Laliberte, S. Spontaneous symmetry breaking in the BFSS model: Analytical results using the Gaussian expansion method. *arXiv* **2022**, arXiv:2209.01255. [CrossRef]
58. Nishimura, J. New perspectives on the emergence of (3+1)D expanding space-time in the Lorentzian type IIB matrix model. *PoS* **2020**, *CORFU2019*, 178. [CrossRef]
59. Hirasawa, M.; Anagnostopoulos, K.; Azuma, T.; Hatakeyama, K.; Ito, Y.; Nishimura, J.; Papadoudis, S.; Tsuchiya, A. A new phase in the Lorentzian type IIB matrix model and the emergence of continuous space-time. *PoS* **2022**, *LATTICE2021*, 428. [CrossRef]

60. Hatakeyama, K.; Anagnostopoulos, K.; Azuma, T.; Hirasawa, M.; Ito, Y.; Nishimura, J.; Papadoudis, S.; Tsuchiya, A. Relationship between the Euclidean and Lorentzian versions of the type IIB matrix model. *PoS* **2022**, *LATTICE2021*, 341. [CrossRef]
61. Hatakeyama, K.; Anagnostopoulos, K.; Azuma, T.; Hirasawa, M.; Ito, Y.; Nishimura, J.; Papadoudis, S.; Tsuchiya, A. Complex Langevin studies of the emergent space-time in the type IIB matrix model. In *Proceedings of the East Asia Joint Symposium on Fields and Strings 2021;* Iso, S., Itoyama, H., Maruyoshi, K., Nishinaka, T., Oota, T., Sakai, K., Tsuchiya, A., Eds.; World Scientific Co. Ltd.: Singapore, 2022; pp. 9–18. [CrossRef]
62. Khoury, J.; Ovrut, B.A.; Steinhardt, P.J.; Turok, N. The ekpyrotic universe: Colliding branes and the origin of the hot big bang. *Phys. Rev. D* **2001**, *64*, 123522. [CrossRef]
63. Khoury, J.; Ovrut, B.A.; Seiberg, N.; Steinhardt, P.J.; Turok, N. From big crunch to big bang. *Phys. Rev. D* **2002**, *65*, 086007. [CrossRef]
64. Kim, S.W.; Nishimura, J.; Tsuchiya, A. Late time behaviors of the expanding universe in the IIB matrix model. *J. High Energy Phys.* **2012**, *10*, 147. [CrossRef]
65. Ito, Y.; Nishimura, J.; Tsuchiya, A. Power-law expansion of the Universe from the bosonic Lorentzian type IIB matrix model. *J. High Energy Phys.* **2015**, *11*, 70. [CrossRef]
66. Hatakeyama, K.; Matsumoto, A.; Nishimura, J.; Tsuchiya, A.; Yosprakob, A. The emergence of expanding space–time and intersecting D-branes from classical solutions in the Lorentzian type IIB matrix model. *Prog. Theor. Exp. Phys.* **2020**, *2020*, 43B10. [CrossRef]
67. Wetterich, C. Cosmology and the fate of dilatation symmetry. *Nucl. Phys. B* **1988**, *302*, 668–696. [CrossRef]
68. Peebles, P.J.E.; Ratra, B. Cosmology with a time variable cosmological constant. *Astrophys. J.* **1988**, *325*, L17–L20. Available online: https://articles.adsabs.harvard.edu/full/1988ApJ...325L..17P (accessed on 5 December 2022). [CrossRef]
69. Ratra, B.; Peebles, P.J.E. Cosmological consequences of a rolling homogeneous scalar field. *Phys. Rev. D* **1988**, *37*, 3406–3427. [CrossRef]
70. Bernardo, H.; Brandenberger, R.; Fröhlich, J. Towards a dark sector model from string theory. *J. Cosmol. Astrpart. Phys.* **2022**, *9*, 40. [CrossRef]
71. Mukhanov, V.F.; Feldman, H.A.; Brandenberger, R.H. Theory of cosmological perturbations. *Phys. Rep.* **1992**, *215*, 203–333. [CrossRef]
72. Brandenberger, R. Lectures on the theory of cosmological perturbations. In *The Early Universe and Observational Cosmology*; Bretón, N., Cervantes-Cota, J.L., Salgado, M., Eds.; Springer: Berlin/Heidelberg, Germany, 2004; pp. 127–167. [CrossRef]
73. Kawahara, N.; Nishimura, J.; Takeuchi, S. High temperature expansion in supersymmetric matrix quantum mechanics. *J. High Energy Phys.* **2007**, *12*, 103. [CrossRef]
74. Brandenberger, R.; Wang, Z. Nonsingular ekpyrotic cosmology with a nearly scale-invariant spectrum of cosmological perturbations and gravitational waves. *Phys. Rev. D* **2020**, *101*, 063522. [CrossRef]

physics

MDPI

Article

Quantum Configuration and Phase Spaces: Finsler and Hamilton Geometries

Saulo Albuquerque [1], Valdir B. Bezerra [1], Iarley P. Lobo [2,3,*], Gabriel Macedo [1], Pedro H. Morais [1], Ernesto Rodrigues [1], Luis C. N. Santos [1] and Gislaine Varão [1]

1 Physics Department, Federal University of Paraíba, Caixa Postal 5008, João Pessoa 58059-900, PB, Brazil
2 Department of Chemistry and Physics, Federal University of Paraíba, Rodovia BR 079-Km 12, Areia 58397-000, PB, Brazil
3 Physics Department, Federal University of Lavras, Caixa Postal 3037, Lavras 37200-000, MG, Brazil
* Correspondence: lobofisca@gmail.com; Tel.: +55-83999882305

Abstract: In this paper, we reviewtwo approaches that can describe, in a geometrical way, the kinematics of particles that are affected by Planck-scale departures, named Finsler and Hamilton geometries. By relying on maps that connect the spaces of velocities and momenta, we discuss the properties of configuration and phase spaces induced by these two distinct geometries. In particular, we exemplify this approach by considering the so-called q-de Sitter-inspired modified dispersion relation as a laboratory for this study. We finalize with some points that we consider as positive and negative ones of each approach for the description of quantum configuration and phases spaces.

Keywords: quantum gravity phenomenology; Finsler geometry; Hamilton geometry

Citation: Albuquerque, S.; Bezerra, V.B.; Lobo, I.P.; Macedo, G.; Morais, P.H.; Rodrigues, E.; Santos, L.C.N.; Varão, G. Quantum Configuration and Phase Spaces: Finsler and Hamilton Geometries. *Physics* **2023**, *5*, 90–115. https://doi.org/10.3390/physics5010008

Received: 1 November 2022
Revised: 14 December 2022
Accepted: 27 December 2022
Published: 19 January 2023

1. Introduction

Since the original works by Bronstein [1] that demonstrated uncertainty in the localization of events when geometrical degrees of freedom are quantized, it has been argued that attempts to formulate quantum gravity in a differentiable manifold endowed with smooth geometric quantities would not be an interesting path to follow if one aims to pursue a fundamental approach to this problem. Attempts in this direction have accumulated over the years, having prominent representatives such as loop quantum gravity (LQG) [2] and causal dynamical triangulation [3]. These approaches to quantum gravity predict or describe several effects that should be manifest at the Planckian regime of length and energy, such as the discretization of geometry, which requires a language that obviously departs from the usual Riemannian construction of general relativity. Despite the elegance of such approaches, with current technology we are far from being able to concretely address the regime in which such discretization would become evident. Nevertheless, the notion that spacetime could effectively behave like a medium formed by "atoms of space" has led to a rich phenomenological approach to quantum gravity, which by encoding generic departures from relativistic equations, can describe common predictions expected to be present at an intermediate stage between classical and quantum gravity. Such an approach is encompassed in the area of quantum gravity phenomenology, which addresses a myriad of effects beyond the one described in this paragraph, as can be seen in Ref. [4], and in particular, has found in multimessenger astronomy a fruitful environment to be explored [5].

Usually, the regime, in which this idea is considered, is the regime, in which the test particle approximation is valid consisting of the approximation, in which one would have simultaneously faint gravitational and quantum effects, described by the limits of the gravitational constant, $G \to 0$, and the reduced Planck's constant, $\hbar \to 0$, however, with the Planck energy, $E_P = \sqrt{c^5\hbar/G}$, being finite, with c the speed of light. This deformed "Minkowski limit", which presents departures from Minskowski spacetime's structure has been suggested by various quantum gravity proposals, such as the linearization of the hypersurface deformation algebra inspired by LQG [6–8] and non-commutative geometry [9–12] (for more details on this Miskowski limit, see Section 3.1.1 of Ref. [4], and for more references on other theoretical approaches, in which such limit emerges, see Section 2.2 of Ref. [5]). It is expected that the path between the differentiable Riemannian description of special (and general) relativity and the complete quantum gravity theory should pass through an intermediate regime, in which one has departures from the Riemannian character of spacetime but still has geometric features that could describe a bottom-up phenomenology.

Furthermore, geometry plays an important role in the description of principles that have guided the developments of relativistic theories; for example, the principle of covariance is manifest through the use of tensorial equations of motion, the local relativity principle is a physical manifestation of having local equations of motion invariant under the Poincaré group (which is the group of isometries of Minkowski space), the equivalence principle of general relativity is manifest in the fact that the motion of free particles is realized through geodesics, and the clock postulate can be expressed by stating that an observer measures its proper time by the arc-length of its own trajectory.

An important part of quantum gravity phenomenology is devoted to the question of whether, in the aforementioned Minkowski limit, the Lorentz invariance, and consequently, the local relativity principle, is preserved or broken due to Planck-scale effects [13]. As is known, a length/energy scale is not invariant under Lorentz transformations, which implies that either a quantum gravity scale breaks the equivalence of inertial frames in the aforementioned Minkowski limit, or the Lorentz or Poincaré group only describes a low energy/large distance approximation of a deeper transformation between inertial frames. The former possibility is known as a Lorentz invariance violation (LIV) scenario [14,15], and the latter is known as doubly (or deformed) special relativity (DSR) [16,17]. As the geometrization of special relativity, due to Minkowski, paved the way to more fundamental descriptions of nature, we shall follow a similar path, but of geometrizing DSR.

Geometric descriptions of DSR through non-commutative geometry are known [9–12], but we revise some continuous, differentiable ways of exploring non-Riemannian degrees of freedom and the possibilities for preserving the aforementioned principles. This way, we critically analyze two extensions of Riemannian geometry that are capable of describing aspects of an emergent "quantum configuration and phase spaces" that preserve the intuition of those principles: they are Finsler and Hamilton geometries. Finsler geometry originally is related to the space of events and velocities (for this reason we refer to a quantum configuration space), and Hamilton geometry originally described the space of events and momenta (for this reason, we call it a quantum phase space). In this paper, we revise the phenomenological opportunities that emerge from these approaches and the interplay between them. We also condensate the utility of each of these geometries and their limitations in the current scenario.

We should also stress that the approaches described in this review, refer to configuration and phase spaces probed by a single particle. The geometry probed by a multi-particle system and its interplay with Finsler and Hamilton languages (or even geometries that go beyond them) should still be further explored, in which, possibly the intuition gained from the relative locality framework [18] would play a prominent role in this approach.

The paper is organized as follows. Section 2 revisits the origin of the idea of describing the effective spacetime probed by a particle that propagates through a modified dispersion relation (MDR) by the proposal of rainbow metrics. Section 3 revisits how this general idea is realized by the use of Finsler geometry in the tangent bundle, whose dual version in the cotangent bundle is discussed in Section 4, which is illustarated by considering the curved non-trivial case of q-de Sitter-inspired Finsler geometry. Section 5 considers the situation of deriving the geometry of the cotangent bundle, and, in Section 6, its dual tangent bundle formalization defined by Hamilton geometry is considered, which is illustrated by the q-de Sitter case. In Section 7, we comparatively discuss these two approaches and highlight points that we consider as useful as well as their limitations. Finally, some important remarks are drawn in Section 8. Throughout the paper, a system of units with $c = \hbar = 1$ is used, so that the Planck length is the inverse of the Planck energy: $\sqrt{G} = \ell = E_{\rm P}^{-1}$.

2. Preliminaries on Rainbow Geometries

As described above, over the years, the intuition that spacetime would behave like material media, where instead of atoms of matter, one would have atoms of spacetime, has been solidified through some approaches of quantum gravity. Just as occurs in matter, in which one does not need to know the specific details of the granular structure of a given medium to study the propagation of particles through it, in spacetime, one can build phenomenology-inspired ways of modeling how elementary particles interact with discrete gravitational degrees of freedom while traveling through space, a so-called "in-vaccuum dispersion". One could say that the most popular way of doing this is through the assumption that particles would obey a modified dispersion relation, whose corrections are given perturbatively by powers of the quantum gravity scale, which we could assume as being in the order of Planck units. (The dispersion relation furnishes the group velocity of waves and defines the trajectory that on-shell particles follow from the Hamilton equations.) Actually, when the interplay between the presence of amplifiers of observables and the uncertainties of observations allows us to constrain this parameter at a level close to its Planckian version, we say that we are at Planck-scale sensitivity [4].

Such behavior also happens in meta-materials [19], in which it is possible to describe the motion of particles through it by geodesics in a given geometry; it also appears in the motion of a charged particle in a pre-metric formulation of electromagnetism [20], in the description of seismic waves [21], etc.; for a review, see Ref. [22]. Additionally, one could wonder if the motion of particles, determined by Planck-scale modified dispersion relations, could also be described by geodesics of a non-Riemannian geometry. Besides, the dispersion relation itself is usually determined by the norm of the 4-momentum measured by a Riemannian metric, which also determines the symmetries observed by measurements in that spacetime.

This intuition was early realized by the so-called "rainbow geometries" [23], idealized by João Magueijo and Lee Smolin which aimed to extend the DSR formulation proposed by them in Ref. [17] to curved spacetimes. In that case, the way found to express local modified dispersion relations through a norm, consisted in absorbing functions of the particle's energy divided by Planck energy, $\epsilon = E/E_{\rm P}$, such as $f(\epsilon)$ and $g(\epsilon)$, which would appear in the MDR that follows:

$$m^2 = f^2(\epsilon)E^2 - g^2(\epsilon)|\vec{p}|^2 , \tag{1}$$

(with the three-momentum $\vec{p}$) into the definition of new spacetime tetrads, $\tilde{e}_{(0)}{}^{\mu}(\epsilon) = f(\epsilon)e_{(0)}{}^{\mu}$ and $\tilde{e}_{(I)}{}^{\mu}(\epsilon) = g(\epsilon)e_{(I)}{}^{\mu}$, such that the MDR reads

$$m^2 = \eta^{AB}\tilde{e}_{(A)}{}^{\mu}\tilde{e}_{(B)}{}^{\nu}p_\mu p_\nu = \tilde{g}^{\mu\nu}(\epsilon)p_\mu p_\nu , \tag{2}$$

where $g^{\mu\nu}(\epsilon) = \eta^{AB}\tilde{e}_{(A)}{}^{\mu}\tilde{e}_{(B)}{}^{\nu}$ is the rainbow metric, η^{AB} is the Minkowski metric diag$(+---)$, Greek letters denote four-dimensional indices and take on the values 0 (time) 1, 2, and 3 (space), low-case Latin letters denote the space indices, and p_μ is the 4-momentum. This description would imply that when an observer uses the motion of a particle with energy E to probe spacetime, then the line element assigned to that spacetime is the following:

$$ds^2 = \tilde{g}_{\mu\nu}dx^\mu dx^\nu = \frac{g_{00}}{f^2(\epsilon)}(dx^0)^2 + \frac{g_{ij}}{g^2(\epsilon)}dx^i dx^j + 2\frac{g_{0i}}{f(\epsilon)g(\epsilon)}dx^0 dx^i , \tag{3}$$

where $g_{\mu\nu}$ is the metric found from undeformed tetrads. Thus, in a nutshell, one identifies the rainbow functions, f and g, from a MDR that is usually inspired by fundamental theories of quantum gravity or by phenomenological intuition; then, one uses $\tilde{g}_{\mu\nu}$ as an input into the classical gravitational field equations. Considering modifications of the stress-energy tensor due to the rainbow functions, one derives what should be $g_{\mu\nu}$ (since f and g are determined a priori). Usually, this procedure gives that $g_{\mu\nu}$ is the Riemannian metric found from the usual gravitational field equations. Therefore, this approach gives basically the usual metric components of a given theory, just modified by factors of the rainbow functions as in Equation (3).

Effective energy-dependent spacetimes have emerged in different approaches to the description of the quantization of gravitational/geometric degrees of freedom [24–26]. Along this line of research, Magueijo-Smolin's proposal has been applied in a myriad of contexts, such as in black hole physics [27,28], cosmology [29], wormholes [30,31], cosmic strings [32], disformal geometries [33,34], and electrostatic self-interaction of charged particles [35]. However, despite its range of applicability and utility in furnishing intuition about extreme scenarios, this approach presents some conceptual and technical limitations that seem unavoidable, such as the lack of a rigorous mathematical framework in which this idea is formulated or the imposition of a preferred vielbein in which the particle's energy is measured, which seems in contradiction with the local DSR intention of this proposal. As shown below, the solution to these problems is actually coincident, and the search for a rigorous mathematical formulation for these ideas will be responsible for giving a framework, in which proper physical questions can be answered and novel phenomenological opportunities to born. The main issue here is what is the proper formulation of a geometry that should not only depend on spacetime points, but also should carry energy dependence of the particle itself that probes this spacetime. This paper deals with the two main proposals—Finsler and Hamilton geometries— solving some of the raised problems and also discusses limitations on their owns.

3. Geometry of the Tangent Bundle: Finsler Geometry

The 1854 Habilitation Dissertation by Bernhard Riemann presents the germ of the idea behind what would later be called Finsler geometry. In the second part of the dissertation, it is said (see Ref. [36], p. 35):

> "For Space, when the position of points is expressed by rectilinear co-ordinates, $ds = \sqrt{\Sigma(dx)^2}$; Space is therefore, included in this simplest case. The next case in simplicity includes those manifoldnesses in which the line-element may be expressed as the fourth root of a quartic differential expression. The investigation of this more general kind would require no really different principles, but would take considerable time and throw little new light on the theory of space, especially as the results cannot be geometrically expressed; I restrict myself, therefore, to those manifoldnesses in which the line-element is expressed as the square root of a quadric differential expression".

The exploration of such more general cases of line elements will be done only 64 years later, in 1918, in the Ph.D. thesis of Paul Finsler [37], where at least from the metric point of view, the distance between points is measured by a 1-homogeneous function (homogeneous with the degree of 1) Such a metric tensor would be defined in the tangent bundle of the base manifold, since it would depend not only on the manifold points, but also on a direction, which is a manifestation of the non-Pythagorean nature of this space. Later on, the issue of non-linear connections was further developed and incorporated as a fundamental structure for the dynamical description of Finsler spaces (for a historical perspective on Finsler geometry, we refer the reader to the Preface of Ref. [38] and references therein). The case of pseudo-Finsler geometries, as an arena for describing spacetime, has been recently discussed [39,40], where, for instance, different definitions are presented and important theorems regarding its causal structure among other issues are being derived [41].

In Section 2, a glimpse of the non-Riemannian nature of spacetime was notified emerging as a manifestation of the quantization of gravitational degrees of freedom. Actually, as one can anticipate, the non-quadratic, i.e., non-Pythagorean nature of a dispersion relation is connected to a possible Finslerian nature of spacetime through an intermediate step that connects the kinematics of particles in a Hamiltonian to a Lagrangian formulation. Actually, the MDR corresponds to a Hamiltonian constraint, which physical particles supposedly obey, the way that the trajectories of free particles, induced by such a deformed Hamiltonian, capture the propagation of a particle through a quantized spacetime. For this reason, the Helmholtz action, associated with such a particle, is naturally given by the functional,

$$S[x,p,\lambda] = \int d\mu(\dot{x}^\alpha p_\alpha - \lambda f(H(x,p),m)), \tag{4}$$

where the dot denotes differentiation with respect to the parameter μ, p_μ is the particle's momenta, f is a function that is null if the dispersion relation is satisfied, namely, $H(x,p) = m$, and λ is a Lagrange multiplier. This is a premetric formulation that is actually defined in the space $T^*M \times \mathbb{R}$, where T^*M is the phase space of analytical mechanics or cotangent bundle. In order to find an arc-length, and consequently, a geometric structure, one needs to calculate an equivalent Lagrangian defined in the configuration space or the tangent bundle TM described by points and velocities (such an observation was firstly presented in Ref. [42]). The algorithm for doing so is as follows [43]:

1. variation with respect to λ enforces the dispersion relation;
2. variation with respect to p_μ yields an equation $\dot{x}^\mu = \dot{x}^\mu(p,\lambda)$, which must be inverted to obtain $p_\mu(x,\dot{x},\lambda)$ to eliminate the momenta p_μ from the action;
3. using $p_\mu(x,\dot{x},\lambda)$ in the dispersion relation, one can solve for $\lambda(x,\dot{x})$; and
4. finally, the desired length measure is obtained as $S[x] = S[x, p(x,\dot{x},\lambda(x,\dot{x})), \lambda(x,\dot{x})]_H$.

This is a Legendre transformation, whose conditions of existence and capability of providing a physical framework are discussed in Refs. [44,45]. These formal conditions are always guaranteed when one considers deformations at the perturbative level. This is crucial because the following algorithm cannot be applied in practice if it is not possible to invert the velocity function to find the momenta as a function of the other variables. In general, this cannot be done, especially for complicated dispersion relations, such as those that depend on sums of hyperbolic functions [46]. Anyway, since quantum gravity phenomenology is usually concerned with first order effects, which are those attainable by experiments nowadays, we shall concentrate on the perturbative level in order to derive our conclusions.

For example, if this algorithm is applied to a Hamiltonian of the form,

$$H(x,p) = g(p,p) + \varepsilon h(x,p), \tag{5}$$

where $g(p,p) = g^{ab}(x)p_a p_b$ is an undeformed dispersion relation, $h(x,p)$ is a function of spacetime points and momenta that depends on the model under consideration, and ε is the perturbation parameter that is usually a function of the energy scale of the deformation (such as the Planck or quantum gravity length scale). As shown in Ref. [43], after the Legendre transformation, the equivalent action takes the form,

$$S[x] = m\int d\mu\sqrt{g(\dot{x},\dot{x})}\left(1 - \varepsilon\frac{h(x,\bar{p}(x,\dot{x}))}{2m^2}\right), \tag{6}$$

where $\bar{p}_a(x,\dot{x}) = m\dot{x}_a/\sqrt{g(\dot{x},\dot{x})}$. In particular, when h is a polynomial function of momenta as (the index is shifted: $n \to n+2$, in comparison with Ref. [43], such that now n corresponds to the power of Planck length in the MDR),

$$h(x,p) = h^{\mu_1\mu_2\ldots\mu_{n+2}}(x)p_{\mu_1}p_{\mu_2}\ldots p_{\mu_{n+2}}, \tag{7}$$

and $\varepsilon = \ell^n$, one finds an action of the form,

$$S[x] = m\int d\mu\sqrt{g(\dot{x},\dot{x})}\left(1 - (\ell m)^n\frac{h_{\mu_1\mu_2\ldots\mu_{n+2}}(x)\dot{x}^{\mu_1}\dot{x}^{\mu_2}\ldots\dot{x}^{\mu_{n+2}}}{2g(\dot{x},\dot{x})^{\frac{n+2}{2}}}\right), \tag{8}$$

where we lowered the indices of h with the components of g. The connection between the mechanics of free particle and geometry takes place when the above expression is identified with the arc length functional, $s[x]$, of a given geometry, i.e., $s[x] = S[x]/m$. Such an identification makes sense if we want to state that the trajectories of free particles are extremizing curves or geodesics in a given geometry, it is related to the preservation of the equivalence principle even in this Planck-scale deformed scenario.

In this case, the spacetime in which a particle propagates by a MDR is described by an arc-length functional that generalizes the one of Riemannian geometry and is given by a function $F(x,\dot{x})$ that is 1-homogeneous in the velocity $\dot{x}$, such that the arc-length is indeed parametrization invariant, as it must be:

$$s[x] = \int F(x,\dot{x})d\mu\,. \tag{9}$$

Actually, this is the kind of scenario envisaged by Riemann in his dissertation, and explored by Finsler, that emerges here quite naturally. There are some definitions of a pseudo-Finsler spacetime in the literature, but we rely on that given in Ref. [39] (the differences in comparison to other definitions are discussed in Ref. [39]). First of all, we are going to work with a smooth manifold, M, endowed with a real valued positive function L that takes values on the tangent bundle TM, described by coordinates (x,y), where $\{x^\mu\}$ are spacetime coordinates and $\{y^\mu\}$ refer to vector or velocity coordinates. Actually, we shall need the slit tangent bundle $\widetilde{TM} = TM/\{0\}$, in which we remove the zero section, and we also need the projection $\pi : TM \to M$. A conic subbundle is a submanifold $\mathcal{D} \subset \widetilde{TM}$ such that $\pi(\mathcal{D}) = M$ and with the conic property that states that if $(x,y) \in \mathcal{D} \Rightarrow (x,\lambda y) \in \mathcal{D}$, $\forall\lambda > 0$.

In a nutshell, a Finsler spacetime is a triple $(M,\mathcal{D},L)$, where $L : \mathcal{D} \to \mathbb{R}$ is a smooth function satisfying the conditions:

1. positive 2-homogeneity: $L(x,\alpha y) = \alpha^2 L(x,y)$, $\forall\alpha > 0$;
2. at any $(x,y) \in \mathcal{D}$ and in any chart of $\widetilde{TM}$, the following Hessian (metric) is non-degenerate:
$$g_{\mu\nu}(x,y) = \frac{1}{2}\frac{\partial^2}{\partial y^\mu \partial y^\nu}L(x,y)\,; \tag{10}$$
3. the metric $g_{\mu\nu}$ has a Lorentzian signature.

The function L is actually the square of the Finsler function, $L(x,y) = F^2(x,y)$, and from it the Finsler arc-length is defined as given in Equation (9). Condition 1 above guarantees that Equation (9) does not depend on the parametrization used to describe the curve and that using Euler's theorem for homogeneous functions, this expression can be cast as

$$s[x] = \int \sqrt{g_{\mu\nu}(x,\dot{x})\dot{x}^\mu\dot{x}^\nu}d\mu\,. \tag{11}$$

From a coordinate transformation,

$$\tilde{x}^\mu = \tilde{x}^\mu(x)\,, \tag{12}$$

$$\tilde{y}^\mu = \frac{\partial\tilde{x}^\mu}{\partial x^\nu}y^\nu\,, \tag{13}$$

the functions $g_{\mu\nu}$ transform according to

$$\tilde{g}_{\mu\nu}(\tilde{x},\tilde{y}) = \frac{\partial x^\alpha}{\partial\tilde{x}^\mu}\frac{\partial x^\beta}{\partial\tilde{x}^\nu}g_{\alpha\beta}(x,y)\,. \tag{14}$$

Due the property (14), $g_{\mu\nu}$ is referred here as the components of a distinguished tensor field (or d-tensor field) on the manifold $\widetilde{TM}$, which follows the notation adopted in Ref. [47]. The extremization of the arc-lenght functional (9) gives the following geodesic equation,

$$\frac{d^2x^\mu}{d\mu^2} + 2G^\mu(x,\dot{x}) = 2\frac{dF}{d\mu}\frac{\partial F}{\partial \dot{x}^\mu}, \tag{15}$$

where $G^\mu = G^\mu(x,\dot{x})$ are the spray coefficients [48] and are given in terms of the Christoffel symbols, $\gamma^\alpha_{\mu\nu}$, of the metric $g_{\mu\nu}$:

$$G^\alpha(x,\dot{x}) = \frac{1}{2}\gamma^\alpha_{\mu\nu}(x,\dot{x})\dot{x}^\mu\dot{x}^\nu, \tag{16}$$

$$\gamma^\alpha_{\mu\nu}(x,\dot{x}) = \frac{1}{2}g^{\alpha\beta}\left(\frac{\partial g_{\mu\beta}}{\partial x^\nu} + \frac{\partial g_{\nu\beta}}{\partial x^\mu} - \frac{\partial g_{\mu\nu}}{\partial x^\beta}\right). \tag{17}$$

If we choose the arc-length parametrization, i.e., the one in which $F = 1$, we have a sourceless geodesic equation. This expression means that the trajectories generated by a MDR of the form $H(x,\dot{x}) = m^2$ are, actually, geodesics of a Finsler metric. The presence of spray coefficients allows us to construct another quite a useful quantity, the so-called Cartan non-linear connection, given by (in this paper, we interchange the notation $\dot{x} \leftrightarrow y$ freely)

$$N^\mu{}_\nu(x,y) = \frac{\partial}{\partial y^\nu}G^\mu(x,y), \tag{18}$$

that transforms according to

$$\tilde{N}^\mu{}_\nu = \frac{\partial \tilde{x}^\mu}{\partial x^\alpha}\frac{\partial x^\beta}{\partial \tilde{x}^\nu}N^\alpha{}_\beta - \frac{\partial^2 \tilde{x}^\mu}{\partial x^\alpha \partial x^\beta}\frac{\partial x^\beta}{\partial \tilde{x}^\nu}y^\alpha. \tag{19}$$

The introduction of this quantity allows us to introduce a useful basis of the tangent space of the tangent bundle at each point. In fact, since according to the coordinate transformation (12) and (13), the usual coordinate basis transforms as

$$\frac{\partial}{\partial \tilde{x}^\mu} = \frac{\partial x^\nu}{\partial \tilde{x}^\mu}\frac{\partial}{\partial x^\nu} + \frac{\partial^2 x^\nu}{\partial \tilde{x}^\mu \partial \tilde{x}^\alpha}\frac{\partial \tilde{x}^\alpha}{\partial x^\beta}y^\beta\frac{\partial}{\partial y^\nu}, \tag{20}$$

$$\frac{\partial}{\partial \tilde{y}^\mu} = \frac{\partial y^\nu}{\partial \tilde{y}^\mu}\frac{\partial}{\partial y^\nu}. \tag{21}$$

In addition, a non-linear connection allows us to define the following frame:

$$\frac{\delta}{\delta x^\mu} = \delta_\mu = \frac{\partial}{\partial x^\mu} - N^\nu{}_\mu\frac{\partial}{\partial y^\nu}, \tag{22}$$

$$\dot{\partial}_\mu = \frac{\partial}{\partial y^\mu}. \tag{23}$$

Due to the transformation properties of the non-linear connection, this basis transforms as

$$\tilde{\delta}_\mu = \frac{\partial x^\nu}{\partial \tilde{x}^\mu}\delta_\nu, \tag{24}$$

$$\tilde{\dot{\partial}}_\mu = \frac{\partial x^\nu}{\partial \tilde{x}^\mu}\dot{\partial}_\nu. \tag{25}$$

This means that one is able to split the tangent space of the tangent bundle into horizontal, $HTM = \text{span}\{\delta_\mu\}$, and vertical, $VTM = \text{span}\{\dot{\partial}_\mu\}$, spaces, such that $T\widetilde{TM} = HTM \oplus VTM$ in each point (x,y). Similarly, the same reasoning applies to the cotangent space; i.e., we split $T^*\widetilde{TM} = H^*TM \oplus V^*TM$ spanned as $H^*TM = \text{span}\{dx^\mu\}$ and $V^*TM = \text{span}\{\delta y^\mu\}$, where

$$\delta y^\mu = dy^\mu + N^\mu{}_\nu dx^\nu \,, \tag{26}$$

which transforms as

$$d\tilde{x}^\mu = \frac{\partial \tilde{x}^\mu}{\partial x^\nu} dx^\nu \,, \tag{27}$$
$$\delta \tilde{y}^\mu = \frac{\partial \tilde{x}^\mu}{\partial x^\nu} \delta y^\nu \,. \tag{28}$$

Such a decomposition of the tangent and cotangent vector spaces implies that a vector X and a 1-form ω with horizontal and vertical terms can read as

$$X = X^\mu \delta_\mu + \dot{X}^\mu \dot{\partial}_\mu = X^H + X^V \,, \tag{29}$$
$$\omega = \omega_\mu dx^\mu + \dot{\omega}_\mu \delta y^\mu = \omega^H + \omega^V \,. \tag{30}$$

Endowed with this basis, the metric $\mathbb{G}(x,y)$ of the configuration space is described by the so-called Sasaki-Matsumoto lift of the metric $g_{\mu\nu}$:

$$\mathbb{G}(x,y) = g_{\mu\nu}(x,y) dx^\mu \otimes dx^\nu + g_{\mu\nu}(x,y) \delta y^\mu \otimes \delta y^\nu \,. \tag{31}$$

Definition 1. *A tensor field T of type $(m+n, p+q)$ on the manifold $\widetilde{TM}$ is called a distinguished tensor field (or d-tensor field) if it has the property*

$$T\left(\overset{1}{\omega},\ldots,\overset{m}{\omega},\overset{1}{\tau},\ldots,\overset{n}{\tau},\underset{1}{X},\ldots,\underset{p}{X},\underset{1}{Y},\ldots,\underset{q}{Y}\right) = T\left(\overset{1}{\omega}{}^H,\ldots,\overset{m}{\omega}{}^H,\overset{1}{\tau}{}^V,\ldots,\overset{n}{\tau}{}^V,\underset{1}{X}{}^H,\ldots,\underset{p}{X}{}^H,\underset{1}{Y}{}^V,\ldots,\underset{q}{Y}{}^V\right). \tag{32}$$

This definition implies that one can write a d-tensor T in the preferred frame as

$$\begin{aligned} T = T^{\mu_1\ldots\mu_m\nu_1\ldots\nu_n}{}_{\alpha_1\ldots\alpha_p\beta_1\ldots\beta_q} \frac{\delta}{\delta x^{\mu_1}} \otimes \cdots \otimes \frac{\delta}{\delta x^{\mu_m}} \otimes \frac{\partial}{\partial y^{\nu_1}} \otimes \cdots \otimes \frac{\partial}{\partial y^{\nu_n}} \\ \otimes dx^{\alpha_1} \otimes \cdots \otimes dx^{\alpha_p} \otimes \delta y^{\beta_1} \otimes \cdots \otimes \delta y^{\beta_q} \,, \end{aligned} \tag{33}$$

and that it transforms according to the rule,

$$\begin{aligned} &\widetilde{T}^{\mu_1\ldots\mu_m\nu_1\ldots\nu_n}{}_{\alpha_1\ldots\alpha_p\beta_1\ldots\beta_q} \\ &= \frac{\partial \tilde{x}^{\mu_1}}{\partial x^{\epsilon_1}} \cdots \frac{\partial \tilde{x}^{\mu_m}}{\partial x^{\epsilon_m}} \frac{\partial \tilde{x}^{\nu_1}}{\partial x^{\lambda_1}} \cdots \frac{\partial \tilde{x}^{\nu_n}}{\partial x^{\lambda_n}} \frac{\partial x^{\gamma_1}}{\partial \tilde{x}^{\alpha_1}} \cdots \frac{\partial x^{\gamma_p}}{\partial \tilde{x}^{\alpha_p}} \frac{\partial x^{\rho_1}}{\partial \tilde{x}^{\beta_1}} \cdots \frac{\partial x^{\rho_q}}{\partial \tilde{x}^{\beta_q}} T^{\epsilon_1\ldots\epsilon_n\lambda_1\ldots\lambda_m}{}_{\gamma_1\ldots\gamma_p\rho_1\ldots\rho_q} \,. \end{aligned} \tag{34}$$

An example of d-tensor field is the metric whose components are given by Equation (14).

3.1. N-Linear Connection

Given a linear connection, D, on the manifold $\widetilde{TM}$, if it preserves the parallelism of the horizontal and vertical spaces, i.e., if it can be written as

$$D_{\delta_\nu} \delta_\mu = L^\alpha_{\mu\nu} \delta_\alpha \,, \qquad D_{\delta_\nu} \dot{\partial}_\alpha = L^\mu_{\alpha\nu} \dot{\partial}_\mu \,, \tag{35}$$

$$D_{\dot{\partial}_\nu} \delta_\mu = C^\alpha_{\mu\nu} \delta_\alpha \,, \qquad D_{\dot{\partial}_\nu} \dot{\partial}_\mu = C^\alpha_{\mu\nu} \dot{\partial}_\alpha \,, \tag{36}$$

then is called an N-linear connection. Let us consider a coordinate change; thus, the coefficients (35) and (36) transform as

$$\tilde{L}^{\alpha}_{\mu\nu} = \frac{\partial \tilde{x}^{\alpha}}{\partial x^{\beta}} \frac{\partial x^{\lambda}}{\partial \tilde{x}^{\mu}} \frac{\partial x^{\epsilon}}{\partial \tilde{x}^{\nu}} L^{\beta}_{\lambda\epsilon} + \frac{\partial^2 x^{\beta}}{\partial \tilde{x}^{\mu} \partial \tilde{x}^{\nu}} \frac{\partial \tilde{x}^{\alpha}}{\partial x^{\beta}}, \tag{37}$$

$$\tilde{C}^{\alpha}_{\mu\nu} = \frac{\partial \tilde{x}^{\alpha}}{\partial x^{\beta}} \frac{\partial x^{\lambda}}{\partial \tilde{x}^{\mu}} \frac{\partial x^{\epsilon}}{\partial \tilde{x}^{\nu}} C^{\beta}_{\lambda\epsilon}. \tag{38}$$

Endowed with these coefficients, the derivative of a d-tensor can be decomposed into a horizontal and a vertical parts, such that one can apply the covariant derivative of a tensor T of type $(m+n, p+q)$ in the direction of a vector X as a direction of a vector X as

$$\begin{aligned} D_X T &= D_{X^H} T + D_{X^V} T \\ &= \left(T^{\mu_1 \dots \mu_m \nu_1 \dots \nu_n}{}_{\alpha_1 \dots \alpha_p \beta_1 \dots \beta_q | \epsilon} X^{\epsilon} + T^{\mu_1 \dots \mu_m \nu_1 \dots \nu_n}{}_{\alpha_1 \dots \alpha_p \beta_1 \dots \beta_q || \epsilon} \dot{X}^{\epsilon} \right) \frac{\delta}{\delta x^{\mu_1}} \otimes \dots \otimes \frac{\delta}{\delta x^{\mu_m}} \\ &\otimes \frac{\partial}{\partial y^{\nu_1}} \otimes \dots \otimes \frac{\partial}{\partial y^{\nu_n}} \otimes dx^{\alpha_1} \otimes \dots \otimes dx^{\alpha_p} \otimes \delta y^{\beta_1} \otimes \dots \otimes \delta y^{\beta_q}, \end{aligned} \tag{39}$$

where

$$\begin{aligned} &T^{\mu_1 \dots \mu_m \nu_1 \dots \nu_n}{}_{\alpha_1 \dots \alpha_p \beta_1 \dots \beta_q | \epsilon} \\ &= \frac{\delta}{\delta x^{\epsilon}} T^{\mu_1 \dots \mu_m \nu_1 \dots \nu_n}{}_{\alpha_1 \dots \alpha_p \beta_1 \dots \beta_q} + L^{\mu_1}_{\gamma\epsilon} T^{\gamma \dots \mu_m \nu_1 \dots \nu_n}{}_{\alpha_1 \dots \alpha_p \beta_1 \dots \beta_q} + \dots - L^{\gamma}_{\alpha_1 \epsilon} T^{\mu_1 \dots \mu_m \nu_1 \dots \nu_n}{}_{\gamma \dots \alpha_p \beta_1 \dots \beta_q}, \end{aligned} \tag{40}$$

$$\begin{aligned} &T^{\mu_1 \dots \mu_m \nu_1 \dots \nu_n}{}_{\alpha_1 \dots \alpha_p \beta_1 \dots \beta_q || \epsilon} \\ &= \frac{\partial}{\partial y^{\epsilon}} T^{\mu_1 \dots \mu_m \nu_1 \dots \nu_n}{}_{\alpha_1 \dots \alpha_p \beta_1 \dots \beta_q} + C^{\mu_1}_{\gamma\epsilon} T^{\gamma \dots \mu_m \nu_1 \dots \nu_n}{}_{\alpha_1 \dots \alpha_p \beta_1 \dots \beta_q} + \dots - C^{\gamma}_{\alpha_1 \epsilon} T^{\mu_1 \dots \mu_m \nu_1 \dots \nu_n}{}_{\gamma \dots \alpha_p \beta_1 \dots \beta_q}, \end{aligned} \tag{41}$$

and the property that the covariant derivative is linear in the direction X is used. The triple $D\Gamma(N, L, C)$ describes the parallel transport and decomposition of the tangent and cotangent spaces of the tangent bundle into horizontal and vertical spaces. At this point, we need to comment on some remarkable N-linear connections that are considered in the literature.

The first connection is the metrical Cartan connection, $C\Gamma(N^{\mu}{}_{\nu}, L^{\alpha}_{\mu\nu}, C^{\alpha}_{\mu\nu})$. In this case, $N^{\mu}{}_{\nu}$ is given by the canonical Cartan non-linear connection, defined by the spray coefficients (18). The coefficients $L^{\alpha}_{\mu\nu}$ and $C^{\alpha}_{\mu\nu}$ are given, respectively, by

$$L^{\alpha}_{\mu\nu} = \frac{1}{2} g^{\alpha\beta} \left(\frac{\delta g_{\mu\beta}}{\delta x^{\nu}} + \frac{\delta g_{\nu\beta}}{\delta x^{\mu}} - \frac{\delta g_{\mu\nu}}{\delta x^{\beta}} \right), \tag{42}$$

$$C^{\alpha}_{\mu\nu} = \frac{1}{2} g^{\alpha\beta} \left(\frac{\delta g_{\mu\beta}}{\delta y^{\nu}} + \frac{\delta g_{\nu\beta}}{\delta y^{\mu}} - \frac{\delta g_{\mu\nu}}{\delta y^{\beta}} \right). \tag{43}$$

This connection is metrical (i.e., without non-metricity tensors) considering both horizontal and vertical covariant derivatives of the Finsler metric.

Besides, the Berwald connection is given by the triple $B\Gamma(N^{\mu}{}_{\nu}, \partial N^{\alpha}{}_{\mu} / \partial y^{\nu}, 0)$ and presents horizontal and vertical non-metricities. The Chern–Rund connection, $R\Gamma(N^{\mu}{}_{\nu}, L^{\alpha}_{\mu\nu}, 0)$, is horizontally metrical, but represents vertical non-metricity. Additionally, the Hashiguchi connection, $H\Gamma(N^{\mu}{}_{\nu}, \partial N^{\alpha}{}_{\mu} / \partial y^{\nu}, C^{\alpha}_{\mu\nu})$, represents horizontal non-metricity, but it is vertically metrical. In these expressions, N is the canonical Cartan non-linear connection (18), L is given by Equation (42), and C is given by Equation (43).

3.2. Symmetries

Geometrical language naturally realizes the concept of symmetry of physical equations. General relativity given in terms of Riemannian geometry encompasses the invariance under general coordinate transformations, and the isometries of the Minkowski space describe the Poincaré transformations (actually, one can further apply this technique for maximally symmetric spaces, including de Sitter and anti-de Sitter ones). Finsler geometry, as we have been using, allows us to go beyond this scope and to define deformed Lorentz/Poincaré transformations that present Planck scale corrections even in the presence of a local modified dispersion relation. One can see how this will naturally emerge, since the invariance of the arc-length (9) is compatible with the invariance of the action in the Hamiltonian formulation (4), from which such an arc-length was derived. This idea was firstly noticed in Ref. [42] and later explicitly explored in Refs. [49,50]. The master equation for this purpose is the one that follows from the invariance of the Finslerian interval ds^2, as done in Appendix A of Ref. [49]. From this invariance, the Finslerian killing equation for the killing vector was found, with components ξ^α, which should be solved in order to derive the deformed symmetries in the DSR context,

$$\xi^\alpha \partial_\alpha g_{\mu\nu} + g_{\alpha\nu}\partial_\mu \xi^\alpha + g_{\mu\alpha}\partial_\nu \xi^\alpha + y^\alpha \partial_\alpha \xi^\beta \dot{\partial}_\beta g_{\mu\nu} = 0\,. \tag{44}$$

3.3. Finsler–q-de Sitter (Tangent Bundle Case)

As an example that presents a non-trivial non-linear connection, we shall consider the case of a Finsler geometry inspired by the so-called q-de Sitter deformed relativity. This case has been previously studied in the literature, e.g., in Refs. [50–53], and can be described by an algebra that deforms the one of Poincaré in a way that gives the de Sitter symmetry when a quantum gravity parameter goes to zero, and on the other hand, gives the so-called κ-Poincaré algebra (that deforms the Poincaré one by an energy scale parameter, supposedly the Planck energy) when the de Sitter curvature parameter goes to zero. Therefore, it corresponds to an authentic realization of a deformed relativity scenario, even in the presence of what can be interpreted as spacetime curvature. In this Subsection, we initially consider results that were originally presented in Ref. [52] in $1+1$ dimensions.

The MDR related to this algebra (in a given basis) can be perturbed to first order in the Planck length and de Sitter curvature parameters ℓ and H, respectively, as

$$\mathcal{H}(x,p) = p_0^2 - p_1^2(1+\ell p_0)(1-2Hx^0)\,. \tag{45}$$

By using the action given by Equation (4) and the algorithm that follows it, the following Finsler function can be obtained:

$$F(x,\dot{x}) = \sqrt{(\dot{x}^0)^2 - (1-2Hx^0)(\dot{x}^1)^2} + \ell\frac{m}{2}\frac{(1-2Hx^0)\dot{x}^0(\dot{x}^1)^2}{(\dot{x}^0)^2 - (1-2Hx^0)(\dot{x}^1)^2}\,, \tag{46}$$

from which the Finsler metric can be found from Equation (10):

$$g^F_{\mu\nu}(x,\dot{x}) = \begin{pmatrix} 1 + \frac{3a^4 m\ell \dot{x}^0(\dot{x}^1)^4}{2[(\dot{x}^0)^2 - a^2(\dot{x}^1)^2]^{5/2}} & \frac{m\ell a^4(\dot{x}^1)^3[a^2(\dot{x}^1)^2 - 4(\dot{x}^0)^2]}{2[(\dot{x}^0)^2 - a^2(\dot{x}^1)^2]^{5/2}} \\ \frac{m\ell a^4(\dot{x}^1)^3[a^2(\dot{x}^1)^2 - 4(\dot{x}^0)^2]}{2[(\dot{x}^0)^2 - a^2(\dot{x}^1)^2]^{5/2}} & -a^2 + \frac{m\ell a^2(\dot{x}^0)^3[2(\dot{x}^0)^2 + a^2(\dot{x}^1)^2]}{2[(\dot{x}^0)^2 - a^2(\dot{x}^1)^2]^{5/2}} \end{pmatrix}, \tag{47}$$

where $a = a(t) = e^{Ht} = 1 + Ht + \mathcal{O}(H^2)$ (in this paper, the terms that grow with higher orders of H and ℓ are discarded). The geodesic equation is found from the extremization of the Finsler arc-length defined by F, from which Christoffel symbols and spray coefficients can be calculated. Actually, the $\gamma^\alpha_{\mu\nu}(x,\dot{x})$ are given, for an arbitrary parametrization, by the set of Equations (44) of Ref. [52], from which the spray coefficients are given by

$$G^0(x,\dot{x}) = \frac{1}{8}a^2 H(\dot{x}^1)^2\left[4 - \frac{\ell m \dot{x}^0}{[(\dot{x}^0)^2 - a^2(\dot{x}^1)^2]^{7/2}}\left(-28a^6(\dot{x}^1)^6 + 12a^2(\dot{x}^0)^4(\dot{x}^1)^2 \right.\right.$$
$$\left.\left. + a^2\left(17a^2 + 28\right)(\dot{x}^0)^2(\dot{x}^1)^4 + 16(\dot{x}^0)^6\right)\right], \tag{48}$$

$$G^1(x,\dot{x}) = H\dot{x}^0\dot{x}^1 + \ell\left[\frac{a^2 Hm(\dot{x}^1)^3(a^6(\dot{x}^1)^6 - 6a^4(\dot{x}^0)^2(\dot{x}^1)^4 + 3a^2(\dot{x}^0)^4(\dot{x}^1)^2 - 28(\dot{x}^0)^6)}{4((\dot{x}^0)^2 - a^2(\dot{x}^1)^2)^{7/2}}\right]. \tag{49}$$

As can be seen, these coefficients are 2-homogeneous in the velocities, as expected. The Cartan non-linear connection coefficients read:

$$N^0{}_0(x,\dot{x}) = \frac{H\ell m(\dot{x}^1)^4\left(-28(\dot{x}^1)^6 - 33(\dot{x}^1)^4(\dot{x}^0)^2 + 240(\dot{x}^1)^2(\dot{x}^0)^4 + 136(\dot{x}^0)^6\right)}{8((\dot{x}^0)^2 - (\dot{x}^1)^2)^{9/2}}, \tag{50}$$

$$N^0{}_1(x,\dot{x}) = H\dot{x}^1 - \frac{H\ell m\dot{x}^1\dot{x}^0}{8((\dot{x}^0)^2 - (\dot{x}^1)^2)^{9/2}}\Big(28(\dot{x}^1)^8 - 179(\dot{x}^1)^6(\dot{x}^0)^2 + 306(\dot{x}^1)^4(\dot{x}^0)^4 + 128(\dot{x}^1)^2(\dot{x}^0)^6 + 32(\dot{x}^0)^8\Big),$$

$$N^1{}_0(x,\dot{x}) = H\dot{x}^1 + \frac{H\ell m(\dot{x}^1)^3\dot{x}^0\left(5(\dot{x}^1)^6 + 18(\dot{x}^1)^4(\dot{x}^0)^2 + 159(\dot{x}^1)^2(\dot{x}^0)^4 + 28(\dot{x}^0)^6\right)}{4((\dot{x}^0)^2 - (\dot{x}^1)^2)^{9/2}}, \tag{51}$$

$$N^1{}_1(x,\dot{x}) = H\dot{x}^0 - \frac{H\ell m(\dot{x}^1)^2\left(2(\dot{x}^1)^8 - 9(\dot{x}^1)^6(\dot{x}^0)^2 + 36(\dot{x}^1)^4(\dot{x}^0)^4 + 97(\dot{x}^1)^2(\dot{x}^0)^6 + 84(\dot{x}^0)^8\right)}{4((\dot{x}^0)^2 - (\dot{x}^1)^2)^{9/2}}, \tag{52}$$

where the worldlines are autoparallel curves of this non-linear connection. Let us note that some terms of the connection are only present due to the coupling between the spacetime curvature parameter, H, and the one that gives a non-trivial velocity space, ℓ. Some curvature-triggered effects in quantum gravity have been recently analyzed [54].

Endowed with these coefficients, the preferred frames that induce the horizontal and vertical decomposition can be immediately found, in addition the N-linear connection coefficients $L^\alpha_{\mu\nu}$ and $C^\alpha_{\mu\nu}$, as discussed in Section 2. Till now, only kinematical properties were discussed, but the choice of the given connection should be given either by physical conditions imposed on the dynamics of the spacetime or by possible effective gravitational field equations for a quantum configuration space.

To finalize this Section, let us discuss the symmetries of the spacetime. A deep analysis of the killing vectors of the $H \to 0$ limit of this Finsler framework was carried out in Ref. [51]. Even in that simplified scenario, the equations are quite lengthy which we omit here. However, some properties should be mentioned. Firstly, the transformations generated by the killing vectors seem to not exactly preserve the line element, but contribute with a term that is given by a total derivative in the action parameter; therefore, the kinematical results of these two line elements coincide. Secondly, the results found are compatible with the κ-Poincaré scenario that inspired this approach. From the Finsler perspective, it is possible to derive more general results, but they reduces to those of the bicrossproduct basis of κ-Poincaré by an appropriate choice of free functions and parameters. The third point is that a finite version of transformations that preserve the κ-Poincaré dispersion relation was recently made in Ref. [55] through an alternative approach, which does not rely on the killing vectors but is determined by the Finsler function and the definition of momentum (explored in Section 4 below); however, a complete integration of the finite isometry and a comparison between these approaches is still missing in the literature. To finalize, the case of $H \neq 0$ was investigated in Ref. [50], but in conformal coordinates (which are not the ones that are considered in this application), and was not done in so much detail as the flat case, but a generator of the corresponding curved boost transformation was made explicit in Equation (25) of Ref. [50].

4. The Cotangent Bundle Version of Finsler Geometry

As was discussed in Ref. [42], by mapping the velocity of the particle to its momentum, it is possible to find the version of the Finsler metric defined in the cotangent bundle or phase space. Already from the definition of the 4-momentum,

$$p_\mu = m\frac{\partial F}{\partial y^\mu}, \tag{53}$$

when it is possible to invert this expression to find $y = y(p)$, one can substitute this result in the Finsler metric as $h^F_{\mu\nu}(x,p) = g^F_{\mu\nu}(x, y(p))$. This metric is defined on the slit cotangent bundle, $\widetilde{T^*M} = T^*M/\{0\}$, where we also remove the zero section in each spacetime point for the same technical reasons as discussed in Section 3 above. Since the quantities are now defined in the cotangent bundle, we need to also address some issues that were raised in Section 3 concerning the tangent bundle. This Section's notation is applied according to Ref. [47]. For instance, under a change of coordinates, the spacetime and momentum variables transformed according to

$$\tilde{x}^\mu = \tilde{x}^\mu(x)\,, \tag{54}$$

$$\tilde{p}_\mu = \frac{\partial x^\nu}{\partial \tilde{x}^\mu} p_\nu\,, \tag{55}$$

which means that the frame $(\partial/\partial x^\mu, \partial/\partial p_\nu)$ transforms as

$$\frac{\partial}{\partial \tilde{x}^\mu} = \frac{\partial x^\nu}{\partial \tilde{x}^\mu}\frac{\partial}{\partial x_\nu} + \frac{\partial p_\nu}{\partial \tilde{x}^\mu}\frac{\partial}{\partial p_\nu}\,, \tag{56}$$

$$\frac{\partial}{\partial \tilde{p}_\mu} = \frac{\partial \tilde{x}^\mu}{\partial x^\nu}\frac{\partial}{\partial p_\nu}\,. \tag{57}$$

On the other hand, the natural coframe (dx^μ, dp_ν) changes as

$$d\tilde{x}^\mu = \frac{\partial \tilde{x}^\mu}{\partial x^\nu} dx^\nu\,, \tag{58}$$

$$d\tilde{p}_\mu = \frac{\partial x^\nu}{\partial \tilde{x}^\mu} dp_\nu + \frac{\partial^2 x^\nu}{\partial \tilde{x}^\mu \partial \tilde{x}^\lambda} p_\nu d\tilde{x}^\lambda\,. \tag{59}$$

Simlarly to that in Section 3, the presence of a nonlinear connection, $O_{\mu\nu}$, allows one to split the cotangent bundle into a horizontal and a vertical subbundle. Inspired by the consideration of the Hamilton case considered in Ref. [56] (discussed below), we propose the following dual non-linear connection (constructed in Appendix A):

$$O_{\mu\nu}(x,p) = -m\left[N^\alpha{}_\mu \frac{(g_{\alpha\nu} - p_\alpha p_\nu/m^2)}{F} - \partial_\mu \dot{\partial}_\nu F\right]\Bigg|_{(x,y(p))}\,, \tag{60}$$

where $p = p(y)$ is the kinematical map defined by Equation (53). By construction, these symbols have the transformation properties of a nonlinear connection,

$$\tilde{O}_{\mu\nu} = \frac{\partial x^\lambda}{\partial \tilde{x}^\mu}\frac{\partial x^\epsilon}{\partial \tilde{x}^\nu} O_{\lambda\epsilon} + \frac{\partial^2 x^\beta}{\partial \tilde{x}^\mu \partial \tilde{x}^\nu} p_\beta\,. \tag{61}$$

Endowed with a nonlinear connection $O_{\mu\nu}$, one can decompose the tangent bundle of the cotangent bundle by the Whitney sum in each point $T_u\widetilde{T^*M} = O_u \oplus V_u$, $\forall u \in \widetilde{T^*M}$. The subbundle O_u is called horizontal space and is spanned by the frame,

$$\frac{\delta}{\delta x^\mu} = \delta_\mu = \frac{\partial}{\partial x^\mu} + O_{\mu\nu}\frac{\partial}{\partial p_\nu}\,, \tag{62}$$

and the subbundle V_u is called vertical space and is spanned by the frame in each point of $\widetilde{T^*M}$:

$$\bar{\partial}^\mu = \frac{\partial}{\partial p_\mu}\,, \tag{63}$$

such that $T_u\widetilde{T^*M} = \text{span}\{\delta_\mu, \bar{\partial}^\nu\}$. The transformation properties of the nonlinear connection are implied in the following rule for transforming this basis:

$$\frac{\delta}{\delta\tilde{x}^\mu} = \tilde{\delta}_\mu = \frac{\partial x^\nu}{\partial\tilde{x}^\mu}\frac{\delta}{\delta x^\nu} = \frac{\partial x^\nu}{\partial\tilde{x}^\mu}\delta_\nu\,, \tag{64}$$

$$\frac{\partial}{\partial\tilde{p}_\mu} = \tilde{\bar{\partial}}^\mu = \frac{\partial\tilde{x}^\mu}{\partial x^\nu}\frac{\partial}{\partial p_\nu} = \frac{\partial\tilde{x}^\mu}{\partial x^\nu}\bar{\partial}^\nu\,. \tag{65}$$

Equivalently, with the nonlinear connection, we can decompose the cotangent space $T^*_u\widetilde{T^*M} = \text{span}\{dx^\mu, \delta p_\nu\}$, where

$$\delta p_\mu = dp_\mu - O_{\nu\mu}dx^\nu\,. \tag{66}$$

Therefore, the dual basis transforms as

$$d\tilde{x}^\mu = \frac{\partial\tilde{x}^\mu}{\partial x^\nu}dx^\nu\,, \tag{67}$$

$$\delta\tilde{p}_\mu = \frac{\partial x^\nu}{\partial\tilde{x}^\mu}\delta p_\nu\,. \tag{68}$$

Similarly to what has been done for the tangent bundle case, such a decomposition allows us to express a vector and a 1-form via horizontal and vertical components, where now, the vertical component is considered along momenta instead of velocities,

$$X = X^\mu\delta_\mu + \bar{X}_\mu\bar{\partial}^\mu = X^H + X^V\,, \tag{69}$$

$$\omega = \omega_\mu dx^\mu + \bar{\omega}^\mu\delta p_\mu = \omega^H + \omega^V\,. \tag{70}$$

Besides, the metric $\mathbb{H}(x,p)$ of the configuration space is defined as follows. Given a metric $h^{\mu\nu}(x,p)$, and the nonlinear connection $O_{\mu\nu}(x,p)$, the quantum phase space presents metrical properties given by the tensor,

$$\mathbb{H}(x,p) = h_{\mu\nu}(x,p)dx^\mu \otimes dx^\nu + h^{\mu\nu}(x,p)\delta p_\mu \otimes \delta p_\nu\,. \tag{71}$$

We refer to the tensor $\mathbb{H}$ as the N-lift to $\widetilde{T^*M}$ of the metric $h_{\mu\nu}$. The map between y and p cannot be done, in general, involving quantities that are parametrization-dependent because p itself is parametrization-invariant, whereas y is not. That is why one can only assume $y(p)$ for the definition of the metric $h^F_{\mu\nu}$.

Endowed with these quantities, one can just extend the definition of d-tensors 1 to the cotangent case, in which one only needs to consider the use of the nonlinear connection $O_{\mu\nu}$ and the adapted basis defined in this Section.

The above implies that a d-tensor T of type $(m+q, n+p)$ can be rewritten in the preferred basis as

$$T = T^{\mu_1\ldots\mu_m}{}_{\nu_1\ldots\nu_n\alpha_1\ldots\alpha_p}{}^{\beta_1\ldots\beta_q}\frac{\delta}{\delta x^{\mu_1}} \otimes \ldots \otimes \frac{\delta}{\delta x^{\mu_m}} \otimes \frac{\partial}{\partial p_{\nu_1}} \otimes \ldots \otimes \frac{\partial}{\partial p_{\nu_n}}$$
$$\otimes\, dx^{\alpha_1} \otimes \ldots \otimes dx^{\alpha_p} \otimes \delta p_{\beta_1} \otimes \ldots \otimes \delta p_{\beta_q}\,, \tag{72}$$

whose components transform according to usual linear transformation rules, as the one of Equation (34).

4.1. N-Linear Connection

Equivalently, the notion of differentiation can be defined in the cotangent bundle through the N-linear connection D, which has the following coefficients in the frame $(\delta_\mu, \bar{\partial}^\nu)$ (see Theorem 4.9.1 in Ref. [47]):

$$D_{\delta_\nu}\delta_\mu = H^\alpha_{\mu\nu}\delta_\alpha\,, \qquad D_{\delta_\nu}\bar{\partial}^\mu = -H^\mu_{\alpha\nu}\bar{\partial}^\alpha\,, \tag{73}$$

$$D_{\bar{\partial}^\nu}\delta_\mu = C^{\alpha\nu}_\mu\delta_\alpha\,, \qquad D_{\bar{\partial}^\nu}\bar{\partial}^\mu = -C^{\mu\nu}_\alpha\bar{\partial}^\alpha\,. \tag{74}$$

Otherwise, in the frame $(dx^\mu, \delta p_\nu)$ one has (see Proposition 4.9.1 in Ref. [47])

$$D_{\delta_\nu}dx^\mu = -H^\mu_{\alpha\nu}dx^\alpha\,, \qquad D_{\delta_\nu}\delta p_\mu = H^\alpha_{\mu\nu}\delta p_\alpha\,, \tag{75}$$

$$D_{\bar{\partial}^\nu}dx^\mu = -C^{\mu\nu}_\alpha dx^\alpha\,, \qquad D_{\bar{\partial}^\nu}\delta p_\mu = C^{\alpha\nu}_\mu\delta p_\alpha\,. \tag{76}$$

Considering a N-linear connection D with set of coefficients, $D\Gamma(N) = (H^\alpha_{\mu\nu}, C^\alpha_{\mu\nu})$, one can add to it a nonlinear connection, $N_{\mu\nu}$, that is in general independent of the coefficients of D, such that the new set is $D\Gamma = (N_{\mu\nu}, H^\alpha_{\mu\nu}, C^\alpha_{\mu\nu})$. For this reason, the derivative of a d-tensor in the cotangent bundle presents similar usual rules for dealing with up and down indices:

$$\begin{aligned} &T^{\mu_1\ldots\mu_m}{}_{\nu_1\ldots\nu_n\alpha_1\ldots\alpha_p}{}^{\beta_1\ldots\beta_q}{}_{|\epsilon} \\ &= \frac{\delta}{\delta x^\epsilon}T^{\mu_1\ldots\mu_m}{}_{\nu_1\ldots\nu_n\alpha_1\ldots\alpha_p}{}^{\beta_1\ldots\beta_q} + H^{\mu_1}_{\gamma\epsilon}T^{\gamma\ldots\mu_m}{}_{\nu_1\ldots\nu_n\alpha_1\ldots\alpha_p}{}^{\beta_1\ldots\beta_q} + \ldots - H^\gamma_{\nu_1\epsilon}T^{\mu_1\ldots\mu_m}{}_{\gamma\ldots\nu_n\alpha_1\ldots\alpha_p}{}^{\beta_1\ldots\beta_q}\,, \end{aligned} \tag{77}$$

$$\begin{aligned} &T^{\mu_1\ldots\mu_m}{}_{\nu_1\ldots\nu_n\alpha_1\ldots\alpha_p}{}^{\beta_1\ldots\beta_q||\epsilon} \\ &= \frac{\partial}{\partial p_\epsilon}T^{\mu_1\ldots\mu_m}{}_{\nu_1\ldots\nu_n\alpha_1\ldots\alpha_p}{}^{\beta_1\ldots\beta_q} + C^{\mu_1\epsilon}_\gamma T^{\gamma\ldots\mu_m}{}_{\nu_1\ldots\nu_n\alpha_1\ldots\alpha_p}{}^{\beta_1\ldots\beta_q} + \ldots - C^{\gamma\epsilon}_{\nu_1}T^{\mu_1\ldots\mu_m}{}_{\gamma\ldots\nu_n\alpha_1\ldots\alpha_p}{}^{\beta_1\ldots\beta_q}\,. \end{aligned} \tag{78}$$

Let us note that from the kinematical map relating velocities and momenta, the coefficients $H^\alpha_{\mu\nu}(x, y(p))$ and $C^\alpha_{\mu\nu}(x, y(p))$ can be found as been parametrization-invariant.

4.2. Finsler–q-de Sitter (Cotangent Bundle Case)

Here, we again consider the q-de Sitter-inspired case. Then, using the Finsler function (46), the momentum is given by Equation: (53)

$$p_0 = \frac{m\dot{x}^0}{\sqrt{(\dot{x}^0)^2 - a^2(\dot{x}^1)^2}} - \ell\frac{m^2a^2(\dot{x}^1)^2(a^2(\dot{x}^1)^2 + (\dot{x}^0)^2)}{2[(\dot{x}^0)^2 - a^2(\dot{x}^1)^2)]^2}\,, \tag{79}$$

$$p_1 = -\frac{ma^2\dot{x}^1}{\sqrt{(\dot{x}^0)^2 - a^2(\dot{x}^1)^2}} + \ell\frac{m^2a^2(\dot{x}^0)^3\dot{x}^1}{((\dot{x}^0)^2 - a^2(\dot{x}^1)^2))^2}\,, \tag{80}$$

which furnishes a helpful expression that is throughout this Section and is a common trick when trying to find momentum-dependent quantities from the Finsler approach:

$$\frac{m\dot{x}^0}{\sqrt{(\dot{x}^0)^2 - a^2(\dot{x}^1)^2}} = p_0 + \ell\frac{a^{-2}(p_1)^2(a^{-2}(p_1)^2 + (p_0)^2)}{2m^2}\,, \tag{81}$$

$$\frac{ma\dot{x}^1}{\sqrt{(\dot{x}^0)^2 - a^2(\dot{x}^1)^2}} = -a^{-1}p_1\left(1 + \ell\frac{(p_0)^3}{m^2}\right). \tag{82}$$

The above expressions allow us to express the Finsler metric through its momentum dependence:

$$g^F_{\mu\nu}(x, \dot{x}(p)) = h^F_{\mu\nu}(x, p) = \begin{pmatrix} 1 + \frac{3\ell p_0(p_1)^4}{m^4} & -\frac{\ell a(p_1)^3[(p_1)^2 - 4(p_0)^2]}{2m^4} \\ -\frac{\ell a(p_1)^3[(p_1)^2 - 4(p_0)^2]}{2m^4} & -a^2 + \frac{\ell a^2(p_0)^3[2(p_0)^2 + (p_1)^2]}{m^4} \end{pmatrix}, \tag{83}$$

which can be called a "Finsler-rainbow metric".

One can also find the induced non-linear connection in the cotangent bundle through the definition (60) to read as

$$\begin{aligned} O_{00}(x,p) =& -\frac{H\ell(p_1)^2}{8m^{10}}\Big[4(p_0)^{10}+44(p_0)^8(p_1)^2+190(p_0)^6(p_1)^4-196(p_0)^4(p_1)^6 \\ &+31(p_0)^2(p_1)^8+32(p_1)^{10}\Big], \end{aligned} \tag{84}$$

$$\begin{aligned} O_{01}(x,p) =& Hp_1-\frac{\ell Hp_0p_1}{8m^{10}}\Big[-4m^8(p_0)^2+8(p_0)^{10}+32(p_0)^8(p_1)^2+206(p_0)^6(p_1)^4 \\ &-212(p_0)^4(p_1)^6+43(p_0)^2(p_1)^8+28(p_1)^{10}\Big], \end{aligned} \tag{85}$$

$$\begin{aligned} O_{10}(x,p) =& Hp_1-\frac{H\ell p_0p_1}{8m^{10}}\Big(-4m^8(p_0)^2+4(p_0)^{10}+140(p_0)^8(p_1)^2+2(p_0)^6(p_1)^4 \\ &-106(p_0)^4(p_1)^6+61(p_0)^2(p_1)^8+4(p_1)^{10}\Big), \end{aligned} \tag{86}$$

$$\begin{aligned} O_{11}(x,p) =& Hp_0+\frac{H\ell}{8m^{10}}\Big(4(p_0)^2(p_1)^2\Big(m^8+3(p_1)^8\Big)+8m^8(p_1)^4-8(p_0)^{12}-124(p_0)^{10}(p_1)^2 \\ &-30(p_0)^8(p_1)^4+138(p_0)^6(p_1)^6-89(p_0)^4(p_1)^8-4(p_1)^{12}\Big). \end{aligned} \tag{87}$$

From these expressions, one can construct the decomposition of the tangent and cotangent spaces of the cotangent bundle into horizontal and vertical parts, accordingly.

5. Geometry of the Cotangent Bundle: Hamilton Geometry

Besides the Finsler geometry, another interesting proposal for building a natural geometry for propagation of particles that probe a modified dispersion relation consists of the so-called Hamilton geometry. In this case, different from the Finsler geometry, we start with a geometric structure defined in the cotangent bundle (the definitions used in this metric follow that in the book [47] and in papers [56–59]).

A Hamilton space is a pair, $(M, H(x,p))$, where M is a smooth manifold and $H: T^*M \to \mathbb{R}$ is a continuous function on the cotangent bundle that satisfies the following properties:

1. H is smooth on the manifold $\widetilde{T^*M}$;
2. the Hamilton metric, h_H, with components,

$$h_H^{\mu\nu}(x,p)=\frac{1}{2}\frac{\partial}{\partial p_\mu}\frac{\partial}{\partial p_\nu}H(x,p), \tag{88}$$

is nondegenerate.

Since one does not have an arc-length functional, worldlines as extremizing curves are an absent concept in this approach. Instead, the equations of motion of a particle that obeys a given Hamiltonian are given by the Hamilton equations of motion:

$$\dot{x}^\mu=\frac{\partial H}{\partial p_\mu}, \tag{89}$$

$$\dot{p}_\mu=-\frac{\partial H}{\partial \dot{x}^\mu}. \tag{90}$$

Since this is just another metric structure defined in the cotangent bundle, the same results regarding the tools for coordinate transformations given by Equation (54) are applicable here. As the case of Hamiltonian mechanics, the definition of Poisson brackets is useful enough for our purposes. For two real valued functions $F(x,p)$ and $G(x,p)$, their Poisson brackets are given in [56] (the geometry of the cotangent bundle with deformed

Hamiltonian can also be described with the language of symplectic geometry, which is reviewed in Ref. [60]):

$$\{F(x,p),G(x,p)\} = \partial_\mu F \bar{\partial}^\mu G - \partial_\mu G \bar{\partial}^\mu F\,. \tag{91}$$

As above, in order to divide the tangent and cotangent spaces of the cotangent bundle into horizontal and vertical spaces, a non-linear connection is necessary, and the canonical choice is given in Theorem 5.5.1 of Ref. [47] and Definition 2 of Ref. [56] as

$$O_{\mu\nu}(x,p) = \frac{1}{4}(\{h^H_{\mu\nu},H\} + h^H_{\mu\alpha}\partial_\nu\bar{\partial}^\alpha H + h^H_{\nu\alpha}\partial_\mu\bar{\partial}^\alpha H)\,, \tag{92}$$

where $h^H_{\mu\nu}$ is the inverse of the metric $h^{\mu\nu}_H$. This non-linear connection allows us to use the basis $\delta_\mu = \partial_\mu + O_{\mu\nu}\bar{\partial}^\nu$ and $\bar{\partial}^\mu$ as a special basis of $T_{(x,p)}\widetilde{T^*M}$, and to use the basis dx^μ and $\delta p_\mu = dp_\mu - O_{\nu\mu}dx^\nu$ as a special basis of $T^*_{(x,p)}\widetilde{T^*M}$, which transforms according to Equations (64), (65) and (67), (68).

Endowed with these coefficients, following Theorem 5.6.1 of Ref. [47], there exists a unique N-linear connection $D\Gamma(O) = (H^\alpha_{\mu\nu}, C^{\mu\nu}_\alpha)$ such that:

1. $O_{\mu\nu}$ is the canonical non-linear connection;
2. the metric $h^{\mu\nu}_H$ is h-covariant constant (no horizontal non-metricity):

$$D_{\delta_\alpha} h^{\mu\nu}_H = 0\,; \tag{93}$$

3. the metric $h^{\mu\nu}_H$ is v-covariant constant (no vertical non-metricity):

$$D\bar{\partial}^\alpha h^{\mu\nu}_H = 0\,; \tag{94}$$

4. $D\Gamma(N)$ is horizontally torsion free:

$$T^\alpha_{\mu\nu} = H^\alpha_{\mu\nu} - H^\alpha_{\nu\mu} = 0\,; \tag{95}$$

5. $D\Gamma(N)$ is vertically torsion free:

$$S^{\mu\nu}_\alpha = C^{\mu\nu}_\alpha - C^{\nu\mu}_\alpha = 0\,; \tag{96}$$

6. the triple $(O_{\mu\nu}, H^\alpha_{\mu\nu}, C^{\mu\nu}_\alpha)$ has coefficients given by

$$O_{\mu\nu}(x,p) = \frac{1}{4}(\{h^H_{\mu\nu},H\} + h^H_{\mu\alpha}\partial_\nu\bar{\partial}^\alpha H + h^H_{\nu\alpha}\partial_\mu\bar{\partial}^\alpha H)\,, \tag{97}$$

$$H^{\mu\nu}_\alpha = \frac{1}{2}h^{\alpha\beta}_H(\delta_\mu h^H_{\beta\nu} + \delta_\nu h^H_{\beta\mu} - \delta_\beta h^H_{\mu\nu})\,, \tag{98}$$

$$C^{\mu\nu}_\alpha = -\frac{1}{2}h^H_{\alpha\beta}\bar{\partial}^\mu h^{\beta\nu}_H\,. \tag{99}$$

This is called a Cartan N-linear covariant derivative. Equivalently, the notion of d-tensors and their derivatives discussed in Section 4.1 are applicable.

5.1. Symmetries

Hamilton geometry also allows one to encompass a DSR language, as was the case for Finsler geometry discussed in Section 3.2. However, its realization does not come from the invariance of an interval ds^2, since one does not have it, but from the invariance of the Hamiltonian function $H(x,p)$. The approach, which we highlight here, was done starting from Definition 4 of Section II-D of Ref. [56]. In a Hamilton space (M,H) with manifold M, and Hamiltonian H, let $X = \xi^\mu\partial_\mu$ be a vector field in the basis manifold M

and $X^C = \xi^\mu \partial_\mu - p_\nu \partial_\mu \xi^\nu \bar{\partial}^\mu$ be the so-called complete lift of X to $\widetilde{T^*M}$. A symmetry of the Hamiltonian is a transformation generated by X^C, whose components satisfy

$$X^C(H)\xi^\mu \partial_\mu H - p_\nu \partial_\mu \xi^\nu \bar{\partial}^\mu H = 0\,. \tag{100}$$

If one derivates this expression twice with respect to momenta, one gets the following result:

$$0 = \frac{1}{2}\bar{\partial}^\mu \bar{\partial}^\nu X^C(H) = \xi^\alpha \partial_\alpha h_H^{\mu\nu} - h_H^{\mu\alpha} \partial_\alpha \xi^\nu - h_H^{\nu\alpha} \partial_\alpha \xi^\mu - p_\beta \partial_\alpha \xi^\beta \bar{\partial}^\alpha h_H^{\mu\nu}\,. \tag{101}$$

This is just the generalization of the killing equation to a general Hamilton space. In general, if h_H does not depend on momenta, then it reduces to the standard Riemannian case. Besides, from the expression of the Poisson brackets (91), it can verified that such symmetries give rise to conserved charges $\xi^\mu p_\mu$; i.e., that Poisson commutes with the Hamiltonian:

$$\{\xi^\mu p_\mu, H\} = 0. \tag{102}$$

These are the charges that, at an algebraic level, can generate translations, boosts, and rotations, for instance.

5.2. Hamilton–q-de Sitter (Cotangent Bundle Case)

As an example, we rely on the results presented in Ref. [56], which are as well inspired by the q-de Sitter Hamiltonian (45). In this case, the Hamilton metric, defined by Equation (88), reads:

$$h_H^{\mu\nu}(x,p) = \begin{pmatrix} 1 & -\ell p_1(1+2Hx^0) \\ -\ell p_1(1+2Hx^0) & -(1+2Hx^0)(1+\ell p_0) \end{pmatrix}, \tag{103}$$

which, as can be seen, acquires a shape much simpler than the rainbow-Finsler one (83) due to the much direct way in which it is calculated.

The non-linear connection can be read from Equation (92) and can be cast in a matrix form due to its simplicity:

$$O_{\mu\nu}(x,p) = \begin{pmatrix} H\ell p_1^2 & Hp_1 \\ Hp_1 & Hp_0(1-\ell p_0) \end{pmatrix}. \tag{104}$$

As expected, it coincides with the case (84) in the Riemannian case, i.e., when $\ell = 0$.

The Hamilton equations of motion can be found from Equation (89) and read:

$$\dot{x}^0 - 2p_0 + \ell p_1^2(1+2Hx^0) = 0\,, \tag{105}$$

$$\dot{x}^1 + 2p_1(1+Hx^0) + 2\ell p_0 p_1(1+2Hx^0) = 0\,, \tag{106}$$

$$\dot{p}_0 - 2Hp_1^2 - 2H\ell p_0 p_1^2 = 0\,, \tag{107}$$

$$\dot{p}_1 = 0\,. \tag{108}$$

The autoparallel (horizontal) curves of the non-linear connection satisfy (see Equation (8.2) in Ref. [47])

$$\dot{p}_\mu - O_{\nu\mu}\dot{x}^\nu = 0\,, \tag{109}$$

and, as can be seen from Equation (104) for $O_{\mu\nu}$, the worldlines, defined from the Hamilton equations of motion, are not autoparallels of the non-linear connection.

The symmetries have also been analyzed in Ref. [56], where it has been noticed that the conserved charges that generate translations and the boost coincide with the results from Ref. [51] that do not rely on the geometrical approach used in this paper.

6. The Tangent-Bundle Version of Hamilton Geometry

Endowed with Hamilton equations of motion (89), one has a map between the momenta and velocities from $\dot{x}^\mu = y^\mu = \partial H/\partial p_\mu$. When it is possible to invert this map to find $p_\mu = p_\mu(y)$ (as done in Appendix B of Ref. [58]), one derives an interesting map between the cotangent and tangent space version of Hamilton geometry. Indeed, using this map, a Hamilton metric defined in the tangent bundle reads:

$$g_H^{\mu\nu}(x,y) \doteq h_H^{\mu\nu}(x,p(y)). \tag{110}$$

The dual non-linear connection in this case has been discussed in Appendix C of Ref. [56], and is given by

$$N(x,y)^\mu{}_\nu = 2O(x,p(y))_{\nu\alpha} h_H^{\alpha\mu}(x,p(y)) - (\partial_\nu \bar{\partial}^\mu H)|_{p=p(y)}. \tag{111}$$

Its main property is the preservation of the horizontal tangent spaces of the cotangent and tangent bundle connected through the kinematical map $y^\mu = \partial H/\partial p_\mu$.

With this map, it is possible to define the dual non-linear and N-linear connections, now defined in the tangent bundle. It should be stressed that although this gives geometrical quantities defined in the tangent bundle, this does not represent a Finsler geometry, since there is no arc-length functional and the Hamilton metric is not, in general, 0-homogeneous to start with.

Hamilton-κ-Poincaré (Tangent Bundle Case)

The kinematical map that allows us to describe $y = y(p)$ is found by inverting the relation $y^\mu = \partial H/\partial p_\mu$ for the q-de Sitter Hamiltonian, given by

$$p_0 = \frac{y^0}{2} + \ell\frac{(y^1)^2}{8}, \tag{112}$$

$$p_1 = -\frac{y^1}{2} + H\frac{x^0 y^1}{2} + \ell\frac{y^0 y^1}{4}. \tag{113}$$

The metric in the tangent bundle reads:

$$g_H^{\mu\nu}(x,y) = \begin{pmatrix} 1 & \ell(Hx^0y^1 + y^1)/2 \\ \ell(Hx^0y^1 + y^1)/2 & -(1+2Hx^0)(1+\ell y^0/2) \end{pmatrix}. \tag{114}$$

The dual non-linear connection reads

$$N^\mu{}_\nu(x,y) = \begin{pmatrix} -H\ell(y^1)^2/2 & \ell h y^0 y^1 + h y^1 \\ H y^1 & -h y^0 - 3\ell h (y^1)^2/4 \end{pmatrix}. \tag{115}$$

In Section 7 below, some key points of each approach are discussed while comparing the descriptions of configuration and phase spaces.

7. Advantages and Difficulties of Each Formalism

The approaches considered—Finsler and Hamilton spaces—present the points that can be considered positive or negative. In this Section, we highlight some of those points which look to be most important from theoretical and phenomenological points of view.

7.1. Finsler Geometry

Let us emphasize that here not a complete list of positive or negative points is given, and, certainly, the points listed represent just our view on the subject under scrutiny and some points we are classifying in one way or another can be seen by others completely differently.

7.1.1. Advantages

Preservation of the equivalence principle. Due to the presence of an arc-length functional, the extremizing geodesics of the Finsler function are the same worldlines of the Hamiltonian, from which the arc-length was derived. This means that, in the Finslerian language, the equivalence principle is satisfied, as soon as the worldlines are trajectories of free particles in this spacetime. There is a fundamental difference in comparison to the special or general relativity formulation, since these trajectories are now mass-dependent, since the Finsler function and the metric carry the mass of the particle due to Planck-scale effects. Intriguingly, although the metric does not present a massless limit (which is discussed below), it is possible to find trajectories of massless particles, which are compatible with the Hamiltonian formulation, by taking the limit $m \to 0$ in the geodesic Equation [49,50]. This finding leads to some effects due to modifications of the trajectories of particles. For instance, one of the most explored avenues of quantum gravity phenomenology (maybe competing with threshold effects) is the time delay until particles with different energies might arrive at a detector after a (almost) simultaneous emission [61,62] (for reviews, see [4,5]). This kind of experimental investigation is not exhausted, and novelties have arrived in the analysis of sets of gamma-ray bursts and candidate neutrinos emitted from them in the multimessenger astronomy approach [63,64].

Preservation of the relativity principle. This formalism allows one to derive and solve the killing equation, which furnishes infinitesimal symmetry transformations of the metric. It has been shown in Ref. [49] that generators of these transformations can be constructed and identified with the transformations that are generally depicted in the doubly special relativity. The latter implies, in a preservation of the relativity principle, that inertial frames should assign the same MDR to a given particle which, in its turn, implies that the deformation scale of quantum gravity is observer-independent, i.e., two observers would not assign different values, in the same system of units, to the quantum gravity scale. This preservation has important phenomenological consequences, such as the point that the threshold constraints on the quantum gravity parameter do not apply in the DSR scenario. The reason is that, accompanied by the deformation of the Lorentz (Poincaré) symmetries, comes a deformation of the composition law of momenta of particles (for instance p and q), such that the nature of interaction vertices to not get modified when transforming from one frame to another:

$$\Lambda(p \oplus q) = \Lambda(p) \oplus \Lambda(q), \tag{116}$$

where Λ is a deformed Lorentz transformation induced by the killing vectors and $\oplus$ represents a modified composition of components of the involved momenta (this covariance condition usually needs a back-reaction on the boost parameter, but we do not dwell on that here; for more details, see [55,65] and references therein). Threshold constraints, such as the one placed in Ref. [66], assumes that the composition of momenta is undeformed, although the dispersion relation is modified in a Lorentz invariance violation (LIV) scenario. When this is the case, processes that are forbidden in special relativity, such as the decay of the photon into an electron–positron pair, becomes kinematicaly allowed for a given threshold energy. The no observation of such decays allows one to place constraints on the quantum gravity parameter. When the dispersion relation is modified as well, what happens is that generally these kinds of processes remain forbidden or modifications in the threshold energies are so minute that they are unobservable for a quantum gravity parameter in the order of the Planck energy [55]. This is an important feature of "deforming" instead of "violating" the Lorentz symmetry.

Preservation of the clock postulate. The availability of an arc-length functional leads to a possibility to analyze the consequences of having the proper time of a given particle given by it. If this is the case, then the worldlines or geodesics are just paths that extremize the proper time an observer measures in spacetime, similar to that in special relativity. One of the consequences of this feature consists of the possibility of connecting the time elapsed in the comoving frame of a particle during its lifetime (which is its lifetime at rest) and the coordinate time, which is the one that is assigned to this phenomenon in the laboratory coordinates. Using this expression, one can investigate the relativistic time dilation (responsible for the "twin paradox") or the so-called first clock effect (for further details on the first and also on the second clock effect, which can appear in theories with a non-metricity tensor, see Ref. [67]), in which, for instance, the lifetime of a particle is dilated in comparison to the one assigned in the laboratory. Due to Finslerian corrections, the lifetime of a particle in the laboratory would receive Planckian corrections, which, actually, is a novel avenue of phenomenological investigation that is being currently carried out [43,55] through the search for signatures in particle accelerators and cosmic rays.

7.1.2. Difficulties

Absence of massless rainbow Finsler metric. The Finsler approach had emerged as an opportunity to describe in a consistent way the intuition that the quantum spacetime probed by a high-energy particle would present some energy-momentum (of the particle itself) corrections, which is justified by different approaches to quantum gravity [24,25]. Since then, proposals of rainbow metrics have considered a smooth transition from massive to massless cases, not only from the point of view of the trajectories, but from the metric itself. This is not the case for the Finsler approach presented here. Although the trajectories and symmetries are defined for both massive and massless cases by considering the $m \to 0$ limit, the rainbow metric of Finsler geometry, given by Equation (83), is certainly not defined for massless particles. The reason for this is the point that when passing from the Hamiltonian to the Lagrangian formalism, we defined an arc-length functional, which is not a legitimate action functional for massless particles. In other words, a crucial step for deriving the Finsler function is the handling of the Lagrange multiplier λ of action (4), which can only be solved if the particle is massive, as can be found in Refs. [43,49,50,53]. A possibility that has been explored consisted of not solving the equation for λ and defining a metric that depends on λ and on velocities from a Polyakov-like action for free particles (instead of the Nambu–Goto one given by the arc-length), which turned out to be out of the Finsler geometry scope [50,53]. However, this possibility has not been further explored beyond preliminary investigations. The issue of the absence of a massless rainbow-Finsler metric could be circumvented by proposing a different kind of geometry, which from the very beginning started from the momenta formulation, like the other possibility described in this paper, namely the Hamilton geometry.

Definition only through perturbations. The Finsler geometry was considered in this paper in this context at most perturbatively around the quantum-gravity-length scale (or inverse of energy scale), which may be considered as a negative point if one aims to make it at a more fundamental or theoretical level. Nevertheless, from the pragmatic perspective of phenomenology, since such effects, if they exist, are minute, then the perturbative approach is enough for proposing new effects that could serve as avenues of experimental investigation.

The handling of finite symmetries. Another issue that can be problematic is the handling of finite symmetries in the Finsler context. Up to today, the connection between Finsler geometry and quantum gravity phenomenology has not faced the issue of integrating the symmetries and finding finite versions of deformed Lorentz transformations. Some initial investigations were carried out in Ref. [55] from the momentum space perspective, but further issues are being currently faced by some authors of the present paper.

7.2. *Hamilton Geometry*

The descriptions of the points here is rather short, without discussion of some universal points already described above, so those universal points can be addressed there when that is the case.

7.2.1. Advantages

Presence of a massless rainbow Hamilton metric. Differently from the Finsler case, the Hamilton geometry does not need an arc-length functional; instead, it only needs a given Hamiltonian, from which the metric, non-linear connection, and symmetries are derived. This means that from the very beginning, the massless limit of geometrical quantities exists.

Does not require perturbative methods. Another positive point about the Hamilton geometry is the finding that one can handle with the exact form of the proposed Hamiltonian, and it does not need to consider perturbations around a certain scale. Instead, independently of the form of the (smooth) dispersion relation that arises from de facto approaches to quantum gravity, the geometry can be handled, as has been considered, e.g., in Refs. [57,58].

Preservation of the relativity principle and the handling of symmetries. Due to the proximity of this approach to the way that the DSR formalism generally handles with Planck scale corrections, i.e., from the point of view of momentum space and Hamilton equations, the handling of symmetries is facilitated in this approach. For instance, it is straightforward to find the conserved charges from the killing vectors, which generate finite transformations that are momenta-dependent without tedious terms in the denominator of the equations when one is working in velocity space, as Finsler geometry is initially formulated (or without mass terms in the denominator in the Finsler version of the phase space).

Generalization to curved spacetimes. This approach is considered in more curved space cases, beyond the q-de Sitter exemplified in this paper; for instance, its spherically symmetric and cosmological versions were explored giving rise to interesting phenomenological opportunities, from the point of view of time delays and gravitational redshift, among others (for some applications of Hamilton geometry in this context, see [59] and references therein).

7.2.2. Difficulties

Non-geodesic trajectory. An issue that may be considered problematic is the point that the worldlines of particles, given by the Hamilton equations, are not geodesics of the non-linear connection that means that there exists a force term in the geodesic equation, which is in contrast with the Finsler case. This is a property of the Hamilton geometry, as has been shown in Ref. [56], and is not specific to the q-de Sitter case analyzed here.

Absence of the arc-length. The Hamilton geometry does not dwell with an arc-length functional that means that the only geodesics present are those of the non-linear or of the N-linear connections and there are no extremizing ones. The absence of a function that allows one to measure distances in spacetime can be seen as a difficulty of this geometry; if the distances cannot be calculated, one could wonder what such a metric means. Even if the norm of a tangent vector can be integrated, this integral would not be, in general, parametrization-independent, which is also a drawback of this tentative. Besides, the absence of an arc-length limits the phenomenology of the preservation of the clock postulate that was discussed in the Finsler case.

8. Final Remarks

We revised two proposals that have been considered as candidates for describing the quantum configuration and phase spaces probed by particles whose kinematics are modified by a length scale identified as the quantum gravity scale.

Finsler geometry starts from a configuration space framework that presents applications on its own in biology, thermodynamics, and modified gravity; and it finds a natural environment in quantum gravity phenomenology due to its power to describe a scenario in which important principles that guided physics in the XXth century, such as the relativity principle, are preserved even at a Planckian regime. Besides its traditional description in terms of the couple spacetime and velocity space (configuration space), we also explored its development in terms of the induced couple spacetime and momentum space (phase space),

References

1. Bronstein, M. Quantum theory of weak gravitational fields. *Gen. Rel. Grav.* **2012**, *44*, 267–283. [CrossRef]
2. Rovelli, C. Loop quantum gravity. *Living Rev. Rel.* **2008**, *11*, 5. [CrossRef] [PubMed]
3. Ambjørn, J.; Jurkiewicz, J.; Loll, R. Causal dynamical triangulations and the quest for quantum gravity. In *Foundations of Space and Time: Reflections on Quantum Gravity*; Murugan, J., Weltman, A., Ellis, G.F.R., Eds.; Cambridge University Press: Cambridge, UK, 2012; pp. 321–337. [CrossRef]
4. Amelino-Camelia, G. Quantum-spacetime phenomenology. *Living Rev. Rel.* **2013**, *16*, 5. [CrossRef] [PubMed]
5. Addazi, A.; Alvarez-Muniz, J.; Batista, R.A.; Amelino-Camelia, G.; Antonelli, V.; Arzano, M.; Asorey, M.; Atteia, J.L.; Bahamonde, S.; Bajardi, F.; et al. Quantum gravity phenomenology at the dawn of the multi-messenger era—A review. *Prog. Part. Nucl. Phys.* **2022**, *125*, 103948, [CrossRef]
6. Amelino-Camelia, G.; da Silva, M.M.; Ronco, M.; Cesarini, L.; Lecian, O.M. Spacetime-noncommutativity regime of loop quantum gravity. *Phys. Rev. D* **2017**, *95*, 024028, [CrossRef]
7. Brahma, S.; Ronco, M.; Amelino-Camelia, G.; Marciano, A. Linking loop quantum gravity quantization ambiguities with phenomenology. *Phys. Rev. D* **2017**, *95*, 044005, [CrossRef]
8. Brahma, S.; Ronco, M. Constraining the loop quantum gravity parameter space from phenomenology. *Phys. Lett. B* **2018**, *778*, 184–189. [CrossRef]
9. Majid, S.; Ruegg, H. Bicrossproduct structure of kappa Poincare group and noncommutative geometry. *Phys. Lett. B* **1994**, *334*, 348–354. . [CrossRef]
10. Lukierski, J.; Ruegg, H.; Nowicki, A.; Tolstoi, V.N. Q deformation of Poincare algebra. *Phys. Lett. B* **1991**, *264*, 331–338. [CrossRef]
11. Lukierski, J.; Nowicki, A.; Ruegg, H. New quantum Poincare algebra and k deformed field theory. *Phys. Lett. B* **1992**, *293*, 344–352. [CrossRef]
12. Majid, S. *Foundations of Quantum Group Theory*; Cambridge University Press: Cambridge, UK, 2010. [CrossRef]
13. Amelino-Camelia, G. Doubly-special relativity: Facts, myths and some key open issues. *Symmetry* **2010**, *2*, 230–271. [CrossRef]
14. Mattingly, D. Modern tests of Lorentz invariance. *Living Rev. Rel.* **2005**, *8*, 5, [CrossRef]
15. Liberati, S. Tests of Lorentz invariance: A 2013 update. *Class. Quant. Grav.* **2013**, *30*, 133001, [CrossRef]
16. Amelino-Camelia, G. Relativity in space-times with short distance structure governed by an observer independent (Planckian) length scale. *Int. J. Mod. Phys. D* **2002**, *11*, 35–60. [CrossRef]
17. Magueijo, J.; Smolin, L. Lorentz invariance with an invariant energy scale. *Phys. Rev. Lett.* **2002**, *88*, 190403, [CrossRef] [PubMed]
18. Amelino-Camelia, G.; Freidel, L.; Kowalski-Glikman, J.; Smolin, L. The principle of relative locality. *Phys. Rev. D* **2011**, *84*, 084010, [CrossRef]
19. Proutorov, E.; Matsuyama, N.; Koibuchi, H. Finsler geometry modeling and Monte Carlo study of liquid crystal elastomers under electric fields. *J. Phys. Condens. Matter* **2018**, *30*, 405101. [CrossRef]
20. Hehl, F.W.; Obukhov, Y.N. *Foundations of Classical Electrodynamics: Charge, Flux, and Metric*; Springer Science+Business Media, LLC/Birkhäuser Boston: New York, NY, USA, 2003. [CrossRef]
21. Červený, V. Fermat's variational principle for anisotropic inhomogeneous media. *Stud. Geophys. Geod.* **2002**, *46*, 567–588. [CrossRef]
22. Pfeifer, C. Finsler spacetime geometry in Physics. *Int. J. Geom. Meth. Mod. Phys.* **2019**, *16*, 1941004, [CrossRef]
23. Magueijo, J.; Smolin, L. Gravity's rainbow. *Class. Quant. Grav.* **2004**, *21*, 1725–1736. [CrossRef]
24. Weinfurtner, S.; Jain, P.; Visser, M.; Gardiner, C.W. Cosmological particle production in emergent rainbow spacetimes. *Class. Quant. Grav.* **2009**, *26*, 065012, [CrossRef]
25. Assanioussi, M.; Dapor, A.; Lewandowski, J. Rainbow metric from quantum gravity. *Phys. Lett. B* **2015**, *751*, 302–305. [CrossRef]
26. Olmo, G.J. Palatini Actions and Quantum Gravity Phenomenology. *J. Cosmol. Astropart. Phys.* **2011**, *10*, 018, [CrossRef]
27. Ling, Y.; Li, X.; Zhang, H.b. Thermodynamics of modified black holes from gravity's rainbow. *Mod. Phys. Lett. A* **2007**, *22*, 2749–2756. [CrossRef]
28. Lobo, I.P.; Santos, L.C.N.; Bezerra, V.B.; Morais Graça, J.P.; Moradpour, H. The extended phase space thermodynamics of Planck-scale-corrected Reissner-Nordström-anti-de Sitter black hole. *Nucl. Phys. B* **2021**, *972*, 115568, [CrossRef]
29. Gorji, M.A.; Nozari, K.; Vakili, B. Gravity's rainbow: A bridge between LQC and DSR. *Phys. Lett. B* **2017**, *765*, 113–119. [CrossRef]
30. Garattini, R.; Lobo, F.S.N. Gravity's rainbow and traversable wormholes. In Proceedings of the Fourteenth Marcel Grossmann Meeting on Recent Developments in Theoretical and Experimental General Relativity, Astrophysics, and Relativistic Field Theories, Rome, Italy, 12–18 July, 2015; World Scientific Co. Ltd.: Singapore, 2017; pp. 1448–1453. [CrossRef]
31. Amirabi, Z.; Halilsoy, M.; Mazharimousavi, S.H. Thin-shell wormholes in rainbow gravity. *Mod. Phys. Lett. A* **2018**, *33*, 1850049. [CrossRef]
32. Bezerra, V.B.; Lobo, I.P.; Mota, H.F.; Muniz, C.R. Landau levels in the presence of a cosmic string in rainbow gravity. *Ann. Phys.* **2019**, *401*, 162–173. [CrossRef]
33. Carvalho, G.G.; Lobo, I.P.; Bittencourt, E. Extended disformal approach in the scenario of Rainbow Gravity. *Phys. Rev. D* **2016**, *93*, 044005, [CrossRef]
34. Lobo, I.P.; Carvalho, G.G. The geometry of null-like disformal transformations. *Int. J. Geom. Meth. Mod. Phys.* **2019**, *16*, 1950180, [CrossRef]

35. Santos, L.C.N.; Bezerra, V.B. Electrostatic self-interaction of charged particles in the space-time of a cosmic string in the context of gravity's rainbow. *Gen. Rel. Grav.* **2019**, *51*, 145, [CrossRef]
36. Riemann, B. *On the Hypotheses Which Lie at the Bases of Geometry;* Jost, J., Ed.; Birkhäuser/Springer International Publishing Switzerland: Cham, Switzerland, 2016. [CrossRef]
37. Finsler, P. *Über Kurven und Flähen in allgemeinen Räumen;* Springer Basel AG/Verlag Birkhäser AG: Basel, Switzerland, 1951. [CrossRef]
38. Bao, D.; Chern, S.S.; Shen, Z. *An Introduction to Riemann-Finsler Geometry;* Springer-Verlag New York, Inc.: New York, NY, USA, 2000. [CrossRef]
39. Hohmann, M.; Pfeifer, C.; Voicu, N. Mathematical foundations for field theories on Finsler spacetimes. *J. Math. Phys.* **2022**, *63*, 032503, [CrossRef]
40. Bernal, A.; Javaloyes, M.A.; Sánchez, M. Foundations of Finsler spacetimes from the observers' viewpoint. *Universe* **2020**, *6*, 55. [CrossRef]
41. Minguzzi, E. Light cones in Finsler spacetime. *Commun. Math. Phys.* **2015**, *334*, 1529–1551. [CrossRef]
42. Girelli, F.; Liberati, S.; Sindoni, L. Planck-scale modified dispersion relations and Finsler geometry. *Phys. Rev. D* **2007**, *75*, 064015, [CrossRef]
43. Lobo, I.P.; Pfeifer, C. Reaching the Planck scale with muon lifetime measurements. *Phys. Rev. D* **2021**, *103*, 106025, [CrossRef]
44. Raetzel, D.; Rivera, S.; Schuller, F.P. Geometry of physical dispersion relations. *Phys. Rev. D* **2011**, *83*, 044047, [CrossRef]
45. Rodrigues, E.; Lobo, I.P. Revisiting Legendre transformations in Finsler geometry. *arXiv* **2022**, arXiv:2208.11406. [CrossRef]
46. Gubitosi, G.; Mercati, F. Relative locality in κ-Poincaré. *Class. Quant. Grav.* **2013**, *30*, 145002. [CrossRef]
47. Miron, R.; Anastasiei, M. *The Geometry of Lagrange Spaces: Theory and Applications;* Springer Scienc+Business Media, B.V./Kluwer Academic Publishers: Dordrecht, The Netherlands, 1994. [CrossRef]
48. Shen, Z. *Differential Geometry of Spray and Finsler Spaces;* Springer Science+Business Media, B.V./Kluwer Academic Publishers: Dordrecht, The Netherlands, 2001. [CrossRef]
49. Amelino-Camelia, G.; Barcaroli, L.; Gubitosi, G.; Liberati, S.; Loret, N. Realization of doubly special relativistic symmetries in Finsler geometries. *Phys. Rev. D* **2014**, *90*, 125030, [CrossRef]
50. Lobo, I.P.; Loret, N.; Nettel, F. Investigation of Finsler geometry as a generalization to curved spacetime of Planck-scale-deformed relativity in the de Sitter case. *Phys. Rev. D* **2017**, *95*, 046015, [CrossRef]
51. Barcaroli, L.; Gubitosi, G. Kinematics of particles with quantum-de Sitter-inspired symmetries. *Phys. Rev. D* **2016**, *93*, 124063, [CrossRef]
52. Letizia, M.; Liberati, S. Deformed relativity symmetries and the local structure of spacetime. *Phys. Rev. D* **2017**, *95*, 046007, [CrossRef]
53. Lobo, I.P.; Loret, N.; Nettel, F. Rainbows without unicorns: Metric structures in theories with modified dispersion relations. *Eur. Phys. J. C* **2017**, *77*, 451, [CrossRef]
54. Amelino-Camelia, G.; Rosati, G.; Bedić, S. Phenomenology of curvature-induced quantum-gravity effects. *Phys. Lett. B* **2021**, *820*, 136595, [CrossRef]
55. Lobo, I.P.; Pfeifer, C.; Morais, P.H.; Batista, R.A.; Bezerra, V.B. Two-body decays in deformed relativity. *J. High Energy Phys.* **2022**, *9*, 003, [CrossRef]
56. Barcaroli, L.; Brunkhorst, L.K.; Gubitosi, G.; Loret, N.; Pfeifer, C. Hamilton geometry: Phase space geometry from modified dispersion relations. *Phys. Rev.* **2015**, *D92*, 084053, [CrossRef]
57. Barcaroli, L.; Brunkhorst, L.K.; Gubitosi, G.; Loret, N.; Pfeifer, C. Planck-scale-modified dispersion relations in homogeneous and isotropic spacetimes. *Phys. Rev.* **2017**, *D95*, 024036, [CrossRef]
58. Barcaroli, L.; Brunkhorst, L.K.; Gubitosi, G.; Loret, N.; Pfeifer, C. Curved spacetimes with local κ-Poincaré dispersion relation. *Phys. Rev. D* **2017**, *96*, 084010, [CrossRef]
59. Pfeifer, C. Redshift and lateshift from homogeneous and isotropic modified dispersion relations. *Phys. Lett. B* **2018**, *780*, 246–250. [CrossRef]
60. Nozari, K.; Gorji, M.A.; Hosseinzadeh, V.; Vakili, B. Natural cutoffs via compact symplectic manifolds. *Class. Quant. Grav.* **2016**, *33*, 025009, [CrossRef]
61. Jacob, U.; Piran, T. Lorentz-violation-induced arrival delays of cosmological particles. *J. Cosmol. Astrpart. Phys.* **2008**, *01*, 031, [CrossRef]
62. Zhu, J.; Ma, B.Q. Lorentz-violation-induced arrival time delay of astroparticles in Finsler spacetime. *Phys. Rev. D* **2022**, *105*, 124069, [CrossRef]
63. Amelino-Camelia, G.; D'Amico, G.; Rosati, G.; Loret, N. In-vacuo-dispersion features for GRB neutrinos and photons. *Nature Astron.* **2017**, *1*, 0139, [CrossRef]
64. Amelino-Camelia, G.; Di Luca, M.G.; Gubitosi, G.; Rosati, G.; D'Amico, G. Could quantum gravity slow down neutrinos? *arXiv* **2022**, arXiv:2209.13726. [CrossRef]
65. Majid, S. Algebraic approach to quantum gravity. II. Noncommutative spacetime. In *Approaches to Quantum Gravity. Toward a New Understanding of Space, Time and Matter;* Oriti, D., Ed.; Cambridge University Press: Cambridge, UK, 2009; pp. 466–492. . [CrossRef]

66. Albert, A.; Alfaro, R.; Alvarez, C.; Camacho, J.A.; Arteaga-Velázquez, J.C.; Arunbabu, K.P.; Rojas, D.A.; Solares, H.A.; Baghmanyan, V.; Belmont-Moreno, E.; et al. Constraints on Lorentz invariance violation from HAWC observations of gamma rays above 100 TeV. *Phys. Rev. Lett.* **2020**, *124*, 131101, [CrossRef]
67. Lobo, I.P.; Romero, C. Experimental constraints on the second clock effect. *Phys. Lett. B* **2018**, *783*, 306–310. [CrossRef]
68. Hossenfelder, S. The soccer-ball problem. *Symm. Integrab. Geom. Meth. Applic. (SIGMA)* **2014**, *10*, 074. [CrossRef]

physics

MDPI

Review
An Introduction to Noncommutative Physics

Shi-Dong Liang [1,2,*] and Matthew J. Lake [1,3,4,5,6,*]

1 School of Physics, Sun Yat-Sen University, Guangzhou 510275, China
2 State Key Laboratory of Optoelectronic Material and Technology, Guangdong Province Key Laboratory of Display Material and Technology, Sun Yat-Sen University, Guangzhou 510275, China
3 National Astronomical Research Institute of Thailand, 260 Moo 4, T. Donkaew, A. Maerim, Chiang Mai 50180, Thailand
4 Department of Physics and Materials Science, Faculty of Science, Chiang Mai University, 239 Huaykaew Road, T. Suthep, A. Muang, Chiang Mai 50200, Thailand
5 Department of Physics, Babeş-Bolyai University, Mihail Kogălniceanu Street 1, 400084 Cluj-Napoca, Romania
6 Office of Research Administration, Chiang Mai University, 239 Huaykaew Rd, T. Suthep, A. Muang, Chiang Mai 50200, Thailand
* Correspondence: stslsd@mail.sysu.edu.cn (S.-D.L.); matthewjlake@narit.or.th (M.J.L.)

Abstract: Noncommutativity in physics has a long history, tracing back to classical mechanics. In recent years, many new developments in theoretical physics, and in practical applications rely on different techniques of noncommutative algebras. In this review, we introduce the basic concepts and techniques of noncommutative physics in a range of areas, including classical physics, condensed matter systems, statistical mechanics, and quantum mechanics, and we present some important examples of noncommutative algebras, including the classical Poisson brackets, the Heisenberg algebra, Lie and Clifford algebras, the Dirac algebra, and the Snyder and Nambu algebras. Potential applications of noncommutative structures in high-energy physics and gravitational theory are also discussed. In particular, we review the formalism of noncommutative quantum mechanics based on the Seiberg–Witten map and propose a parameterization scheme to associate the noncommutative parameters with the Planck length and the cosmological constant. We show that noncommutativity gives rise to an effective gauge field, in the Schrödinger and Pauli equations. This term breaks translation and rotational symmetries in the noncommutative phase space, generating intrinsic quantum fluctuations of the velocity and acceleration, even for free particles. This review is intended as an introduction to noncommutative phenomenology for physicists, as well as a basic introduction to the mathematical formalisms underlying these effects.

Keywords: noncommutative quantum mechanics; noncommutative phase space; quantum Hall effect; Seiberg-Witten map

Citation: Liang, S.-D.; Lake, M.J. An Introduction to Noncommutative Physics. *Physics* **2023**, *5*, 436–460. https://doi.org/10.3390/physics5020031

Received: 29 January 2023
Revised: 14 March 2023
Accepted: 20 March 2023
Published: 18 April 2023

1. Introduction

Different physical theories depend on different mathematical structures. For example, classical Hamiltonian mechanics is defined on a symplectic manifold, and relativistic gravity theories describe the dynamics of pseudo-Riemann geometries, while quantum theories are defined by using complex vector spaces [1]. Noncommutativity naturally arises in various formalisms, tracing back to the descriptions of angular momentum and work in classical mechanics, through to the Heisenberg algebra and its associated uncertainty relations in canonical quantum mechanics, and on to more speculative recent theories regarding the noncommutative nature of spacetime at the Planck scale [2–5]. The latter include the Snyder algebra [6], Heisenberg–Weyl algebra, various types of Lie algebra, and the so-called noncommutative phase space algebra [2,7].

Snyder proposed a five-dimensional noncommutative spacetime, with global Lorentz invariance, to remove singularities in particle physics without renormalization techniques [6].

His model was generalized to curved spacetimes by C. N. Yang, in order to include gravitational effects [8]. These early studies were among the first to suggest that quantized spacetime should be described by a kind of noncommutative geometry (NCG), which, it is hoped, can provide a self-consistent formalism to unify quantum theory with gravity [9–11]. Thus, it is believed that NCG could play a vital role in removing the infinities and singularities that unavoidably emerge in both particle physics and cosmology.

In addition, the recent discoveries of dark energy and dark matter in cosmological observations lead to many puzzles, and the fundamental physics behind these phenomena is not well understood. There are many candidate theories, and, even though dark energy can be interpreted phenomenologically as a cosmological constant, there is still no direct experimental evidence capable of determining the physical mechanism that gives rise to it [12,13]. One possibility is that the small but nonzero vacuum energy emerges from the quantum fluctuations of the spacetime background at the Planck scale, which is expected in the framework of noncommutative field theory and quantum gravity models based on noncommutative geometry [14–16].

In less speculative fields, noncommutativity also plays a vital role in describing a range of important physical phenomena. In condensed matter physics, it may be shown that a two-dimensional electronic system, immersed in a strong magnetic field, is equivalent to its free electron counterpart, formulated in a noncommutative phase space [17–22]. Thus, by generalizing the canonical Heisenberg algebra to include nontrivial space–space and/or momentum–momentum commutation relations, this so-called noncommutative quantum mechanics (NCQM) can successfully model known phenomena, such as the Aharonov–Bohm effect [23–26], the quantum Hall effect [22,27,28], the existence of magnetic monopoles [17] and the Berry phase [29,30], by using a different language and mathematical formalism.

Currently, there are several schemes used to implement such phase space noncommutativity [7,27,31–33], including the Seiberg–Witten (SW) map [34–36], the Moyal product formalism [3] and the Wigner–Weyl phase space approach [37–40]. Based on these noncommutative algebras, the Heisenberg uncertainty principle can be extended to include new space–space and momentum–momentum uncertainty relations, which are clearly of great phenomenological interest to physicists. These physical phenomena also stimulate mathematical interest in noncommutative algebras and noncommutative geometry [14–16].

An example of a new mathematical formalism, with potential applications in high-energy physics and gravitational theory, is the Nambu generalization of symplectic geometry, in which the Poisson algebra is generalized to the so-called Nambu bracket, yielding generalized Hamiltonian [41–43] and Lagrangian mechanics [44,45]. The quantisation of the classical Nambu brackets generates a deformed Heisenberg algebra [46–49], which leads to new physical phenomenology [50–52], and to a new phase space structure, which is also of interest to mathematicians [53–55].

In this paper, we present an accessible introduction to noncommutative physics in a pedagogical format with a strong focus on phenomenology. For completeness, and so that the text can be read as a self-contained reference, we also include brief summaries of the basic mathematical formalisms needed to implement noncommutative structures in a range of example systems. We begin by reviewing important examples of noncommutative phenomena in physics in Section 2, including the canonical Poisson brackets of classical mechanics, permutation symmetries in statistical mechanics, and the classical and quantum Hall effects.

In Section 3, we give an overview of some known noncommutative algebras, including the Poisson algebra in classical symplectic geometry, the Heisenberg, Lie, Clifford, and Dirac algebras in canonical quantum theory, the Snyder and Nambu algebras in theoretical physics, and the deformed Heisenberg algebra used to generate models of NCQM [7].

In Section 4, we give a detailed presentation of NCQM, based on the deformed Heisenberg algebra, and review its phenomenological implications for gravitational theories and high-energy particle physics. Based on the SW map, we give the Heisenberg representation of the Schrödinger, Heisenberg, and Pauli equations, and consider some basic properties of the model, including the existence of anomalous velocity and acceleration terms in the free-particle dynamics induced by the noncommutativity of the background. We propose a parameterization scheme that associates the noncommutative parameters of the space–space and momentum–momentum commutators, respectively, with the Planck length and the cosmological constant and discuss its implications for particle physics and cosmology.

However, in this pedagogical introduction to noncommutative physics we do not discuss Conne's approach to noncommutative geometry, K theory, noncommutative field theory, the Moyal star product technique, or $\kappa-$deformed symmetries of spacetime. For details of these (more mathematical) approaches, readers are referred to [14–16,56–59] and references therein. We summarise our conclusions, and offer a few opinions on the outlook for noncommutative geometry in physics research in Section 5.

2. Noncommutative Phenomena in Physics: Important Examples

In this Section, we review important examples of noncommutative phenomena in classical and quantum physics.

2.1. Noncommutativity in Classical Physics

The best known example of noncommutativity in classical physics comes from Hamiltonian mechanics, in which the Poisson bracket is defined as

$$\{f,g\} := \frac{\partial f}{\partial q^i}\frac{\partial g}{\partial p_i} - \frac{\partial f}{\partial p_i}\frac{\partial g}{\partial q^i}. \tag{1}$$

This relation implies that the generalised coordinates, q^i, $i \in \{1,2,\ldots d\}$, where d is the dimensionality of the system, and their corresponding canonical momenta, $p_i = \partial L/\partial q^i$, obey the commutation relations [60]

$$\left\{q^i, q^j\right\} = 0, \tag{2}$$

$$\left\{p_i, p_j\right\} = 0, \tag{3}$$

$$\left\{q^i, p_j\right\} = \delta^i{}_j, \tag{4}$$

where $\delta^i{}_j$ is the Kronecker delta. These relations are called the fundamental Poisson brackets. Together, the coordinates and canonical momenta define the phase space of the system. All physical quantities are expressed in terms of maps on the phase space and the evolution of the system follows the integral curves generated by the flow of the Hamiltonian vector field,

$$\dot{\gamma} = \{\gamma, \mathcal{H}\}. \tag{5}$$

The physical quantity γ is conserved if $\{\gamma, \mathcal{H}\} = 0$ [60].

2.2. Permutation Symmetries in Statistical Mechanics

In statistical physics, the statistical behavior of multiparticle systems depends sensitively on the permutation symmetries of elementary particles. The mean occupation numbers for free particles in thermal equilibrium states are given by

$$n = e^{-(\varepsilon-\mu)/k_B T} \tag{6}$$

for classical particles, and

$$n = \begin{cases} \frac{1}{e^{(\varepsilon-\mu)/k_B T}-1} & \text{for bosons,} \\ \frac{1}{e^{(\varepsilon-\mu)/k_B T}+1} & \text{for fermions} \end{cases} \tag{7}$$

in quantum physics, where T is the temperature of the system, k_B is Boltzmann's constant, and μ is the chemical potential. The differing bosonic and fermionic distributions, known as the Bose–Einstein (BE) distribution and the Fermi–Dirac (FD) distribution, respectively, are governed by different algebras. In the particle number representation, the particle creation and annihilation operators for bosons and fermions obey, respectively,

$$\left[\hat{b}_i, \hat{b}_j^\dagger\right] = \delta_{ij}\hat{\mathbb{1}}, \left[\hat{b}_i, \hat{b}_j\right] = \left[\hat{b}_i^\dagger, \hat{b}_j^\dagger\right] = 0\,(\text{BE}), \tag{8}$$

$$\left\{\hat{a}_i, \hat{a}_j^\dagger\right\} = \delta_{ij}\hat{\mathbb{1}}, \{\hat{a}_i, \hat{a}_j\} = \left\{\hat{a}_i^\dagger, \hat{a}_j^\dagger\right\} = 0\,(\text{FD}), \tag{9}$$

where $\hat{\mathbb{1}}$ is the identity operator, $[A, B] \equiv AB - BA$ denotes the commutator of AB and B, and $\{A, B\} \equiv AB + BA$ is the anticommutator. The anticommutative property of fermion operators ($\hat{a}_i\hat{a}_j = -\hat{a}_j\hat{a}_i, \hat{a}_i^2 = 0$) is known as a Grassmann algebra. The relationship between the Fermi–Dirac distribution, spin, and the Grassmann algebra, is called the spin-statistics theorem [54].

2.3. The Heisenberg Algebra

To understand the discrete atomic spectrum of hydrogen, Bohr proposed an orbital model of the atom in which discrete spectra are generated by the transitions of electrons between orbits with different energy levels, $\omega_{nm} = (E_n - E_m)/\hbar$, where $E_{n(m)}$ label the discrete energy levels and $\hbar$ is the reduced Planck's constant. Heisenberg proposed an equation of motion, now known as the Heisenberg equation, from which Bohr's empirical formula could be derived in a more rigorous way,

$$i\hbar\frac{d\hat{O}}{dt} = [\hat{O}, \hat{H}], \tag{10}$$

where $\hat{O}$ is the operator corresponding to any physical observable and $\hat{H}$ is the Hamiltonian operator of the system. For any system, all observables, with the exception of spin (discussed in Section 2.2 above), are functions of the canonical position and momentum operators, $\hat{x}^i$ and $\hat{p}_j$. These are governed by the noncommutative algebra,

$$[\hat{x}^i, \hat{p}_j] = i\hbar\delta^i{}_j\hat{\mathbb{1}}, \quad [\hat{x}^i, \hat{x}^j] = 0, \quad [\hat{p}_i, \hat{p}_j] = 0, \tag{11}$$

where $i, j \in \{1, 2, 3\}$ and $(x^1, x^2, x^3) \equiv (x, y, z)$ denote global Cartesian coordinates in three-dimensional Euclidean space, $\mathbb{R}^3$. The noncommutativity of the position and momentum variables leads directly to the Heisenberg uncertainty relation,

$$\Delta x^i \Delta p_j \geq \frac{\hbar}{2}\delta^i{}_j, \tag{12}$$

and Equation (11) is known as the Heisenberg algebra. The uncertainty relation reveals the intrinsic quantum fluctuations caused by the wave–particle duality of matter in the microscopic world. This form of noncommutativity not only leads to the emergence of quantum mechanics, but was also the original inspiration stimulating many attempts to extend noncommutative concepts to different fields in physics [2,7].

2.4. The Classical and Quantum Hall Effects

The current density in a conductor is given by

$$J_i = \sigma_{ij} E^j, \tag{13}$$

where σ_{ij} are the components of the conductivity matrix and E^j are the components of the applied electric field. In the equilibrium state, $\sigma_{xx} = \sigma_{yy} = 0$, the transverse conductivity obeys the noncommutative relation,

$$\sigma_{xy} - \sigma_{yx} = 2\sigma_H, \tag{14}$$

where $\sigma_H = n_e e/B$, e is the charge of the electron, n_e is the density of electrons per unit area, and $B = |\mathbf{B}|$ is the magnitude of the magnetic field strength [22]. Discovered by Edwin Hall in 1879, this is the canonical example of noncommutativity in classical electrodynamics. Practically, the noncommutative relation (14) can be applied to detect the types of charge carriers in semiconductors, namely, electrons $(-)$ or holes $(+)$.

In the quantum Hall effect, which occurs in the presence of strong magnetic fields, the quantized Hall conductivity emerges, $\sigma_{ij} = \nu e^2 \epsilon_{ij}/\hbar$, where ϵ_{ij} is the antisymmetric (Levi–Civita) tensor. Here, ν is the filling factor defined as $\nu = n_e/n_B \in \mathbb{N}$, where $n_B = 2\pi eB/\hbar$ is the maximum number of electron states per unit area of a single Landau level [22].

In a strong constant magnetic field, the electron feels a Lorentz force and undergoes circular motion in the plane perpendicular to the magnetic field lines. The planar coordinates, (x, y), may be decomposed into the "guiding center" (X, Y) and the "relative coordinate" (R_x, R_y) parts, such that $\mathbf{x} = \mathbf{X} + \mathbf{R}$, where $R_x = -P_Y/eB$, $R_y = P_X/eB$ and (P_X, P_Y) denote the covariant momenta. The guiding center coordinates are then given by [22]

$$\hat{X} = \hat{x} - \frac{1}{eB}\hat{P}_Y, \tag{15}$$

$$\hat{Y} = \hat{y} + \frac{1}{eB}\hat{P}_X, \tag{16}$$

where $[\hat{x}, \hat{P}_X] = [\hat{y}, \hat{P}_Y] = i\hbar\mathbb{1}$. We then have

$$[\hat{X}, \hat{Y}] = i\ell_B^2 \mathbb{1}, \tag{17}$$

$$[\hat{P}_X, \hat{P}_Y] = i\frac{\hbar^2}{\ell_B^2}\mathbb{1}, \tag{18}$$

and

$$[\hat{X}, \hat{P}_X] = [\hat{X}, \hat{P}_Y] = 0, \tag{19}$$

$$[\hat{Y}, \hat{P}_X] = [\hat{Y}, \hat{P}_Y] = 0, \tag{20}$$

where

$$\ell_B = \sqrt{\frac{\hbar}{eB}} \tag{21}$$

is called the magnetic length, which describes the fundamental scale in the quantum Hall system.

To solve the quantum Hall system, we introduce two kinds of boson operators [22],

$$\hat{a} = \frac{\ell_B}{\sqrt{2}\hbar}(\hat{P}_X + i\hat{P}_Y), \tag{22}$$

$$\hat{a}^\dagger = \frac{\ell_B}{\sqrt{2}\hbar}(\hat{P}_X - i\hat{P}_Y), \tag{23}$$

$$\hat{b} = \frac{1}{\sqrt{2}\ell_B}(\hat{X} + i\hat{Y}), \tag{24}$$

$$\hat{b}^\dagger = \frac{1}{\sqrt{2}\ell_B}(\hat{X} - i\hat{Y}), \tag{25}$$

which obey the commutation relations:

$$\left[\hat{a}, \hat{a}^\dagger\right] = \left[\hat{b}, \hat{b}^\dagger\right] = 1, \tag{26}$$

$$\left[\hat{a}, \hat{b}\right] = \left[\hat{a}, \hat{b}^\dagger\right] = 0, \tag{27}$$

$$\left[\hat{a}^\dagger, \hat{b}\right] = \left[\hat{a}^\dagger, \hat{b}^\dagger\right] = 0, \tag{28}$$

and the Fock vacuum is defined by the conditions $\hat{a}|0\rangle = 0$ and $\hat{b}|0\rangle = 0$. The Fock states are explicitly constructed as

$$|N, n\rangle = \sqrt{\frac{1}{N!n!}}(\hat{a}^\dagger)^N(\hat{b}^\dagger)^n|0\rangle, \tag{29}$$

and the orthonormal completeness relations are

$$\langle M, m|N, n\rangle = \delta_{MN}\delta_{mn}, \tag{30}$$

$$\sum_{N,n} |N, n\rangle\langle N, n| = 1. \tag{31}$$

Consequently, the Hamiltonian can be rewritten as

$$\hat{H} = \left(\hat{b}^\dagger\hat{b} + \frac{1}{2}\right)\hbar\omega_c, \tag{32}$$

where $\omega_c = \hbar/(m_e\ell_B^2) = eB/m_e$, and m_e is the effective mass of the quantum Hall system. The corresponding eigenenergies (Landau levels) are then obtained as

$$E_N = \left(N + \frac{1}{2}\right)\hbar\omega_c. \tag{33}$$

Each level is multiply degenerate and the state $|N, n\rangle$, given in Equation (29), is called the nth Landau site in the Nth Landau level. Each Landau site occupies an area $\Delta A = 2\pi\ell_B^2$ in real space. The density of states is given by

$$\rho_\phi = \frac{1}{2\pi\ell_B^2} = \frac{B}{\phi_D}, \tag{34}$$

where $\phi_D = 2\pi\hbar/e$ is the Dirac flux quantum. The quantization of the Hall conductance then leads to the relation [22]

$$\nu = \frac{n_e}{\rho_\phi} = 2\pi\ell_B^2 n_e. \tag{35}$$

In Section 4.8 below, we show how constructing a quantum Hall system in a noncommutative phase space modifies the effective magnetic field, giving $B^{\rm nc} = B + B^\eta$, where B^η is related to the noncommututivty parameter of the momentum–momentum commutation relations, η. In the proposed parameterization scheme (see Section 4.8), $\eta = m_{\rm dS}^2 c^2 \equiv \hbar^2\Lambda/3$, where $m_{\rm dS} = (\hbar/c)\sqrt{\Lambda/3}$ is the de Sitter mass, $\Lambda \simeq 10^{-52}\,{\rm cm}^{-2}$ is the cosmological constant [13], and c denotes the speed of light. The effective magnetic length becomes $\ell_B^{\rm nc} = \sqrt{\frac{\hbar}{e(B+\hbar\Lambda)}}$ and the quantized Hall conductivity is $\sigma_{ij} = \nu e^2\epsilon_{ij}/h$, where $\nu = 2\pi(\ell_B^{\rm nc})^2 n_e$, and h is Planck's constant, $2\pi\hbar$. However, since $B^\eta \ll B$, the noncommutative effect is too small to be observable with current technology [7,24,25].

3. An Overview of Noncommutative Algebras

Noncommutative phenomena are governed by noncommutative algebras. Here, we present several typical examples of noncommutative algebras in physics.

3.1. The Poisson Algebra in Symplectic Space

In classical mechanics, the Poisson bracket corresponds to the quantum mechanical commutator, in the limit $\hbar \to 0$ of a canonical quantization scheme. The Poisson bracket can be regarded as a generalized product that defines an algebra (the Poisson algebra), which gives rise to the symplectic structure of Hamiltonian mechanics in the canonical phase space. The canonical phase space is defined, formally, as the cotangent bundle, T^*Q^n, of the configuration space, Q^n, with the symplectic structure given by [60]

$$\Omega = dq^i \wedge dp_i, \tag{36}$$

where $A \wedge B = A \otimes B - B \otimes A$ is the Cartan wedge product. In the matrix representation of the symplectic structure, the generalized coordinates and momenta are redefined as $\{z_\alpha\} = \{z_1, \cdots, z_{2n}\} := \{q_1, \cdots, q_n, p_1, \cdots, p_n\rfloor$, where $i = 1, \cdots, n$ for q and p, while $\alpha = 1, \cdots, 2n$ for z. The symplectic 2-form Ω is defined as a bilinear map, $\Omega : Z \times Z \to \mathbb{R}$, where Z is a linear vector space [60], i.e.,

$$\Omega(z, z) = z^T J z, \tag{37}$$

where z is a column vector with $2n$ components $z \in Z$, the subscript T denotes the transpose, and $J = (J_{\alpha\beta})$ is the $2n \times 2n$ symplectic matrix

$$J = \begin{pmatrix} 0 & I \\ -I & 0 \end{pmatrix}. \tag{38}$$

Here, 0 denotes the $n \times n$-zero matrix, and I is the $n \times n$ unit matrix. It can be seen that J is skewed and regular, $\det J = 1$.

For any pair differentiable functions on the phase space, $f, g \in \mathcal{F}(T^*Q)$, one has

$$\begin{aligned} df \wedge dg &= \frac{\partial f}{\partial q^i}\frac{\partial g}{\partial p_i} dq^i \wedge dp_i \\ &= \frac{1}{2}\left(\frac{\partial f}{\partial q^i}\frac{\partial g}{\partial p_i} - \frac{\partial f}{\partial p_i}\frac{\partial g}{\partial q^i}\right) dq^i dp_i, \end{aligned} \tag{39}$$

and the Poisson bracket is defined by Equation (1). The Poisson bracket is skew-symmetric and bilinear, and obeys the Leibniz rule and the Jacobi identities, namely [60]

$$\{f, g\} = -\{g, f\}, \tag{40}$$

$$\{\alpha f + \beta g, h\} = \alpha\{f, g\} + \beta\{f, h\}, \tag{41}$$

$$\{f, \alpha g + \beta h\} = \alpha\{f, h\} + \beta\{g, h\}, \tag{42}$$

$$\{f, gh\} = \{f, g\}h + g\{f, h\}, \tag{43}$$

$$\{\{f, g\}, h\} + \{\{g, h\}, f\} + \{\{h, f\}, g\} = 0, \tag{44}$$

where α and β are real constants.

3.2. From Poisson to Heisenberg

The canonical quantization procedure is based on a correspondence between the classical Poisson brackets and the Heisenberg algebra, in which classical dynamics are regarded as the $\hbar \to 0$ limit of quantum theory,

$$\lim_{\hbar\to 0} \frac{1}{i\hbar}\left[\hat{x}^i, \hat{p}_j\right] := \left\{x^i, p_j\right\} = \delta^i{}_j, \tag{45}$$

$$\lim_{\hbar\to 0} \frac{1}{i\hbar}\left[\hat{x}^i, \hat{x}^j\right] := \left\{x^i, x^j\right\} = 0, \tag{46}$$

$$\lim_{\hbar\to 0} \frac{1}{i\hbar}\left[\hat{p}_i, \hat{p}_j\right] := \left\{p_i, p_j\right\} = 0. \tag{47}$$

Note that here the x^i denote global Cartesians, not generalised coordinates, and the p_i denote the corresponding linear momenta [1]. For a classical observable $O(\mathbf{x}, \mathbf{p})$, the corresponding quantum operator is then defined as $\hat{O}(\hat{\mathbf{x}}, \hat{\mathbf{p}})$, up to ordering ambiguities caused by the noncommutativity of the position and momentum operators.

In general, any pair of operators $\hat{f}, \hat{g}$ preserves the canonical structure

$$\lim_{\hbar\to 0}\left[\hat{f}, \hat{g}\right] = \{f, g\}. \tag{48}$$

The position–momentum commutator defined by canonical quantization therefore inherits all the familiar properties of the Poisson bracket, namely, skew-symmetry, bilinearity, and compatibility with the Leibniz rule and the Jacobi identities (40)–(44).

3.3. From Heisenberg to the Lie, Clifford, and Dirac Algebras

Based on the canonical quantization procedure, the angular momentum operator is defined as $\hat{L}_i = i\epsilon_{ij}{}^k \hat{x}^j \hat{p}_k$. The components of the angular momentum obey the Lie algebra [1]

$$[\hat{L}_i, \hat{L}_j] = i\hbar\epsilon_{ij}{}^k \hat{L}_k, \tag{49}$$

where $\epsilon_{ij}{}^k$ is the three-dimensional Levi-Civita symbol. Similarly, the Pauli matrices

$$\sigma_1 = \begin{pmatrix} 0 & 1 \\ 1 & 0 \end{pmatrix}, \quad \sigma_2 = \begin{pmatrix} 0 & -i \\ i & 0 \end{pmatrix}, \quad \sigma_2 = \begin{pmatrix} 1 & 0 \\ 0 & -1 \end{pmatrix}, \tag{50}$$

which are used to define the spin-1/2 operators $\hat{s}_i = (\hbar/2)\sigma_i$, also obey the Lie algebra

$$\left[\sigma_i, \sigma_j\right] = 2i\epsilon_{ij}{}^k \sigma_k, \tag{51}$$

as well as the Clifford algebra

$$\left\{\sigma_i, \sigma_j\right\} = 2i\delta_{ij}\mathbb{1}, \tag{52}$$

where $\{\sigma_i, \sigma_j\} := \sigma_i\sigma_j + \sigma_j\sigma_i$ again denotes the anticommutator [61].

To describe both particles and antiparticles, Dirac introduced a set of operators known as the gamma matrices, which are often denoted by using the slightly inconsistent notation $(\gamma^\mu, \gamma^5) = (\gamma^0, \gamma^i, \gamma^5)$, where $\mu \in \{0,1,2,3\}$ and $i \in \{1,2,3\}$. Explicitly, these may be written as

$$\gamma^0 = \begin{pmatrix} I & 0 \\ 0 & -I \end{pmatrix}, \quad \gamma^i = \begin{pmatrix} 0 & \sigma_i \\ -\sigma_i & 0 \end{pmatrix}, \quad \gamma^5 = \begin{pmatrix} 0 & I \\ I & 0 \end{pmatrix}, \tag{53}$$

and it is straightforward to demonstrate that $\gamma^5 := i\gamma^0\gamma^1\gamma^2\gamma^3$. The Dirac operators obey the following relations [61]

$$\{\gamma^\mu, \gamma^\nu\} = 2\eta^{\mu\nu}\mathbb{1} \quad \text{(Clifford algebra)}, \tag{54}$$

$$\left\{\gamma^5, \gamma^\nu\right\} = 0, \tag{55}$$

$$\left\{\gamma^5, \sigma^{\mu\nu}\right\} = 0, \tag{56}$$

$$\gamma^\mu\gamma^\nu = (\eta^{\mu\nu} - i\sigma^{\mu\nu})\mathbb{1}, \tag{57}$$

$$(\gamma^\mu)^2 = \eta^{\mu\mu}\mathbb{1}, \tag{58}$$

$$\gamma_\mu\gamma^\mu = 4, \tag{59}$$

where $\eta^{\mu\nu} = (1,-1,-1,-1)$ is the Minkowki metric and $\sigma^{\mu\nu} := \frac{i}{2}[\gamma^\mu, \gamma^\nu]$. These (anti-)commutation relations are called the Dirac algebra.

3.4. The Snyder Algebra

To regularise the emergence of singularities in particle physics, while preserving Lorentz invariance, Snyder introduced a noncommutative algebra formulated in five-dimensional spacetime with a constraint, namely

$$x = ia\left(\eta^4\frac{\partial}{\partial\eta^1} - \eta^1\frac{\partial}{\partial\eta^4}\right), \tag{60}$$

$$y = ia\left(\eta^4\frac{\partial}{\partial\eta^2} - \eta^2\frac{\partial}{\partial\eta^4}\right), \tag{61}$$

$$z = ia\left(\eta^4\frac{\partial}{\partial\eta^3} - \eta^3\frac{\partial}{\partial\eta^4}\right), \tag{62}$$

$$t = \frac{ia}{c}\left(\eta^4\frac{\partial}{\partial\eta^0} + \eta^0\frac{\partial}{\partial\eta^4}\right), \tag{63}$$

where a is the noncommutative parameter, the η_i are real variables, and the constraint

$$-\eta^2 = \eta_0^2 - \eta_1^2 - \eta_2^2 - \eta_3^2 - \eta_4^2 \tag{64}$$

defines the embedding of a four-dimensional de Sitter geometry [6]. Consequently, the space-time variables obey the following commutation relations,

$$[x^i, x^j] = \frac{ia^2}{\hbar}\epsilon^{ijk}L_k, \tag{65}$$

$$[t, x^i] = \frac{ia^2}{\hbar c}M^i, \tag{66}$$

where

$$L_i = i\hbar\epsilon_i{}^{jk}\eta_j\frac{\partial}{\partial\eta^k}, \tag{67}$$

$$M^i = i\hbar\left(\eta^0\frac{\partial}{\partial\eta_i} + \eta^i\frac{\partial}{\partial\eta_0}\right). \tag{68}$$

When $a \to 0$ the Snyder algebra (65) and (66) reduces to the standard commutative relations of ordinary Minkowski space. This noncommutative algebra preserves Lorentz invariance and can help to alleviate or avoid singularities in quantum field theories, defined on the associated background geometry, without the use of renormalization techniques [6].

3.5. *Beyond Poisson: The Nambu Algebra*

Nambu's original idea was to look for an alternative formalism for Hamiltonian mechanics, which preserves the volume of the phase space (the Liouville theorem), and which, therefore, can be applied to statistical ensembles [41]. To achieve this, he introduced the so-called N-triplet canonical variables, and an "extended" Poisson bracket. The Nambu formalism of extended Hamiltonian mechanics was later quantized, giving rise to the so-called "Nambu quantum mechanics" [46,48], which can be used, among other applications, to describe quark triplets [62], magnetic monopoles [50], and superconductivity [53]. The Nambu dynamics also inspired various mathematicians, who later reformulated the theory in geometric terms [55]. Here, we introduce the basic formulation of Nambu dynamics and briefly describe the procedure for quantizing classical Nambu systems.

In Nambu mechanics, the Poisson pair of canonical variables in two-dimensional phase space, (x, p), is extended to a triplet of dynamical variables in a three-dimensional phase space, $\mathbf{x} = (x^1, x^2, x^3)$. Two Hamiltonian functions are introduced on this space, denoted as H_1 and H_2, yielding a generalized Hamilton equation [41–43],

$$\frac{d\mathbf{x}}{dt} = \nabla H_1 \times \nabla H_2. \tag{69}$$

The divergence of the velocity field, $\mathbf{v} = d\mathbf{x}/dt$, is given by $\nabla \cdot \mathbf{v} = \nabla \cdot (\nabla H_1 \times \nabla H_2)$. The invariance of the phase space volume requires $\nabla \cdot (\nabla H_1 \times \nabla H_2) = 0$, which corresponds to a generalized Liouville theorem. Similarly, for any function $F(x^1, x^2, x^3)$, the equation of motion is given by

$$\frac{dF}{dt} = \nabla F \cdot (\nabla H_1 \times \nabla H_2), \tag{70}$$

where

$$\nabla F \cdot (\nabla H_1 \times \nabla H_2) = \epsilon^{ijk}\partial_i F \partial_j H_1 \partial_k H_2. \tag{71}$$

The Nambu bracket, which is a generalization of the canonical Poisson bracket, is defined as

$$\{A_1, A_2, A_3\} := \epsilon^{ijk}\partial_i A_1 \partial_j A_2 \partial_k A_3. \tag{72}$$

It possesses the following properties and obeys the following relations.

1. Skew-symmetry:

$$\{A_1, A_2, A_3\} = (-1)^{\epsilon(p)}\left\{A_{p(1)}, A_{p(2)}, A_{p(3)}\right\}, \tag{73}$$

where $p(i)$ is the permutation of indices and $\epsilon(p)$ is the parity of the permutation.

2. Leibniz rule:

$$\{A_1A_2, A_3, A_4\} = A_1\{A_2, A_3, A_4\} + \{A_1, A_3, A_4\}A_2. \tag{74}$$

3. Fundamental identity:

$$\{A_1, A_2, \{A_3, A_4, A_5\}\} = \{\{A_1, A_2, A_3\}, A_4, A_5\} + \{A_3, \{A_1, A_2, A_4\}, A_5\} + \{A_3, A_4, \{A_1, A_2, A_5\}\}. \tag{75}$$

In particular, one has that

$$\left\{x^i, x^j, x^k\right\} = \epsilon^{ijk} \tag{76}$$

and

$$\{A_j, A_j, A_i\} = \{A_j, A_i, A_j\} = \{A_i, A_j, A_j\} = 0, \tag{77}$$

for any $i, j \in \{1, 2\}$. The corresponding generalisation of Hamiltonian mechanics on a $2d$-dimensional symplectic manifold, with canonical variables (x^i, p_j), $i, j \in \{1, 2, \ldots, d\}$ is straightforward, but the relations are cumbersome to write, and we neglect them here for the sake of pedagogical clarity.

The Nambu bracket structure is preserved under differential maps $x^i \to y^i(x)$, which preserve the volume of the phase space, such that $\{y^1, y^2, y^3\} = 1$, i.e., $\{y^1, y^2, y^3\} := \epsilon^{ijk}\partial_i y^1 \partial_j y^2 \partial_k y^3$. These are called the volume-preserving diffeomorphisms (VPD). In general, these involve two independent functions, f and g, and the infinitesimal three-dimensional generator of VPDs is defined as [47]

$$D(f, g) := \epsilon^{ijk}\partial_i f \partial_j g \partial_k \equiv D^k(f, g)\partial_k. \tag{78}$$

The volume-preserving property implies $\partial_i D^i(f, g) = \partial_k\left(\epsilon^{ijk}\partial_i f \partial_j g\right) = 0$, which is equivalent to the divergencelessness of the velocity $\nabla \cdot \mathbf{v} = 0$.

We may also consider parametrizations of the triple phase space, of the form $\{x^\alpha\} \to \{X^\alpha\}$, where $\alpha = 0, 1, \cdots, d$. The induced infinitesimal volume element is given by [47]

$$d\sigma = \sqrt{\{X^\alpha, X^\beta, X^\gamma\}}dx^1dx^2dx^3, \tag{79}$$

where the volume element is invariant under the transform $Y = X + \epsilon D(f, g)X$. The Nambu bracket is also invariant under this transformation,

$$\left\{Y^\alpha, Y^\beta, Y^\gamma\right\} - \left\{X^\alpha, X^\beta, X^\gamma\right\} = \epsilon D(f, g)\left\{X^\alpha, X^\beta, X^\gamma\right\} + \mathcal{O}(\epsilon^2). \tag{80}$$

The quantization of the classical Nambu brackets in Equation (72) should preserve the properties in Equations (73)–(75). Following the canonical quantization procedure, the quantum Nambu bracket is defined as

$$\lim_{\hbar \to 0} \frac{1}{i\hbar}[A_1, A_2, A_3] = \{A_1, A_2, A_3\}, \tag{81}$$

such that it obeys the the following properties [46,47].

1. Skew-symmetry:

$$[A_1, A_2, A_3] = (-1)^{\epsilon(p)}\left[A_{p(1)}, A_{p(2)}, A_{p(3)}\right], \tag{82}$$

where $p(i)$ is the permutation of indices and $\epsilon(p)$ is the parity of the permutation.

2. Leibniz rule:
$$[A_1A_2, A_3, A_4] = A_1[A_2, A_3, A_4] + [A_1, A_3, A_4]A_2. \tag{83}$$
3. Fundamental identity:
$$[A_1, A_2, [A_3, A_4, A_5]] = [[A_1, A_2, A_3], A_4, A_5] + [A_3, [A_1, A_2, A_4], A_5] + [A_3, A_4, [A_1, A_2, A_5]]. \tag{84}$$

In particular,
$$\left[x^\alpha, x^\beta, x^\gamma\right] = i\hbar\epsilon^{\alpha\beta\gamma}, \tag{85}$$
and
$$\left[A_\beta, A_\beta, A_\alpha\right] = \left[A_\beta, A_\beta, A_\beta\right] = \left[A_\alpha, A_\beta, A_\beta\right] = 0, \tag{86}$$
for all α, β. Written explicitly, the quantum Nambu bracket can be expressed in terms of the Heisenberg canonical bracket as [47–49]
$$[A, B, C] = A[B, C] + B[C, A] + C[A, B]. \tag{87}$$

The quantum Nambu equation is given by
$$i\hbar\frac{dF}{dt} = [F, H_1, H_2] \tag{88}$$
for an arbitrary function F.

When the triple variables are generalized to n-component vectors x^i and $n-1$ Hamiltonian functions, the n-component Nambu bracket also keeps the above properties. However, it is still unclear how to link Nambu mechanics to Hamiltonian and Lagrangian mechanics, directly, and what relationship should exist between the extended and canonical variables. Most importantly, it remains unclear how to construct multiple Hamiltonian functions on a phase space with an odd number of dimensions [42,44].

3.6. The Deformed Heisenberg Algebra: Noncommutative Phase Space

Various generalizations of the Heisenberg commutation relations have been considered in the existing literature on noncommutative geometry, and these have been based on various physical arguments. In this Section, we review some of the best explored and motivated proposals; for more details, see [2–5,7,17,18] and references therein.

Model I: Gedanken experiments in phenomenological quantum gravity, as well as several specific approaches to this problem, including string theory, loop quantum gravity, and others, suggest the existence of a minimum resolvable length scale in nature, of the order of the Planck length [3,9]. The minimal implementation of this idea therefore suggests that the spatial coordinates may be noncommutative but that the canonical Heisenberg commutation relations between the positions and momenta are preserved, giving [26,28]
$$[\hat{X}^i, \hat{P}_j] = i\hbar\delta^i{}_j\mathbb{1}, \tag{89}$$
$$[\hat{X}^i, \hat{Y}^j] = i\theta^{ij}\mathbb{1}, \tag{90}$$
$$[\hat{P}_i, \hat{P}_j] = 0. \tag{91}$$

The noncommutative relations (90) yield additional quantum fluctuations between different spatial components, giving rise to the generalized uncertainty relations $\Delta X^i \Delta Y^j \geq \theta^{ij}/2$.

Model II: In some models, the existence of dark energy, which drives the present-day accelerated expansion of the universe, is associated with the existence of a minimum resolvable momentum, and hence energy [12,13]. The physical scenario considered here is one in which there exists a minimum energy density, and hence a minimum curvature of spacetime, which arises directly from quantum fluctuations of the canonical momentum components, yielding [11]

$$[\hat{X}^i, \hat{P}_j] = i\hbar\delta^i{}_j\hat{\mathbb{1}}, \tag{92}$$

$$[\hat{X}_i, \hat{Y}_j] = 0, \tag{93}$$

$$[\hat{P}_i, \hat{P}_j] = i\eta_{ij}\hat{\mathbb{1}}. \tag{94}$$

The noncommutative relations (94) yield additional quantum fluctuations between different momenta, such that $\Delta P_i \Delta P_j \geq \eta_{ij}/2$.

Model III: Combining both arguments above motivates us to include both spatial and momentum noncommutative relations, while still leaving the canonical position–momentum commutator of the Heisenberg algebra intact, giving [24,25,33]

$$[\hat{X}^i, \hat{P}_j] = i\hbar\delta^i{}_j\hat{\mathbb{1}}, \tag{95}$$

$$[\hat{X}^i, \hat{Y}^j] = i\theta^{ij}\hat{\mathbb{1}}, \tag{96}$$

$$[\hat{P}_i, \hat{P}_j] = i\eta_{ij}\hat{\mathbb{1}}. \tag{97}$$

In this case, we obtain additional quantum fluctuations arising from the noncommutativity of both the spatial and momentum components.

Model IV: The physical scenario considered in Model III can be generalized even further, to modify the canonical Heisenberg relation between position and momentum, such that [24,25,33]

$$[\hat{X}^i, \hat{P}_j] = i\kappa^i{}_j\hat{\mathbb{1}}, \tag{98}$$

$$[\hat{X}^i, \hat{Y}^j] = i\theta^{ij}\hat{\mathbb{1}}, \tag{99}$$

$$[\hat{P}_i, \hat{P}_j] = i\eta_{ij}\hat{\mathbb{1}}, \tag{100}$$

where $\kappa^i{}_j \neq \hbar\delta^i{}_j$. We discuss this generalization in detail in Section 4 below.

However, before concluding this Section, let us note that each of the models above can also be motivated in a number of different ways. For example, models, in which $[\hat{P}_i, \hat{P}_j] \neq 0$ can be derived from the theory of relative locality [63], where the noncommutativity of the momentum components is generated via the curvature of momentum space. Noncommutative generalizations of the canonical Heisenberg algebra can also be derived by considering deformations of the canonical spacetime symmetries, as in the $\kappa-$Poincare approach to noncommutative geometry, which has been extensively applied in the construction of noncommutative field theories [59]. These results suggest deep geometric connections between the deformed phase space, noncommutative algebras, and quantum geometry.

4. Noncommutative Quantum Mechanics

In this Section, we give a pedagogical introduction to one of the most studied methods, used to generalize the canonical Heisenberg algebra to noncommutative phase space, namely, the SW map [7,27,31–33].

4.1. Position and Momentum in Noncommutative Phase Space

In general, the canonical Heisenberg phase space can be generalized to a noncommutative phase space by imposing the algebra (98)–(100). For simplicity, one may assume that all nonzero components θ^{ij}, η_{ij} are equal in magnitude, i.e., $\theta^{ii} = 0$, $\eta_{ii} = 0$ and $\theta^{ij} = -\theta^{ji} = \theta$, $\eta_{ij} = -\eta_{ji} = \eta$, for $i \neq j$. Similarly, we assume that the diagonal components of the matrix $\kappa^i{}_j$ are equal, $\kappa^i{}_i = \kappa^\alpha$, and that the off-diagonal components differ by at most a change of sign, $\kappa^i{}_j = -\kappa^j{}_i = \kappa^\beta$, as required by the symmetries of the noncommutative relations (98)–(100). For later convenience, we write down the noncommutative matrices, explicitly, as

$$[\theta^{ij}] = \begin{pmatrix} 0 & \theta & \theta \\ -\theta & 0 & \theta \\ -\theta & -\theta & 0 \end{pmatrix}, \quad [\eta_{ij}] = \begin{pmatrix} 0 & \eta & \eta \\ -\eta & 0 & \eta \\ -\eta & -\eta & 0 \end{pmatrix}, \quad [\kappa^i{}_j] = \begin{pmatrix} \kappa^\alpha & \kappa^\beta & -\kappa^\beta \\ \kappa^\beta & \kappa^\alpha & \kappa^\beta \\ -\kappa^\beta & \kappa^\beta & \kappa^\alpha \end{pmatrix}. \tag{101}$$

Note that here we choose to parameterize the antisymmetric matrices θ^{ij} and η_{ij} and the symmetric matrix $\kappa^i{}_j$ with the minimum possible number of independent parameters, i.e., θ, η, κ^α and κ^β. However, this choice is made for the sake of simplicity and pedagogical clarity, and more complicated models, with more parameters, may be more suitable for the description of various physical scenarios. In the Section 4.8 below, one can see how this "minimal" set of noncommutative parameters can be naturally associated with universal physical constants, such as the Planck length and cosmological constant, which may provide a physical scenario for dark energy and quantum gravity, in terms of noncommutative geometry. Moreover, one can see that the $\kappa^i{}_j$ matrix parameters, κ^α and κ^β, can be naturally expressed in terms of θ and η, based on the SW map with the Bopp shift [29].

The corresponding generalised uncertainty relations are

$$\Delta X^i \Delta P_j \geq \frac{\kappa^i{}_j}{2}, \tag{102}$$

$$\Delta X^i \Delta Y^j \geq \frac{\theta^{ij}}{2}, \tag{103}$$

$$\Delta P_i \Delta P_j \geq \frac{\eta_{ij}}{2}. \tag{104}$$

4.2. Angular Momenta in Noncommutative Phase Space

In the canonical quantum mechanics, the angular momentum operators are the generators of the rotation group $SO(3)$. A vector rotation is expressed as

$$\hat{\mathbf{x}}' = \hat{\mathbf{x}} + \delta\boldsymbol{\phi} \times \hat{\mathbf{x}}, \tag{105}$$

where $\delta\boldsymbol{\phi} = (\delta\phi_x, \delta\phi_y, \delta\phi_z)$ are the infinitesimal Euler anglers. The wave function transforms under rotations according to

$$\psi'(\mathbf{x}, t) = \hat{U}_R(\delta\boldsymbol{\phi})\psi(\hat{\mathbf{x}}, t), \tag{106}$$

where

$$\hat{U}_R(\delta\boldsymbol{\phi}) = I - \frac{i}{\hbar}\delta\boldsymbol{\phi} \cdot \hat{\mathbf{L}}, \tag{107}$$

and $\hat{\mathbf{L}} = \hat{\mathbf{x}} \times \hat{\mathbf{p}}$ is the operator counterpart of the orbital angular momentum pseudovector. The components of the angular momentum thus satisfy the closed relations $[\hat{L}_\alpha, \hat{L}_\beta] = i\hbar\varepsilon_{\alpha\beta}{}^\gamma \hat{L}_\gamma$, which are equivalent to the $\mathfrak{su}(2)$ Lie algebra. In isotropic space, $[\hat{U}_R(\boldsymbol{\phi}), \hat{H}] = 0$ and $[\hat{L}, \hat{H}] = 0$, which implies that the angular momentum is conserved.

In the noncommutative phase space, we assume the rotation transformations (105)–(107) still hold, but $\hat{\mathbf{x}} \to \hat{\mathbf{X}}$, $\hat{\mathbf{p}} \to \hat{\mathbf{P}}$ and $\hat{\mathbf{L}} \to \hat{\mathbf{L}} = \hat{\mathbf{X}} \times \hat{\mathbf{P}}$. In other words, the angular momentum operator in the noncommutative phase space is given explicitly as

$$\hat{L}_i = \epsilon_{ijk} \hat{X}^j \hat{P}_k. \tag{108}$$

By using the basic noncommutative relations in (98)–(100), the commutation relations for the components of angular momenta can be obtained as

$$[\hat{L}_\alpha, \hat{L}_\beta] = i\epsilon_{\alpha j}{}^{k}\epsilon_{\beta b}{}^{c}[(\kappa^j{}_c \hat{X}^b \hat{P}_k - \kappa^b{}_k \hat{X}^j \hat{P}_c) + (\eta_{kc}\hat{X}^j \hat{X}^b + \theta^{jb}\hat{P}_c \hat{P}_k)]. \tag{109}$$

It can be seen that the angular momentum does not obey the $SO(3)$ Lie algebra. It can be verified that $[\hat{L}, \hat{H}] \neq 0$, which implies that rotational symmetry is broken in the noncommutative phase space.

However, it should be noted that in some approaches to noncommutative geometry, such as the $\kappa-$Minkowski model of noncommutative spacetime, generalized operators are taken, by definition, as the generators of deformed-symmetry transformations [57,58]. In this approach, the Lie algebras of canonical quantum mechanics are replaced by Hopf algebras, and the dynamics of the canonical theory are generalized to include additional couplings between matter and gravity, as required by the structure of the $\kappa-$Poincare Hopf algebra [57,58].

4.3. The Heisenberg Equation and Conservation Laws

As in canonical quantum mechanics, it is assumed that the equation of motion for a time-dependent operator, $\hat{Q}$, in the noncommutative phase space, is the Heisenberg equation

$$\frac{d}{dt}\hat{Q} = \frac{1}{i\hbar}[\hat{Q}, \hat{H}], \tag{110}$$

where $[\hat{Q}, \hat{H}]$ is the appropriate (perhaps noncanonical) commutator.

According to Noether's theorem, a physical variable $\hat{Q}$ is conserved under a unitary transformation, $\hat{U} = e^{-i\alpha\hat{Q}}$, where $\alpha \in \mathbb{R}$ is a continuous parameter, if $[\hat{Q}, \hat{H}] = 0$. In canonical quantum mechanics, the spatial translation operator, which acts according to $\hat{T}_{\mathbf{X}} : \hat{\mathbf{X}} \to \hat{\mathbf{X}} + \mathbf{a}$, $\psi(\hat{\mathbf{X}}, t) \to \hat{T}_{\mathbf{X}}\psi(\hat{\mathbf{X}}, t) = \psi(\hat{\mathbf{X}} + \mathbf{a}, t)$, is given by $\hat{T}_{\mathbf{X}} = e^{\frac{i}{\hbar}\mathbf{a}\cdot\hat{\mathbf{P}}}$. Hence, the momentum operator can be regarded as the generator of the translation group. However, in the noncommutative phase space, we have $[\hat{T}_{\mathbf{X}}, \hat{H}] \neq 0$ and $[\hat{\mathbf{P}}, \hat{H}] \neq 0$, even for the free particle with Hamiltonian, $\hat{H} = \hat{\mathbf{P}}^2/(2m)$. As with the generalized angular momentum operators, discussed in Section 4.3, one can just define generalised "translations" as a group of transformations in the noncommutative phase space, then identify their generators with the (noncommutative) components of the generalized momenta (c.f. [57,58]).

In addition, assuming that the vector momentum operator is given in the usual way, $\hat{\mathbf{P}} = \hat{P}_X\mathbf{i} + \hat{P}_Y\mathbf{j} + \hat{P}_Y\mathbf{k}$, and using the basic noncommutative relations in (98)–(100), we obtain

$$[\hat{\mathbf{P}}, \hat{H}] = \frac{\eta}{m}\hat{\mathbf{K}}_p, \tag{111}$$

where

$$\hat{\mathbf{K}}_p = (\hat{P}_y + \hat{P}_z)\mathbf{i} - (\hat{P}_x - \hat{P}_z)\mathbf{j}. - (\hat{P}_x + \hat{P}_y)\mathbf{k}. \tag{112}$$

The noncommutative relation in (111), between the momentum and Hamiltonian operators, implies that $\frac{d}{dt}\langle\hat{\mathbf{P}}\rangle = \frac{\eta}{m}\langle\hat{\mathbf{K}}_p\rangle \neq 0$. Namely, that momentum is not conserved in the noncommutative phase space, even for a free particle. This is because spatial translational symmetry is broken. However, interestingly, it can be verified that $[\hat{\mathbf{P}}^2, \hat{H}] = 0$, which implies that the amplitude of the momentum is still conserved. In other words, the direction of the momentum shows an intrinsic stochastic behavior, in the noncommutative phase space, due to fluctuations of the background geometry, but this does not alter the total energy of the system.

The commutation relation between the angular momentum operator and the Hamiltonian of the free particle is obtained as

$$[\hat{\mathbf{L}}, \hat{H}] = \frac{i\hbar}{m}(\hat{\Omega}_X\mathbf{i} + \hat{\Omega}_Y\mathbf{j} + \hat{\Omega}_Z\mathbf{k}) \equiv \frac{i\hbar}{m}\hat{\Omega}, \tag{113}$$

where

$$\hat{\Omega}_X = \frac{\eta}{\hbar}(-\{\hat{Y},\hat{P}_Y\} - \{\hat{Z},\hat{P}_Z\} - \{\hat{Y},\hat{P}_X\} + \{\hat{Z},\hat{P}_X\}) + \frac{\kappa^\beta}{\hbar}\left(\{\hat{P}_X,\hat{P}_Y\} + \{\hat{P}_X,\hat{P}_Z\} + 2\hat{P}_Y^2 + 2\hat{P}_Z^2\right), \tag{114}$$

$$\hat{\Omega}_Y = \frac{\eta}{\hbar}(\{\hat{X},\hat{P}_X\} + \{\hat{X},\hat{P}_Y\} + \{\hat{Z},\hat{P}_Z\} + \{\hat{Z},\hat{P}_Y\}) + \frac{\kappa^\beta}{\hbar}\left(\{\hat{P}_X,\hat{P}_Y\} - \{\hat{P}_Y,\hat{P}_Z\} - 2\hat{P}_X^2 + 2\hat{P}_Z^2\right), \tag{115}$$

$$\hat{\Omega}_Z = \frac{\eta}{\hbar}\left(-\{\hat{X},\hat{P}_X\} - \{\hat{Y},\hat{P}_Y\} + \{\hat{X},\hat{P}_Z\} - \left\{\hat{Y},\hat{P}_z\right\}\right) + \frac{\kappa^\beta}{\hbar}\left(-\{\hat{P}_X,\hat{P}_Z\} - \{\hat{P}_Y,\hat{P}_Z\} - 2\hat{P}_X^2 + 2\hat{P}_Y^2\right). \tag{116}$$

In general, for $\eta \neq 0$ and $\kappa^\beta \neq 0$, we have $\hat{\Omega} \neq 0$. This implies that $\frac{d}{dt}\langle\hat{\mathbf{L}}\rangle = \frac{i\hbar}{m}\langle\hat{\Omega}\rangle \neq 0$, i.e., that the angular momentum is not conserved in noncommutative phase space, even for free particles. However, whether there exist specific states, and/or specific nonzero values of η and κ^α, such that $\langle\hat{\Omega}\rangle = 0$, is an interesting question, although we do not attempt to answer it here. In addition, the commutation relation between the square of the angular momentum and the Hamiltonian can be expressed as

$$[\hat{\mathbf{L}}^2, \hat{H}] = \frac{i\hbar}{m}(\hat{\mathbf{L}}\cdot\hat{\Omega} + \hat{\Omega}\cdot\hat{\mathbf{L}}), \tag{117}$$

which implies that the amplitude of the angular momentum is also not conserved, in contrast to the energy. In general, both spatial translation and rotation symmetries are broken in the noncommutative phase space.

By contrast, time translations are defined implicitly through the Heisenberg equation, as the transformations generated by the Hamiltonian, $\hat{H}$. Hence, the preservation of time-translation symmetry is a direct consequence of the Heisenberg equation, which is assumed to also hold in the noncommutative phase space model. More specifically, the time-translation operator, which acts according to $\hat{T}_t : t \to t + \tau$ and $\psi(\hat{\mathbf{X}}, t) \to \hat{T}_t\psi(\hat{\mathbf{X}}, t) = \psi(\hat{\mathbf{X}}, t + \tau)$, is given by $\hat{T}_t = e^{\tau\frac{\partial}{\partial t}} = e^{-\frac{i}{\hbar}\tau\hat{H}}$, where $\hat{H} = i\hbar\frac{\partial}{\partial t}$. We then have $\left[\hat{T}_t, \hat{H}\right] = 0$ and $[\hat{H}, \hat{H}] = 0$, i.e., time-translation symmetry still holds in the noncommutative phase space, which corresponds to the conservation of energy.

However, since spatial translation symmetry is broken in the noncommutative phase space, let us investigate the velocity and acceleration of a free particle. The velocity of the particle is defined as $\hat{\mathbf{v}} := \frac{d}{dt}\hat{\mathbf{X}}$, so that by using the Heisenberg equation (110) one has:

$$\hat{\mathbf{v}} = \frac{1}{i\hbar}[\hat{\mathbf{X}}, \hat{H}]. \tag{118}$$

For the free particle, the velocity is obtained, explicitly, as

$$\hat{\mathbf{v}} = \frac{1}{m}\left(\kappa^\alpha\hat{\mathbf{P}} + \kappa^\beta\hat{\mathbf{K}}_v\right), \tag{119}$$

where

$$\hat{\mathbf{K}}_v = (\hat{P}_Y - \hat{P}_Z)\mathbf{i} + (\hat{P}_X + \hat{P}_Z)\mathbf{j} + (-\hat{P}_X + \hat{P}_Y)\mathbf{k}. \tag{120}$$

It can be seen that there exists an intrinsic velocity, associated with the noncommutative parameters, and driven by the stochastic fluctuations of the background geometry. However, when the canonical momentum of the particle vanishes, $\langle\hat{\mathbf{P}}\rangle = 0$, we obtain

$\langle \hat{\mathbf{K}}_v \rangle = 0$. In Section 4.4 below, we give the Heisenberg representation of the velocity based on the SW map. This raises interesting issues regarding the nature of the principle of relativity in noncommutative space, but an in-depth discussion of these points lies outside the scope of the present review;. for further literature, see references in Ref. [64].

Consistent with the definition of velocity, the acceleration is defined as $\hat{\mathbf{a}} := \frac{d}{dt}\hat{\mathbf{v}}$. Again, by using the Heisenberg equation, this gives

$$\hat{\mathbf{a}} = \frac{1}{i\hbar}[\hat{\mathbf{v}}, \hat{H}]. \tag{121}$$

For the free particle, the acceleration is obtained, explicitly, as

$$\hat{\mathbf{a}} = \frac{\eta}{m^2\hbar}(\kappa^{\alpha} + \kappa^{\beta})\hat{\mathbf{K}}_p, \tag{122}$$

where

$$\hat{\mathbf{K}}_p = \left(\widehat{P}_y + \widehat{P}_z\right)\mathbf{i} - \left(\widehat{P}_x - \widehat{P}_z\right)\mathbf{j} - \left(\widehat{P}_x + \widehat{P}_y\right)\mathbf{k}. \tag{123}$$

Hence, there also exists an intrinsic stochastic acceleration. We can interpret this intrinsic acceleration as arising from quantum fluctuations of the noncommutative phase space background. This is associated with fluctuations in the components of the momentum, but we note that the direction of the acceleration is not, in general, the same as the direction of the momentum fluctuations.

4.4. Seiberg–Witten Map and the Heisenberg Representation

The noncommutative phase space provides a new operator algebra, beyond the canonical Heisenberg algebra, with which to explore unsolved puzzles in physics. To establish a rigorous mathematical basis for the model, and to allow clearer comparison with the canonical quantum formalism, several methods have been proposed in the literature to implement the noncommutative algebra (98)–(100) via a map on the canonical quantum operators. The three most used methods are the SW map, the Moyal star product, and the Wigner–Weyl phase space formulation [19–21]. Here, we outline the method of Seiberg and Witten (SW), which constructs a map linking the noncommutative phase space relations to the canonical Heisenberg algebra [29,34,35].

Formally, the SW map is defined as a map from the noncommutative phase space to the canonical Heisenberg phase space, which preserves the form of functions of the canonical variables, i.e.,

$$\phi_{\mathrm{SW}} : \begin{array}{ccc} (\hat{\mathbf{X}}, \hat{\mathbf{P}}) & \rightarrow & (\hat{\mathbf{x}}, \hat{\mathbf{p}}), \\ \hat{O}(\hat{\mathbf{X}}, \hat{\mathbf{P}}) & \rightarrow & \hat{O}(\hat{\mathbf{x}}, \hat{\mathbf{p}}), \end{array} \tag{124}$$

where $(\hat{\mathbf{X}}, \hat{\mathbf{P}})$ obey the algebra (98)–(100) and $(\hat{\mathbf{x}}, \hat{\mathbf{p}})$ obey Equation (11).

In other words, the noncommutative relations between the operators in Equations (98)–(100) can be implemented equivalently by the Heisenberg commutation relations, together with the SW map. The SW map may be implemented, explicitly, via the so-called Bopp shift [29], which is expressed as

$$\hat{X}^i = \hat{x}^i + \alpha^{ik}\hat{p}_k, \tag{125}$$

$$\hat{P}_j = \hat{p}_j + \beta_{jl}\hat{x}^l, \tag{126}$$

where

$$\theta^{ij} := \hbar(\alpha^{ji} - \alpha^{ij}), \quad \eta_{ij} := \hbar(\beta_{ji} - \beta_{ij}) \tag{127}$$

and

$$\kappa^i{}_j := \hbar(\delta^i{}_j - \alpha^{ik}\beta_{jk}). \tag{128}$$

By using the matrices (101), this gives

$$\begin{pmatrix} \hat{X} \\ \hat{Y} \\ \hat{Z} \end{pmatrix} = \begin{pmatrix} x - \frac{\theta}{2\hbar}(\hat{p}_y + \hat{p}_z) \\ y + \frac{\theta}{2\hbar}(\hat{p}_x - \hat{p}_z) \\ z + \frac{\theta}{2\hbar}(\hat{p}_x + \hat{p}_y) \end{pmatrix}, \tag{129}$$

$$\begin{pmatrix} \hat{P}_X \\ \hat{P}_Y \\ \hat{P}_Z \end{pmatrix} = \begin{pmatrix} \hat{p}_x + \frac{\eta}{2\hbar}(y + z) \\ \hat{p}_y - \frac{\eta}{2\hbar}(x - z) \\ \hat{p}_z - \frac{\eta}{2\hbar}(x + y) \end{pmatrix}, \tag{130}$$

which implies

$$\left[\alpha^{ij}\right] = \begin{pmatrix} 0 & -\frac{\theta}{2\hbar} & -\frac{\theta}{2\hbar} \\ \frac{\theta}{2\hbar} & 0 & -\frac{\theta}{2\hbar} \\ \frac{\theta}{2\hbar} & \frac{\theta}{3\hbar} & 0 \end{pmatrix} \tag{131}$$

and

$$\left[\beta_{ij}\right] = \begin{pmatrix} 0 & \frac{\eta}{2\hbar} & \frac{\eta}{2\hbar} \\ -\frac{\eta}{2\hbar} & 0 & \frac{\eta}{2\hbar} \\ -\frac{\eta}{2\hbar} & -\frac{\eta}{2\hbar} & 0 \end{pmatrix}. \tag{132}$$

This representation of the position and momentum operators can be regarded as the Heisenberg representation of noncommutative quantum mechanics. The SW map provides an efficient way to modify the Heisenberg algebra, giving rise to a noncommutative phase space, even though the map is not unitary or canonical.

By using the SW map (125) and (126) together with the noncommutative matrices (101), we obtain the explicit form of the noncommutative parameters that generalises the position–momentum commutator as

$$\kappa^{\alpha} = \hbar\left(1 + \frac{\theta\eta}{2\hbar^2}\right), \quad \kappa^{\beta} = \frac{\theta\eta}{4\hbar}, \tag{133}$$

which implies that the noncommutative phase space can be described by two independent parameters, θ and η.

However, it should be remarked that there is actually no unique SW map between the noncommutative phase space and the canonical Heisenberg phase space. The SW map with the Bopp shift, defined by Equations (131) and (132), provides a straightforward way to capture, approximately, the essential features of the noncommutative algebra in terms of the usual canonical Heisenberg algebra [29]. Moreover, for two general operators $\hat{A}(\hat{X}, \hat{P})$ and $\hat{B}(\hat{X}, \hat{P})$, the SW map does not preserve the form of the canonical commutator, namely,

$$\phi_{SW}[\hat{A}(\hat{X}, \hat{P}), \hat{B}(\hat{X}, \hat{P})] \neq [\phi_{SW}\hat{A}(\hat{X}, \hat{P}), \phi_{SW}\hat{B}(\hat{X}, \hat{P})]. \tag{134}$$

Hence, there exists additional ordering ambiguity.

Before concluding this section, we again stress that the choice of independent parameters for the antisymmetric matrices, θ and η, and for the symmetric matrix, κ^{α} and κ^{β}, is the simplest "natural" choice available. Nonetheless, in principle, the SW map and noncommutative algebras do not require constant matrices. Furthermore, for the SW map with the Bopp shift, (131) and (132), this relatively simple choice is truly "minimal", since even κ^{α} and κ^{β} can be expressed in terms of θ and η, according to Equation (133). In Section 4.8 below, we associate the two noncommutative parameters, θ and η, with the Planck length and the cosmological constant, respectively, which, it may be hoped, could provide a physical scenario for the emergence of dark energy from noncommutative geometry.

4.5. The Schrödinger Equation in Noncommutative Phase Space

Based on the SW map, the momentum operators in Equations (129) and (130) can be rewritten as $\hat{\mathbf{P}} = \hat{\mathbf{p}} - \mathbf{A}^{\eta}$, where

$$A_X^{\eta} = -\frac{\eta}{2\hbar}(y+z), \tag{135a}$$

$$A_Y^{\eta} = \frac{\eta}{2\hbar}(x-z), \tag{135b}$$

$$A_Z^{\eta} = \frac{\eta}{2\hbar}(x+y), \tag{135c}$$

and the A_i^{η} can be regarded as the components of an effective gauge potential. Thus, the Schrödinger equation for the free particle can be expressed as

$$i\hbar\frac{\partial\psi}{\partial t} = \frac{1}{2m}(\hat{\mathbf{p}} - \mathbf{A}^{\eta})^2\psi. \tag{136}$$

It can be seen that the Schrödinger equation for a free particle in noncommutative phase space is analogous to that obtained, in the canonical theory, for a charged particle in an electromagnetic potential. The continuity equation is given by

$$\frac{\partial\rho}{\partial t} + \nabla\cdot\mathbf{J} = 0, \tag{137}$$

where $\rho = |\psi|^2$ is the probability density and the probability current density is defined as $\mathbf{J} = \frac{1}{m}\langle\psi|\hat{\mathbf{p}} - \mathbf{A}^{\eta}|\psi\rangle$. It should be noted that the current density is not equal to the velocity given by Equation (119).

4.6. Noncommutative Gauge Fields and the Pauli Equation

We suppose that spin in noncommutative phase space has the same form as in canonical quantum mechanics, because spin is an intrinsic property of the particle, which is independent of the spacetime background [61]. Thus, the Pauli equation for a particle with nonzero spin and charge q, in the noncommutative space, may be written as

$$i\hbar\frac{\partial\psi}{\partial t} = \left[\frac{1}{2m}[\sigma\cdot(\hat{\mathbf{p}} - \mathbf{A}^{\mathrm{nc}})]^2 + q\phi\right]\psi, \tag{138}$$

where $\sigma = \sigma_x\mathbf{i} + \sigma_y\mathbf{j} + \sigma_z\mathbf{k}$ and ϕ is the electric potential. The effective vector potential contains two terms,

$$\mathbf{A}^{\mathrm{nc}} = \mathbf{A} + \mathbf{A}^{\eta}, \tag{139}$$

where $\mathbf{A}$ is the canonical vector potential and $\mathbf{A}^{\eta}$ is the effective contribution arising from the noncommutativity of the momentum components. The total effective magnetic field is $\mathbf{B}^{\mathrm{nc}} = \nabla\times\mathbf{A}^{\mathrm{nc}}$, where $\mathbf{B}^{\mathrm{nc}} = \mathbf{B} + \mathbf{B}^{\eta}$ with $\mathbf{B} = \nabla\times\mathbf{A}$ and $\mathbf{B}^{\eta} = \nabla\times\mathbf{A}^{\eta}$. Thus, the Pauli equation can be rewritten as

$$i\hbar\frac{\partial\psi}{\partial t} = \left[\frac{1}{2m}(\hat{\mathbf{p}} - \mathbf{A}^{\mathrm{nc}})^2 - \frac{\hbar q}{2m}\sigma\cdot\mathbf{B}^{\mathrm{nc}} + q\phi\right]\psi. \tag{140}$$

By analogy with canonical electrodynamics, we also define the effective electromagnetic field tensor, generated from the effective gauge potential in the noncommutative phase space, as

$$F_{\mu\nu}^{\eta} = \partial_\mu A_\nu^{\eta} - \partial_\nu A_\mu^{\eta}. \tag{141}$$

Substituting the gauge potential (135) into Equation (141), the explicit form of $F_{\mu\nu}^{\eta}$ is obtained as

$$\left[F_{\mu\nu}^{\eta}\right] = \begin{pmatrix} 0 & -\eta/\hbar & -\eta/\hbar \\ \eta/\hbar & 0 & -\eta/\hbar \\ \eta/\hbar & \eta/\hbar & 0 \end{pmatrix}. \tag{142}$$

Interestingly, the effective gauge field is induced by the noncommutativity between different directional components of momentum, in the noncommutative phase space. This noncommutative effect can be interpreted as a spacetime curvature, so that the resulting gauge field can be naturally related to the value of the scalar curvature. For this reason, we relate η to the observed value of the cosmological constant, in Section 4.8.

Following Equation (124), any observable operator in the noncommutative phase space can be mapped to the Heisenberg representation, based on the SW map. Actually, the wave function in noncommutative phase space is also given by $\Psi(\mathbf{X}) = \langle \mathbf{X}|\Psi\rangle$ or $\tilde{\Psi}(\mathbf{P}) = \langle \mathbf{P}|\Psi\rangle$, by analogy with the canonical theory. However, as a first-order approximation, it is convenient to assume that [24,25,33]

$$\Psi(\mathbf{X}) \simeq \psi(\mathbf{x}) + \mathcal{O}(\theta), \tag{143}$$

$$\Psi(\mathbf{P}) \simeq \psi(\mathbf{p}) + \mathcal{O}(\eta). \tag{144}$$

4.7. Noncommutativity-Induced Anomalous Velocity and Acceleration

For the free particle, using the SW map in Equations (124)–(126), the expectation value of the velocity in the Heisenberg representation can be expressed as

$$\langle \hat{v}^i \rangle = \frac{1}{\hbar m}\left(\kappa^i_j \langle \hat{p}_j \rangle + \frac{\eta}{2}\gamma^i_j \langle x^j \rangle\right), \tag{145}$$

where

$$\left[\gamma^i_j\right] = \begin{pmatrix} 0 & \left(1+\frac{3\theta\eta}{4\hbar^2}\right) & \left(1+\frac{3\theta\eta}{4\hbar^2}\right) \\ -\left(1+\frac{3\theta\eta}{4\hbar^2}\right) & 0 & \left(1+\frac{3\theta\eta}{4\hbar^2}\right) \\ -\left(1+\frac{3\theta\eta}{4\hbar^2}\right) & -\left(1+\frac{3\theta\eta}{4\hbar^2}\right) & 0 \end{pmatrix}. \tag{146}$$

Similarly, the expectation value of the acceleration is obtained as

$$\langle \hat{a}_i \rangle = \frac{1}{m^2\hbar}\left(1+\frac{3\theta\eta}{4\hbar^2}\right)\left(\eta_{ij}\langle \hat{p}_j \rangle - \frac{\eta^2}{\hbar}\chi_{ij}\langle x^j \rangle\right), \tag{147}$$

where

$$[\chi_{ij}] = \begin{pmatrix} 1 & 1/2 & -1/2 \\ 1/2 & 1 & 1/2 \\ -1/2 & 1/2 & 1 \end{pmatrix}. \tag{148}$$

As noted previously, in Section 4.3, the velocity and acceleration contain an intrinsic stochastic perturbation arising from the noncommutativity of momenta in different directions of the phase space. Here, these are expressed in terms of the two independent parameters of our specific implementation of the SW map, θ and η. This anomalous acceleration can be interpreted as a quantum effect, since it is induced by the noncommutative algebra. In Section 4.8 below, we express the noncommutative parameters in terms of the Planck length and the cosmological constant, linking the noncommutative algebra with the spacetime background and nonzero minimal energy density.

4.8. Physical Interpretations of the Noncommutative Parameters

In canonical quantum mechanics, Planck's constant plays an essential role in quantizing the phase space of elementary particles, such that quantum states are described by state vectors in a Hilbert space. Roughly speaking, $\hbar$ represents the minimum (incompressible) volume of a phase space fluid element, $\Delta X^i \Delta P_j$ [9,10]. In noncommutative phase space, the noncommutative parameters θ and η play analogous roles with respect to physical space and momentum space, respectively, inducing minimum bounds on the volumes $\Delta X^i \Delta X^j$ and $\Delta P_i \Delta P_j$, which cannot be further compressed below their minimum values.

Actually, one can adopt different parametrization schemes for the noncommutative parameters, for various physical problems, associated with different energy and spacetime scales. In other words, how one endows the noncommutative parameters with physical meanings (and scales) depends on what physical problems are being addressed. Therefore, let us note that many studies suggest that spacetime should be quantized, inducing a minimum length scale of the order of the Planck length [11]. Similarly, the emergence of dark energy implies that there exists a minimum curvature of spacetime, expressed in terms of the cosmological constant [12,13], which can be naturally associated with the nonzero energy/momentum density obtained in a noncommutative phase space model. (We recall that a minimum positive curvature is equivalent, according to the gravitational field equations, to a minimum positive energy density.)

In particular, gedanken experiments in quantum gravity, together with various theoretical approaches, such as string theory [11], loop quantum gravity [9,10], and others, suggest spacetime quantization at the Planck scale. While the status of a minimum possible momentum is less clear, several models also propose this [12,13], and it is worth noting that in a universe governed by dark energy there exists a finite de Sitter horizon [12,13]. This places on upper bound on the value of a particle's de Broglie wavelength and, hence, a lower value on its momentum uncertainty, of the order of the de Sitter momentum, $m_{dS}c = \hbar\sqrt{\Lambda/3}$, mentioned previously in Section 2.3.

Thus, we propose a parametrization scheme for the noncommutative parameters, which is associated with the Planck length and the cosmological constant, namely

$$\theta = \ell_{\rm Pl}^2, \quad \eta = m_{\rm dS}^2 c^2, \tag{149}$$

where we define the Planck and de Sitter mass scales explicitly as

$$\ell_{\rm Pl} = \sqrt{\frac{\hbar G}{c^3}} \simeq 10^{-33}\,{\rm cm}, \quad m_{\rm Pl} = \sqrt{\frac{\hbar c}{G}} \simeq 10^{-5}\,{\rm g}, \tag{150}$$

$$\ell_{\rm dS} = \sqrt{\frac{3}{\Lambda}} \simeq 10^{28}\,{\rm cm}, \quad m_{\rm dS} = \frac{\hbar}{c}\sqrt{\frac{\Lambda}{3}} \simeq 10^{-66}\,{\rm g}. \tag{151}$$

Combining the relations (149) and (150), we obtain the independent components of the generalised position–momentum commutator, in the noncommutative phase space, as

$$\kappa^\alpha = \hbar(1+\Delta), \quad \kappa^\beta = \frac{\hbar\Delta}{2}, \tag{152}$$

where

$$\Delta := \frac{\rho_\Lambda}{\rho_{\rm Pl}} \simeq 10^{-122}, \tag{153}$$

and

$$\rho_\Lambda = \frac{\Lambda c^2}{8\pi G}, \quad \rho_{\rm Pl} = \frac{3}{4\pi}\frac{c^5}{\hbar G^2}, \tag{154}$$

are the dark energy and Planck densities, respectively, with G the gravitational constant. Interestingly, a new model of generalised uncertainty relations was developed [65,66], in which the action scale $\beta := 2\hbar\sqrt{\Delta} \simeq \hbar \times \mathcal{O}(10^{-61})$ was proposed as the fundamental quantum of action for spacetime, as opposed to matter. In this model, the generalised uncertainties were obtained without modifying the canonical space–space or momentum–momentum commutators, but the parameterization above suggests that an extension of this approach could be used to provide a physical mechanism for noncommutative geometry. This intriguing possibility will be analysed in detail in a future work.

5. Conclusions and Outlook

The interplay between physics and mathematics stimulates new ideas to resolve unsolved puzzles in nature. In this review, we have given a brief introduction to the interplay between physics and mathematics in the fields of noncommutative geometry and noncommutative phase space, encompassing topics in both classical physics and quantum mechanics.

Although noncommutative phenomena were discovered, even in classical mechanics, their true significance became apparent only with the advent of quantum theory in the 1920s. Since then, noncommutative structures have inspired bold new attempts to solve unsolved problems in gravity and high-energy theory, and, arguably, their significance to modern theoretical physics has only increased since the heady days of the early twentieth century.

From classical Poisson brackets to the Heisenberg commutation relations and the quantum Hall effect, physicists have found real-world phenomena described by noncommutative algebras and geometries. Furthermore, as we attempt to extend our current physical theories into previously unprobed regimes at the Planck scale, or dark energy scale, infinities and singularities unavoidably emerge, suggesting new types of noncommutative structures which may be able to cure them. Increasingly, a large number of researchers believe that the infinities and singularities which break all known physical laws may be cured if there exist quantum fluctuations of the spacetime background, governed by noncommutative algebras, which are able to prevent such divergences. Thus, noncommutative phenomena inspire many attempts to construct a unified framework for both gravity and high-energy particle physics [2–4].

In this review, we have introduced the basic concepts underlying noncommutative phenomena in classical and quantum mechanics and presented some important examples in condensed matter and statistical physics. We discussed the basic noncommutative algebras that arise in physical theories, including the classical Poisson brackets in symplectic geometry, the Heisenberg algebra of fundamental operators, acting on the Hilbert space of canonical quantum mechanics, and the Lie, Clifford, and Dirac algebras associated with rotational symmetry and spin. On a more theoretical note, we also gave brief expositions of the Snyder and Nambu algebras, which have been proposed as extensions of existing physical theories, and are intended to help cure the emergence of the singularities mentioned above [12,13].

Based on the SW map, we outlined the basic properties and novel phenomena that occur in the noncommutative extension of the Heisenberg phase space, incorporating both space–space and momentum–momentum noncommutativity. These include the breaking of translation and rotational symmetries, as well as important phenomenological predictions like the existence of anomalous, stochastic perturbations to the velocity and acceleration of free particles, induced by noncommutativity. The stochastic perturbations can be viewed as an additional quantum force, driving particle motion due to quantum fluctuations of the background geometry, and we showed that the noncommutative terms give rise to an effective gauge field in the Schrödinger and Pauli equations. With this in mind, we proposed a parametrization scheme for the noncommutative parameters, which associated them with both the Planck length and the dark energy density, where the latter is expressed in terms of the cosmological constant.

Based on this parametrization scheme, the effective gauge field that arises from the noncommutativity of the phase space can also be interpreted in terms of the minimum length, and minimum energy density of the universe. We showed that this gives rise to phenomenologically interesting effects on the dynamics of free particles, which are subjected to intrinsic stochastic velocity and acceleration perturbations. These perturbations depend on the initial momentum and position, and can be regarded as a quantum effects induced by the noncommutative phase space. The quantum anomalous acceleration of free particles could actually provide a microphysical model for dark energy.

However, we note that noncommutative models in physics have been developed in many different ways, including Conne's approach to noncommutative geometry [14,15], noncommutative M-theory [4,56], noncommutative field theory based on the Moyal star product [3], the principle of relative locality, and $\kappa-$deformed spacetime symmetries based on Hopf algebras [57–59]. These interesting results suggest that noncommutative phase space may provide a deep connection between the dynamics of microscopic particles and the quantum theory of gravity, including dark energy; for details of these and other approaches to noncommutative phenomena, that were not covered in the present pedagogical introduction, see the bibliography and references therein.

Author Contributions: Conceptualization, S.-D.L.; methodology, S.-D.L. and M.J.L.; validation, S.-D.L. and M.J.L.; formal analysis, S.-D.L. and M.J.L.; resources, S.-D.L. and M.J.L.; writing—original draft preparation, S.-D.L.; writing—review and editing, S.-D.L. and M.J.L.; visualization, funding acquisition, S.-D.L. and M.J.L. All authors have read and agreed to the published version of the manuscript.

Funding: The authors thank the Grant of Scientific and Technological Projection of Guangdong Province (China), no. 2021A1515010036. The research of M.J.L. was supported by the Natural Science Foundation of Guangdong Province (China), Grant no. 008120251030.

Data Availability Statement: Not applicable.

Acknowledgments: M.J.L. wishes to thank the Department of Physics and Materials Science, Faculty of Science, and the Office of Research Administration, Chiang Mai University (Thailand), for providing research facilities.

Conflicts of Interest: The authors declare no conflict of interest.

References

1. Takhtajan, L.A. *Quantum Mechanics for Mathematicians*; American Mathematical Society: Providence, RI, USA, 2008.
2. Douglas, M.R.; Nekrasov, N.A. Noncommutative field theory. *Rev. Mod. Phys.* **2001**, *73*, 977–1029. [CrossRef]
3. Szabo, R.J. Quantum field theory on noncommutative spaces. *Phys. Rep.* **2003**, *378*, 207–299. [CrossRef]
4. Konechny, A.; Schwarz, A. Introduction to M(atrix) theory and noncommutative geometry. *Phys. Rep.* **2002**, *360*, 353–465. [CrossRef]
5. Rosenbaum, M.; Vergara, J.D.; Juarez, L.R. Noncommutative field theory from quantum mechanical space-space noncommutativity. *Phys. Lett. A* **2007**, *367*, 1–10. [CrossRef]
6. Snyder, H.S. Quantized space-time. *Phys. Rev.* **1947**, *71*, 38–41. [CrossRef]
7. Gouba, L. A comparative review of four formulations of noncommutative quantum mechanics. *Int. J. Mod. Phys. A* **2016**, *31*, 1630025. [CrossRef]
8. Yang, C.N. On Quantized space-time. *Phys. Rev.* **1947**, *72*, 874. [CrossRef]
9. Rovelli, C. *Quantum Gravity*; Cambridge University Press: Cambridge, UK, 2004. [CrossRef]
10. Thiemann, T. *Modern Canonical Quantum General Relativity*; Cambridge University Press: Cambridge, UK, 2007. [CrossRef]
11. Fredenhagen, K. Gravity induced noncommutative spacetime. *Rev. Math. Phys.* **1955**, *7*, 559–565. [CrossRef]
12. Matarrese, S.; Colpi, M.; Gorini, V.; Moschella, U. *Dark Matter and Dark Energy*; Springer Science+Business Media/Canopus Academic Publishing Limited: Dordrecht, The Netherlands, 2011. [CrossRef]
13. Peebles, P.J.E. The cosmological constant and dark energy. *Rev. Mod. Phys.* **2003**, *75*, 559–606. [CrossRef]
14. Connes, A.; Marcolli, M. A walk in the noncommutative garden. In *An Introduction to Noncommutative Geometry*; Khalkhali, M., Marcolli, M., Eds.; World Scientific: Singapore, 2008; pp. 1–128. [CrossRef]
15. Connes, A. Non-commutative differential geometry. *Int. Hautes Etudes Sci. Publ. Math.* **1985**, *62*, 257–360. [CrossRef]
16. Doplicher, S.; Fredenhagen, K.; Roberts, J.E. The quantum structure of spacetime at the Planck scale and quantum fields. *Commun. Math. Phys.* **1995**, *172*, 187–220. [CrossRef]
17. Delduc, F.; Duret, Q.; Gieres, F.; Lefrancois, M. Magnetic fields in noncommutative quantum mechanics. *J. Phys. Conf. Series* **2008**, *103*, 012020. [CrossRef]
18. Kovačik, S.; Prešnajder, P. Magnetic monopoles in noncommutative quantum mechanics 2. *J. Math. Phys.* **2018**, *59*, 082107. [CrossRef]
19. Bellucci, S.; Nersessian, A.; Sochichiu, C. Two phases of the noncommutative quantum mechanics. *Phys. Lett. B* **2001**, *52*, 345–349. [CrossRef]
20. Gamboa, J.; Loewe, M.; Rojas, J.C. Noncommutative quantum mechanics. *Phys. Rev. D* **2001**, *64*, 067901. [CrossRef]
21. Gamboa, J.; Loewe, M.; Mendez, F.; Rojas, J.C. Noncommutative quantum mechanics: The two dimensional central field. *Int. J. Mod. Phys. A* **2002**, *17*, 2555–2565. [CrossRef]
22. Ezawa, Z.F. *Quantum Hall Effects: Field theoretical Approach and Related Topics*; World Scietific: Singapore, 2008. [CrossRef]

23. Das, A.; Falomir, H.; Gamboa, J.; Mendez, F.; Nieto, M. Aharonov-Bohm effect in a class of noncommutative theories. *Phys. Rev. D* **2011**, *84*, 045002. [CrossRef]
24. Liang, S.-D.; Li, H.; Huang, G.-Y. Detecting noncommutative phase space by the Aharonov-Bohm effect. *Phys. Rev. A* **2014**, *90*, 010102. [CrossRef]
25. Liang, S.-D.; Harko, T. Towards an observable test of noncommutative quantum mechanics. *Ukr. J. Phys.* **2019** *64*, 11. [CrossRef]
26. Chaichiian, M.; Langvik, M.; Sasaki, S.; Tureanu, A. Gauge covariance of the Aharonov-Bohm phase in noncommutative quantum mechancs. *Phys. Lett. A* **2008**, *666*, 199–204. [CrossRef]
27. Kokado, A.; Okamura, T.; Saito, T. Noncommutative quantum mechanics and the Seiberg-Witten map. *Phys. Rev. D* **2004**, *69*, 125007. [CrossRef]
28. Basu, B.; Chowdhury, D.; Ghosh, S. Inertial spin Hall effect in noncommutative space. *Phys. Lett. A* **2013**, *377*, 1661–1667. [CrossRef]
29. Bertolami, O.; Rosa, J.G.; de Aragao, C.M.L.; Castorina, P.; Zappala, D. Noncommutative gravitational quantum well. *Phys. Rev. D* **2005**, *72*, 025010. [CrossRef]
30. Falomir, H.; Gamboa, J.; Lopez-Sarrion, J.; Mendez, F.; Pisani, P.A.G. Magnetic-dipole spin effects in noncommutative quantum mechanics. *Phys. Lett. B* **2009**, *680*, 384–386. [CrossRef]
31. Sivasubramanian, S.; Srivastava, Y.N.; Vitiello, G.; Widom, A. Quantum dissipation induced noncommutative geometry. *Phys. Lett. A* **2003**, *11*, 97–105. [CrossRef]
32. Vilela Mendes, R. Quantum mechanics and non-commutative space-time. *Phys. Lett. A* **1996**, *210*, 232–240. [CrossRef]
33. Harko, T.; Liang, S.-D. Energy-dependent noncommutative quantum mechanics. *Eur. Phys. J. C* **2019**, *79*, 300. [CrossRef]
34. Seiberg, N.; Witten, E. Electric-magnetic duality, monopole condensation, and confinement in $N = 2$ supersymmetric Yang-Mills theory. *Nucl. Phys. B* **1994**, 426, 19–52. [CrossRef]
35. Seiberg, N.; Witten, E. String theory and noncommutative geometry. *J. High Energy Phys.* **1999**, *1999*, 032. [CrossRef]
36. Bastos, C.; Bertolami, O. Berry phase in the gravitational quantum well and the Seiberg-Witten map. *Phys. Lett. A* **2008**, *372*, 5556–5559. [CrossRef]
37. Bastos, C.; Bertolami, O.; Costa Dias, N.; Nuno Prata, J. Violation of the Robertson-Schrodinger uncertainty principle and noncommutative quantum mechanics. *Phys. Rev. D* **2012**, *86*, 105030. [CrossRef]
38. Bastos, C.; Bernardini, A.E.; Bertolami, O.; Costa Dias, N.; Nuno Prata, J. Robertson-Schrödinger-type formulation of Ozawa's noise-disturbance uncertainty principle. *Phys. Rev. D* **2014**, *89*, 042112. [CrossRef]
39. Bastos, C.; Bernardini, A.E.; Bertolami, O.; Costa Dias, N.; Nuno Prata, J. Bell operator and Gaussian squeezed states in noncommutative quantum mechanics. *Phys. Rev. D* 2016, *93*, 104055. [CrossRef]
40. Ho, P.-M.; Kao, H.-C. Noncommutative quantum mechanics from noncommutative quantum field theory. *Phys. Rev. Lett.* **2002**, *88*, 151602. [CrossRef]
41. Nambu, Y. Generalized Hamiltonian dynamicse. *Phys. Rev. D* **1973**, *7*, 2405–2412. [CrossRef]
42. Bayen, F.; Flato, M. Remarks concerning Nambu's generalized mechanics. *Phys. Rev. D* **1975**, *11*, 3049–3053. [CrossRef]
43. Estabrook, F.B. Comments on generalized Hamiltonian dynamics. *Phys. Rev. D* **1973**, *8*, 2741. [CrossRef]
44. Mukunda, N.; Sudarshan, E.C.G. Relation between Nambu and Hamiltonian mechanics. *Phys. Rev. D* **1976**, *13*, 2846–2850. [CrossRef]
45. Kálnay, A.J.; Tascón, R. Lagrange, Hamilton-Dirac, and Nambu mechanics. *Phys. Rev. D* **1978**, *17*, 1552–1562. [CrossRef]
46. Curtright, T.; Zachos, C. Classical and quantum Nambu mechanics. *Phys. Rev. D* **2003**, *68*, 085001. [CrossRef]
47. Awata, H.; Li, M.; Minic, D.; Yoneya, T. On the quantization of Nambu brackets. *J. High Eng. Phys.* **2001**, *02*, 013. [CrossRef]
48. Sahoo, D.; Valsakumar, M.C. Nambu mechanics and its quantization. *Phys. Rev. A* **1992**, *46*, 4410–4412. [CrossRef] [PubMed]
49. Zachos, C.K.; Curtright, T.L. Deformation quantization of Nambu mechanics. *AIP Conf. Proc.* **2003**, *672*, 183–196.
50. Hirayama, M. Realization of Nambu mechanics: A particle interacting with an $SU(2)$ monopole. *Phys. Rev. D* **1977**, *16*, 530–532. [CrossRef]
51. Hirayama, M.; Tze, H.-C. Nambu mechanics of non-Abelian dyons. *Phys. Rev. D* **1978**, *17*, 546. [CrossRef]
52. Minic, D.; Takeuchi, T.; Tze, C.-H. Interference and oscillation in Nambu quantum mechanics. *Phys. Rev. D* **1978**, *17*, 546–556. [CrossRef]
53. Angulo, R.; Codriansky, S.; Gonzalez-Bernardo, C.A.; Kalnay, A.; Rodriguez-Bernardo, F.P.J.; Rodriguez-Nunez, J.J.; Tello-llanos, R.A. Superconductivity and the existence of Nambu's three-dimensional phase space mechanics. *Phys. Lett. A* **1984**, *104*, 106–108. [CrossRef]
54. Takhtajan, L. On foundation of the generalized Nambu mechanics. *Commun. Math. Plays.* **1994**, **160**, 295–315. [CrossRef]
55. Fecko, M. On a geometrical formulation of the Nambu dynamics. *J. Math. Phys.* **1992**, *33*, 926–929. [CrossRef]
56. Prodan, E.; Schulz-Baldes, H. *Bulk and Boundary Invariants for Complex Topological Insulators: From K-theory to Physics*; Springer International Publishing: Cham, Switzerland, 2016. [CrossRef]
57. Agostini, A.; Amelino-Camelia, G.; D'Andrea, F. Hopf-algebra description of noncommutative-space-time symmetries. *Int. J. Mod. Phys. A* **2004**, *19*, 5187–5220. [CrossRef]
58. Arzano, M.; J. Kowalski-Glikman, J. Deformations of Spacetime Symmetries, Gravity: Group-Valued Momenta, and Non-Commutative Fields; Springer: Berlin/Heidelberg, Germany, 2021. [CrossRef]

59. Arzano, M.; Kowalski-Glikman, J. An introduction to κ-deformed symmetries, phase spaces and field theory. *Symmetry* **2021**, *13*, 946. [CrossRef]
60. Deriglazov, A. *Classical Mechanics, Hamiltonian and Lagranian Formulation*; Springer: Berlin/Heidelberg, Germany, 2010.
61. Wachter, A. *Relativistic Quantum Mechanics*; Springer Science+Business Media B.V.: Dordrecht, The Netherlands, 2011. [CrossRef]
62. Sucre, M.G.; Kalnay, A. On the statistics consistent with Nambu's new quantum rules: Quarks? *Int. J. Theor. Phys.* **1975**, *12*, 149–151. [CrossRef]
63. Amelino-Camelia, G.; Freidel, L.; Kowalski-Glikman, J.; Smolin, L. The principle of relative locality. *Phys. Rev. D* **2011**, *84*, 084010. [CrossRef]
64. Liang, S.-D. Klein-Gordon theory in noncommutative phase space. *Symmetry* **2023**, *15*, 367. [CrossRef]
65. Lake, M.J.; Miller, M.; Ganardi, R.F.; Liu, Z.; Liang, S.-D.; Paterek, T. Generalised uncertainty relations from superpositions of geometries. *Class. Quantum Grav.* **2019**, *36*, 155012. [CrossRef]
66. Lake, M.J.; Miller, M.; Liang, S.-D. Generalised uncertainty relations for angular momentum and spin in quantum geometry. *Universe* **2020**, *6*, 56. [CrossRef]

MDPI
St. Alban-Anlage 66
4052 Basel
Switzerland
www.mdpi.com

Physics Editorial Office
E-mail: physics@mdpi.com
www.mdpi.com/journal/physics

www.ingramcontent.com/pod-product-compliance
Lightning Source LLC
Chambersburg PA
CBHW040822250926
51850CB00007BA/1241
9783036587707